4/14 no
update
keep

World Atlas of Biodiversity

Published in association with UNEP-WCMC by the University of California Press

University of California Press
Berkeley and Los Angeles, California
University of California Press, Ltd.
London, England

© 2002 UNEP World Conservation Monitoring Centre

UNEP-WCMC
219 Huntingdon Road
Cambridge CB3 0DL, UK
Tel: +44 (0) 1223 277 314
Fax: +44 (0) 1223 277 136
E-mail: info@unep-wcmc.org
Website: www.unep-wcmc.org

World Atlas of Biodiversity:
Earth's Living Resources in the 21st Century
is a revised and updated edition of
Global Biodiversity:
Earth's Living Resources in the 21st Century

Cloth edition ISBN
0-520-23668-8

Cataloging-in-publication data is on file with the Library of Congress

Citation Groombridge B. and Jenkins M.D. (2002) *World Atlas of Biodiversity*. Prepared by the UNEP World Conservation Monitoring Centre. University of California Press, Berkeley, USA.

UNEP WCMC

World Atlas of Biodiversity

Earth's Living Resources in the 21st Century

Brian Groombridge & Martin D. Jenkins

UNIVERSITY OF CALIFORNIA PRESS

BERKELEY LOS ANGELES LONDON

World Atlas of Biodiversity

Prepared by
UNEP World Conservation
Monitoring Centre
219 Huntingdon Road
Cambridge CB3 0DL, UK
Tel: +44 (0) 1223 277 314
Fax: +44 (0) 1223 277 136
E-mail: info@unep-wcmc.org
Website: www.unep-wcmc.org

Director
Mark Collins

Authors
Brian Groombridge
Martin D. Jenkins

Additional contributors
Adrian C. Newton (Project manager)
Rachel Cook
Neil Cox
Victoria Gaillard
Edmund Green
Janina Jakubowska
Thomas Kaissl
Valerie Kapos
Charlotte Lusty
Anna Morton
Mark Spalding
Christoph Zöckler

Production of maps
Simon Blyth
with the assistance of
Igor Lysenko
Corinna Ravilious
Jonathan Rhind

Layout
Yves Messer

UNEP WCMC

The UNEP World Conservation Monitoring
Centre is the biodiversity information and
assessment arm of the United Nations
Environment Programme, the world's
foremost intergovernmental environmental
organization. UNEP-WCMC aims to help
decision-makers recognize the value of
biodiversity to people everywhere, and to
apply this knowledge to all that they do. The
Centre's challenge is to transform complex
data into policy-relevant information, to build
tools and systems for analysis and
integration, and to support the needs of
nations and the international community as
they engage in joint programs of action.

A Banson production
27 Devonshire Road
Cambridge CB1 2BH, UK

Color separations
Swaingrove

Printed in the UK

Acknowledgments

First and foremost we would like to express our deepest thanks to the Aventis Foundation, without whose generous funding the research and production work for this book could not have been undertaken. Preparation of the book was also generously supported by the Department of Environment, Food and Rural Affairs (DEFRA) of the UK Government. The Owen Family Trust is also acknowledged for financial support to the first edition of this text.

We also acknowledge with thanks the generous assistance extended by the following, listed approximately in the same sequence as the chapters in which their material appears:

Christopher Field and George Merchant, Department of Global Ecology, Carnegie Institution of Washington for use of data from a global model of net primary production.

Robert Lesslie of the Department of Geography, Australian National University, Canberra, for allowing us to use data resulting from his global wilderness analysis.

BirdLife International, of Cambridge, UK, for allowing use of spatial data on endemic bird areas and on threatened bird species.

Gene Carl Feldman, Oceanographer at NASA/Goddard Space Flight Center, Greenbelt, Maryland, for approving use of material from the SeaWiFS Project of NASA/Goddard Space Flight Center and ORBIMAGE.

The University of Maryland Global Land Cover Facility, for facilitating use of land cover data.

Professor Wilhelm Barthlott of the Botanisches Institut und Botanischer Garten, Rheinischen Friedrich-Wilhelms-Universität, Bonn, for kindly allowing use of a map showing contours of global plant species diversity.

Jonathan Loh, responsible for the WWF Living Planet Report, for kindly approving use of global trend indices from the *Living Planet Report 2000*.

John E.N. Veron, Chief Scientist at the Australian Institute of Marine Sciences, Townsville, Queensland for allowing use of coral generic diversity data.

Several biologists associated with IUCN/SSC specialist groups on fishes, mollusks and inland water crustacea, for providing data and expertise on important areas for freshwater biodiversity collated in an earlier publication: Gerald R. Allen (Western Australian Museum); the late Denton Belk (Texas); Philippe Bouchet (Laboratoire de Biologie des invertébrés marins et malacologie, Muséum National d'Histoire Naturelle, Paris); Keith Crandall (Department of Zoology, Brigham Young University); Neil Cumberlidge (Department of

Biology, Northern Michigan University); Olivier Gargominy (Laboratoire de Biologie des invertébrés marins et malacologie, Muséum National d'Histoire Naturelle, Paris); Maurice Kottelat (Cornol, Switzerland); Sven O. Kullander (Department of Vertebrate Zoology, Swedish Museum of Natural History, Stockholm); Christian Lévêque (ORSTOM, Paris); R. von Sternberg (Center for Intelligent Systems, State University of New York at Binghamton); Guy Teugels (Laboratoire Ichthyologie, Musée Royal de l'Afrique Centrale, Tervuren).

Ben ten Brink, Jan Bakkes and Jaap van Woerden for facilitating use of material illustrating work on scenarios carried out at the Rijksinstituut voor Volksgezondheid en Milieu (RIVM), Bilthoven, the Netherlands.

Christian Nellemann (Norwegian Institute for Nature Research), Hugo Ahlenius (UNEP GRID-Arendal) and the Secretariat of GLOBIO (Global methodology for mapping human impacts on the biosphere), for material applying this approach to scenario development.

PHOTOGRAPHS
Pages: 6, L. Olesen/UNEP/Still Pictures; 7, L.L. Hock/UNEP/Topham;
14, B. Groombridge; 15, M. Wakabayashi/UNEP/Topham; 37, Giotto Castelli;
38, M. Friedlander/UNEP/Topham; 39, UNEP/Topham; 72, G. Bluhm/UNEP/Topham;
73, K. Kaznaki/UNEP/Topham; 86, M. Schneider/UNEP/Topham;
93, M.R. Andrianavalona/UNEP/Still Pictures; 96, H. Mundell/UNEP/Topham;
98, R. Faria/UNEP/Topham; 99, UNEP/Topham; 103 J. Nuab/UNEP/Topham;
110 top, R. del Rosarion/UNEP/Topham;110 bottom, Mazinsky/UNEP/Topham;
118, E. Green; 143 top, E. Green; bottom, M. Spalding; 144, M. Garcia Blanco/UNEP/Topham;
151, E. Green; 152, D. Nayak/UNEP/Still Pictures; 154, E. Green;
155, D. Seifert/UNEP/Still Pictures; 165, UNEP/Topham; 169, S.W. Ming/UNEP/Still Pictures;
174, F. Colombini/UNEP/Topham; 178, C.K. Au/UNEP/Still Pictures;
185, K. Lohua/UNEP/Still Pictures; 194, C. Petersen/UNEP/Topham;
212, P. Garside/UNEP/Topham; 216, C. Senanunsakul/UNEP/Still Pictures

Contents

Foreword

Klaus Töpfer, Executive Director, United Nations Environment Programme

It is a great pleasure for me to introduce this important new book from the UNEP World Conservation Monitoring Centre. Building on the analyses it carried out for the Earth Summit in Rio de Janeiro in 1992 and for the new millennium just two years ago, UNEP-WCMC has once again updated and revised its important overview of life on Earth in time for a major global event.

In tune with the message of the Johannesburg World Summit on Sustainable Development, this new atlas places humankind firmly in the context of the species and ecosystems upon which we all depend for our livelihoods. We are a part of biodiversity and as such we should treasure it, use it wisely and share its benefits in our own enlightened self-interest.

I commend this book to all who seek a greater appreciation of the inter-dependency between our own future and that of global biodiversity.

In closing I should like on behalf of UNEP to thank most warmly the Aventis Foundation and the UK Department for Environment, Food and Rural Affairs for their support in the preparation of the *World Atlas of Biodiversity*.

Preface

Mark Collins, Director, UNEP World Conservation Monitoring Centre

The diversity of life is the defining feature of planet Earth. It is unique – as far as we know – in the infinity of the universe. For 11 000 years since agriculture began, humankind has increasingly appropriated the biological resources and natural productivity of lands and seas to support the expansion of civilizations and technologies. Everything that we have achieved has its origins in living animals, plants and the communities and ecosystems of which they are a part. But it is only in the past 30 years, since the United Nations Conference on the Human Environment in Stockholm in 1972, that we have begun to recognize the limits to nature's gifts. We now know that our own success is placing strain on nature's ability to evolve, diversify, cleanse our air and water and provide us with the raw materials we need for food, fuel, fiber and health.

Just ten years ago we began to take integrated and holistic action to ensure conservation and sustainable use of biological resources. The UN Conference on Environment and Development (UNCED) saw the signing of the Convention on Biological Diversity, the first global agreement on biodiversity that clearly positioned humankind as an integral part of the complex of life on Earth, rather than a special case somehow separate from nature and immune from its laws. The 'ecosystem approach' espoused by the Biodiversity Convention acknowledges that our relations with the rest of the living world are truly interactive, and that what we do to nature will in turn reflect on nature's ability to respond to our own needs. The Convention foresaw a careful balance in the management of the Earth's living wealth through conservation, sustainable use and equitable sharing of costs and benefits.

There could be no better time to launch a fresh assessment of the living world. This *World Atlas of Biodiversity* is published to coincide with the World Summit on Sustainable Development in Johannesburg, Republic of South Africa. The focus of the meeting is once again on sustainable development, but this time the emphasis is clearly on poverty alleviation. The message is clear: harmonized economic, social and environmental development will be but a dream while so many of the world's people have no choices and no opportunities to take a planned approach to their lives. What is the relevance of this book in the context of the Johannesburg message?

The reality is that this *World Atlas of Biodiversity* is of greater relevance now than at any time in the past. The world's living wealth remains the cornerstone of sustainable livelihoods and quality lifestyles in both the industrialized and developing worlds. Recognition of this fact is spreading, and the value of biodiversity in people's lives, socially, economically and environmentally, has never been more apparent than it is today.

This is not a textbook, it is a resource pack and a survival kit for the future. I hope that all who read it will find new insights into the significance of life on Earth to their own lives. And that they will take steps within their homes, communities and nations to utilize and enjoy living resources wisely, share the benefits and hold the capital in trust for future generations.

Introduction

OBJECTIVES

The past ten years have seen a remarkable growth in concern for wildlife and the environment, with an increased appreciation of the links between the state of ecosystems and the state of humankind. Many analysts have concluded that achieving sustainable and equitable human development will require, among other measures, taking a more effective approach to managing human impacts on the biosphere. This was reinforced by the 1992 United Nations Conference on Environment and Development (the Earth Summit), at which the Convention on Biological Diversity (CBD) was opened for signature. Many conservation and management initiatives worldwide have arisen from efforts to meet the objectives framed by the CBD text.

In principle, any kind of variation at any level of biological organization – encompassing genes, populations, species and communities – is biological diversity. The text of the CBD defines biological diversity as 'the variability among living organisms from all sources including, *inter alia*, terrestrial, marine and other aquatic ecosystems and the ecological complexes of which they are part; this includes diversity within species, between species and of ecosystems'. In practice, the term is often contracted to 'biodiversity', and used to refer collectively to all such variation: in effect, as a convenient shorthand for the total complex of life in some given area, or on the Earth as a whole.

In the present volume we aim to use the data now available to provide an overview of the current state of global biodiversity, using maps where helpful, and to ensure that this information is accessible to a wide readership. While biodiversity has many dimensions, attention is here focused on the diversity of living organisms and their populations, and on major aquatic and terrestrial ecosystem types. Far more space is given to the macro-scale organisms and landscape elements that may be subject to planning and management intervention than to microorganisms, despite the immense metabolic diversity of the latter, and their pivotal role in driving biosphere cycles.

STRUCTURE OF THIS BOOK

The eight chapters fall informally into four thematic sections. The first section opens with an outline of some fundamental aspects of material cycles and energy flow in the biosphere (Chapter 1). This is followed by a synopsis of the diversity of living organisms (Chapter 2) and of change in this diversity through geological time (Chapter 3). The second section (Chapter 4) is largely concerned with relationships between humankind and biodiversity, noting the increasing human impact on the environment from early modern humans onward, the use of biodiversity in human nutrition, and reviewing trends in recent time, focusing on depletion and extinction of species. The third section aims to characterize communities and biodiversity trends in the three basic biome types: terrestrial, marine and inland waters (Chapters 5, 6 and 7, respectively). Finally, Chapter 8 introduces some of the management and planning responses that have been implemented with a view to maintaining ecosystem health and putting human development on a sustainable foundation.

1 The biosphere

THE BIOSPHERE IS THE THIN AND IRREGULAR ENVELOPE around and including the Earth's surface that contains all living organisms and the elements they exchange with the non-living environment. Water makes up about two thirds of an average living cell, and organic molecules based on hydrogen, carbon, nitrogen and oxygen make up the remaining one third. These and other elements of living cells cycle repeatedly between the soil, sediment, air and water of the environment and the transient substance of living organisms.

The energy to maintain the structure of organisms enters the biosphere when sunlight is used by bacteria, algae and plants to produce organic molecules by photosynthesis, and all energy eventually leaves the biosphere again in the form of heat. Photosynthetic organisms themselves use a proportion of the organic material they synthesize; net primary production is the amount of energy-rich material left to sustain all other life on Earth.

Humans now appropriate a large proportion of global net primary production, and have caused planetary-scale perturbations in cycling of carbon, nitrogen and other elements.

THE LIVING PLANET

The defining characteristic of the planet Earth is that it supports life, and has done so for at least 70 percent of its history (see Chapter 3). The position of the Earth relative to the sun, its size and composition appear to be the main factors that have allowed life to develop here. Most importantly, these factors have combined to ensure the permanent presence of a large amount of liquid water on the planet's surface, and this is the fundamental prerequisite of life as we know it.

The space occupied by living organisms and the part of the planet that supports them is called the biosphere[1]. The non-living biosphere comprises the hydrosphere (the waters), the soil and upper part of the lithosphere (the solid matter that forms the rocky crust of the Earth), and the lower part of the atmosphere (the thin layer of gas coating the planet's surface). These domains interact in ways critical to the operation of the biosphere, and are linked in particular by the properties of water as a solvent and medium that fosters the chemical reactions basic to life.

While providing the conditions necessary for life, the structure and composition of the non-living parts of the biosphere have themselves been profoundly affected through time by living organisms. Most clearly, and from the human viewpoint most importantly, the presence of significant quantities of free oxygen in the atmosphere is entirely the product of oxygen-releasing photosynthesis by cyanobacteria starting more than 2 000 million years ago. The idea that living organisms do not merely influence conditions in the biosphere but in some way regulate them to maintain the conditions conducive to life has received considerable attention in recent years, chiefly in terms of the Gaia hypothesis[2] proposed in the 1970s.

The extent of the biosphere

At planetary scale, the biosphere can be pictured as a thin and irregular envelope around the Earth's surface, just a few kilometers deep on the globe's 6 371-kilometer radius. Because most living organisms depend directly or indirectly on sunlight, the regions reached by sunlight form the core of the

**Map 1.1
Physical geography of
the Earth**

The relative areas occupied
by dry land and by water,
and the general distribution
of areas of extreme height
or depth.

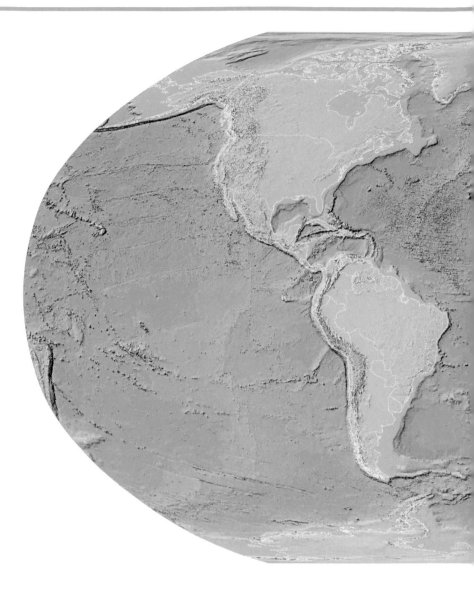

biosphere: i.e. the land surface, the top few millimeters of the soil, and the upper waters of lakes and the ocean through which sunlight can penetrate.

The biosphere is not homogenous, because actively metabolizing living organisms are sparse or absent where liquid water is absent, such as in the permanent ice at the poles and on the very highest mountain peaks, but abundant where conditions are favorable. Nor are its boundaries sharply defined, because bacterial spores and other dormant forms of life passively disperse virtually everywhere, from polar icecaps to several tens of kilometers above the surface of the Earth (approaching the upper limit of the stratosphere), and living microorganisms occur[3] within rocks more than 3 kilometers deep in the lithosphere.

The whole of the sea is theoretically capable of supporting active life and constitutes therefore the vast majority of the volume of the biosphere (Figure 1.1). Depending on water clarity, the sunlit (photic) zone may reach just a few centimeters to a few hundred meters in depth, but the marine biosphere is extended into regions of total darkness, down to more than 10 000 meters in the ocean depths, by organisms that subsist on the rain of organic debris falling from the upper waters. In addition, there are animal communities on the sea floor based on microorganisms deriving their energy from hydrogen sulfide emitted from hydrothermal vents. Overall, however, the amount of living material in most of the sea – that part of the open ocean below the upper hundred or so meters – is relatively low.

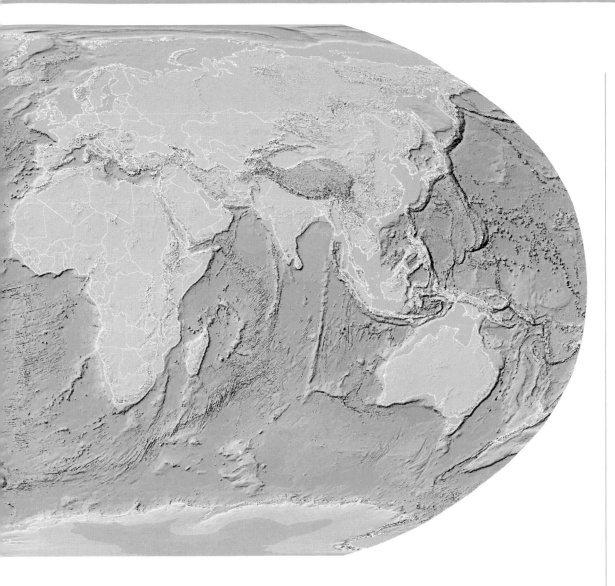

Figure 1.1
Hypsographic curve

The horizontal baseline in
this figure represents the
Earth's total surface area
of 510 million km².
The figure shows that 71%
of this surface is covered by
marine waters and 29% is
dry land. It also shows the
mean land elevation and
mean ocean depth, and the
amount of Earth's surface,
in percentage terms,
standing at any given
elevation or depth.

The atmosphere plays a vital role in the biosphere, not only in providing a source of essential gases, but also in buffering conditions at ground level, by regulating temperature and providing a shield against excessive ultraviolet radiation. Many organisms, from microscopic bacteria to bats and birds, spend part of their lives suspended in the atmosphere; however, no organism is known that passes its complete life cycle in the air, and living biomass per unit volume above the Earth's solid or liquid surface is extremely low.

Photosynthesis and the biosphere

Life on Earth is based essentially on the chemistry of water and carbon. Indeed, in biochemical terms, living organisms are simply elaborate systems of organic macro-

molecules dispersed in an aqueous medium. The average cell is about 70 percent water by weight; the remainder consists very largely of carbon-containing (organic) compounds composed mainly of the four elements hydrogen, carbon, nitrogen and oxygen. These com-

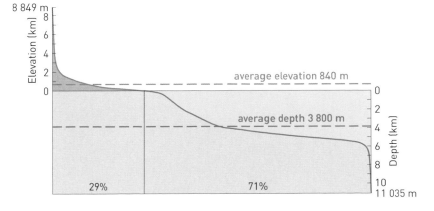

pounds include four major types of large organic molecule – proteins, carbohydrates, lipids and nucleic acids – and about 100 different small organic molecules. A number of other elements are required in much smaller, though still vital, quantities. These include phosphorus, sulfur, iron and magnesium. All these elements cycle through the biosphere in a variety of forms, both organic and inorganic, following complex and interlinked pathways many of which are yet to be fully elucidated. Except for some micro-organisms that use energy derived from inorganic chemicals, the engine that drives the organic part of this turnover is photo-synthesis – the capture by living tissues of energy from the sun.

Photosynthesis essentially involves the use of energy from sunlight to reduce carbon dioxide (CO_2) with a source of electrons (almost invariably hydrogen) to produce carbohydrates, water (H_2O) and, generally, a by-product from the hydrogen donor. In some bacteria the hydrogen donor is hydrogen gas, in others it is hydrogen sulfide; but, in cyanobacteria, algae and plants, water is the hydrogen donor and gaseous elemental oxygen (O_2) is the by-product. This is over-whelmingly the predominant and most impor-tant form of photosynthesis on the planet, and is described by the following equation:

$$2nH_2O + nCO_2 + light \rightarrow nH_2O + nCH_2O + 2nO$$

The initial products of photosynthesis in plants are simple sugars such as glucose ($C_6H_{12}O_6$). Larger carbohydrate molecules made from glucose include cellulose, the main component of plant cell walls and woody tissues, and starch, a key storage carbohydrate found in roots and tubers. Energy is needed to make the chemical bonds within these organic molecules, and energy is released when the bonds are broken down again. The controlled breakdown of these molecules within cells is the mechanism by which all cells obtain energy to do useful work.

All organisms can break down sugars very directly without the need for oxygen. Many bacteria that live in aerobic conditions use only this method. Virtually all eukaryotes (see Chapter 2) have evolved a more complex additional pathway that requires oxygen but yields much more energy. This latter pathway – aerobic respiration – essentially reverses the basic photosynthetic reaction shown above.

The major cycling process of the biosphere, therefore, consists of the photosynthetic fixing of carbon dioxide with water to produce organic compounds, in which energy is stored, and oxygen; this is followed by respiration of these compounds, in which the stored energy is released and carbon dioxide and water are produced. Photosynthesis therefore is not only responsible for the vast majority of organic production, but also for the maintenance of free oxygen in the atmosphere, without which aerobic organisms (the great majority of eukaryotic organisms, including humans) could not survive.

Although photosynthesis is the primary engine of the biosphere, in the sense that it injects energy into the system and creates basic organic molecules, production of the full range of organic molecules on which life depends requires additional elements. Of the four key elements, nitrogen is often the one in limited supply, but it is an essential com-ponent of nucleic acids and proteins. Although the atmosphere consists of 79 percent nitrogen, this inert gaseous form of the element cannot be used by plants or most other organisms until combined (fixed) with other elements. In the biosphere, atmos-pheric nitrogen is fixed by a range of bacteria, including cyanobacteria, some free-living soil

Energy from the sun drives photosynthesis, responsible for the vast majority of organic production.

bacteria, and most importantly by specialized bacteria that live symbiotically in the root nodules of leguminous plants (peas, beans, etc.) Some nitrogen is also fixed by lightning in electric storms and, in the modern world, industrially in the production of fertilizer. Fixed nitrogen is made available to plant roots through association with fungi (mychorrhizas) and as nitrogen-fixing organisms decay. From plant roots, it is transported to metabolizing plant cells. On death, these steps are reversed, and the fixed nitrogen may be immediately recycled or revert to elemental nitrogen.

PRODUCTIVITY AND THE CARBON CYCLE

About half of the solar energy reaching the upper atmosphere of the Earth is immediately reflected. Most of the remainder interacts with the atmosphere, ocean or land, where it evaporates water and heats air, so driving atmospheric and ocean circulation. Much less than 1 percent of the incoming energy is intercepted and absorbed by photosynthetic organisms. On land these photosynthesizers are overwhelmingly green plants, although cyanobacteria and algae are also present, the latter particularly in the symbiotic associations with fungi known as lichens. In aquatic habitats, particularly the sea, virtually all photosynthesis is carried out by cyanobacteria and algae, although green plants are also present in shallow coastal and inland waters.

Photosynthesizers fix carbon and therefore accumulate organic mass or biomass (often measured in dry form – that is, the once-living tissues of an organism with the water extracted). These organisms are the primary producers. The amount of carbon fixed is referred to as gross primary production and is typically measured in grams (g) of carbon (C) per unit of space (area or volume) per unit of time.

The photosynthetic producers also respire to meet their own energetic needs. Under some circumstances, respiration of photosynthesizers over a given period may balance their carbon fixation, so that there is no net accumulation of organic carbon. More normally, however, there is a surplus of fixation over respiration, so that organic matter is accumulated over time. This accumulation is referred to as net primary production (NPP). The accumulated matter is available to the vast suite of organisms of all sizes, including humans, that cannot synthesize their own organic compounds from an inorganic base or harness energy from inorganic sources. Such organisms are referred to as heterotrophs, while photosynthesizers and the few kinds of microorganisms that use other energy sources to synthesize organic compounds are referred to as autotrophs.

Organic products pass through the food chain, through processes such as predation.

Food webs

An organic product produced by a photosynthesizer may pass through a number of heterotrophs before finally being broken down again to its inorganic constituents. Conventionally this can be viewed as a food chain. At macroscopic level, a green plant may be eaten by a herbivore – a grasshopper, say – which is eaten by a lizard, which is itself eaten by a hawk, which dies and is disassembled and partially consumed by animal scavengers, with the remainder decomposed by bacteria and fungi.

In reality, this is an enormous oversimplification. The plant will almost certainly have a complex network of symbiotic fungi associated with its roots, which make use of some of the gross production of the plant but which also provide it with some essential nutrients. The plant itself may shed leaves which are directly broken down by other fungi, protoctists such as slime molds,

**Map 1.2
Primary production in the biosphere**

Global spatial variation in annual net primary production (NPP), in g C per m² per year, calculated from an integrated model of production based on satellite indices of absorbed solar radiation.

Source: Map created from data supplied by Chris Field and George Merchant, Department of Global Ecology, Carnegie Institution of Washington. See http://jasper.stanford.edu/chrisweb/flab/flab.html, and Field et al.[4].

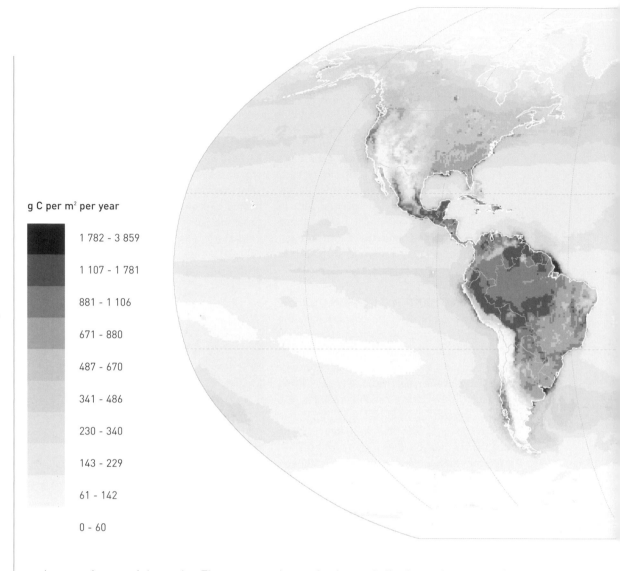

g C per m² per year

	1 782 – 3 859
	1 107 – 1 781
	881 – 1 106
	671 – 880
	487 – 670
	341 – 486
	230 – 340
	143 – 229
	61 – 142
	0 – 60

and many forms of bacteria. The grasshopper is likely to be parasitized by a host of smaller organisms, some of which are themselves in turn parasitized. It will also support a host of benign microorganisms in its intestine that are themselves constantly growing and reproducing. The lizard may die and decompose and the hawk may eat the grasshopper directly. The overall pattern of feeding relationships thus forms a web of immense complexity in any but the simplest ecosystems.

Each organism in the food web respires, releasing energy which is eventually dissipated in the form of heat, carbon dioxide and water. At each stage, therefore, some carbon is returned to the inorganic part of the carbon cycle. In addition, all living organisms produce waste products, some of which are

incompletely metabolized organic compounds. Heterotrophic organisms are also not completely efficient in their appropriation of the organic material they consume, so that some proportion of this is excreted as waste product. These organic wastes are theoretically available to other organisms in the food web. The assimilation efficiency of heterotrophic organisms may be anything from 20 percent (in the case of some terrestrial herbivores) to 90 percent (in the case of some carnivores), with the remainder excreted.

Of the amount assimilated, a high proportion is expended as respiration, with the remainder available to add biomass, i.e. to enable the organism to grow and reproduce. The proportion available to add biomass is dependent on the organisms involved as well

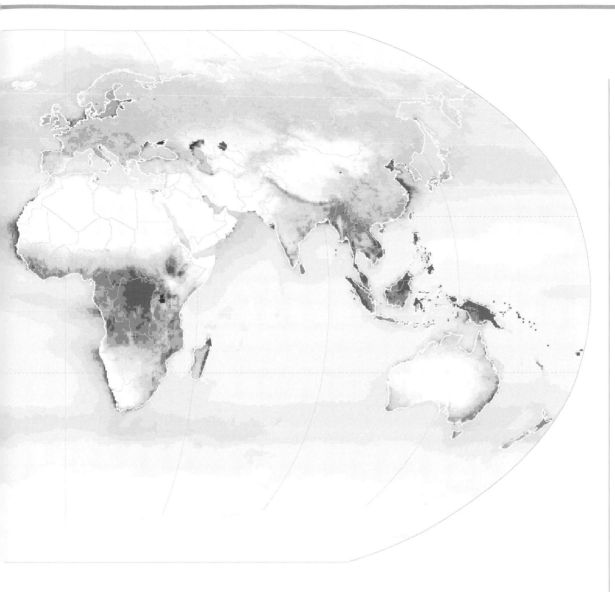

as a range of other factors. It can be as low as 10 percent or less and as high as 50 percent or more. This proportion is a measure of the net growth efficiency of the organism.

For purposes of ecological analysis, particularly involving productivity estimates, the gross growth efficiency is the most commonly used measure. This is simply the product of the assimilation efficiency and the net growth efficiency of a particular heterotroph and is a measure of the proportion of food consumed by that organism that, after excretion and respiration, is ultimately available for its growth. As a very coarse generalization, a value of 10 percent is widely used, although it is acknowledged that in terrestrial herbivores the figure is likely to be lower and in planktonic communities and terrestrial

carnivores it is likely to be higher. Using the figure of 10 percent in the example above, for every kilo of plant matter eaten by the grasshopper, the latter would add 10 grams to its body weight. When the grasshopper was eaten by the lizard, this would add 1 gram to the lizard's body weight, and when the lizard was eaten by the hawk, this would add 0.1 grams to the hawk's weight. This explains why, at the species level, so-called higher predators are rarer than herbivores and in any given area have a lower biomass, while the biomass of primary producers exceeds that of all heterotrophs combined.

Measures of local and global productivity

Primary productivity varies enormously, both spatially and temporally, at all scales. Most obviously, under natural conditions productivity

Table 1.1
Global annual net primary production

Note: This represents one among several attempts to estimate global production; see text for further details. Ocean data averaged 1978-83, land 1982-90, units in petagrams (1 Pg = 10^{15} g).

Source: Adapted from Field *et al* [4].

Biosphere units	NPP (x 10^{15} g C)
Ocean	48.5
Terrestrial	56.4
Tropical rainforest	17.8
Deciduous broadleaf forest	1.5
Broadleaf and needleleaf forest	3.1
Evergreen needleleaf forest	3.1
Deciduous needleleaf forest	1.4
Savannah	16.8
Perennial grassland	2.4
Broadleaf shrubs with bare soil	1.0
Tundra	0.8
Desert	0.5
Cultivation	8.0

effectively ceases every night. Seasonal variations in most parts of the world are also marked. Productivity is, however, difficult to measure, so that estimates at all scales are subject to considerable uncertainty.

On land one major source of uncertainty is below ground productivity: in natural ecosystems, less than 20 percent of plant production is typically consumed by herbivores. The remainder enters the soil system, either through the plant roots or as leaf litter. Measuring this portion of terrestrial productivity – probably over 80 percent of the total – is particularly problematic. In the past there has been a marked tendency to underestimate it. Similarly, it had long been assumed that the nutrient-poor waters of the open ocean were extremely unproductive, but it is now known that large populations of extremely small photosynthesizing unicellular organisms – the so-called picoplankton – form the basis of a surprisingly productive ecosystem in these regions.

One approach to estimating global productivity is based on measurements in particular ecosystems and extrapolation from these using estimates of the global extent of those ecosystems. This suggests that net primary production on land is of the order of 45-65 x 10^{15} g C per year (that is, 45-65 petagrams or thousand million metric tons), and at sea is around 51 x 10^{15} g C per year. This

gives a global estimate of annual net primary production in the order of 100 x 10^{15} g C. Gross primary production is estimated to be about twice this.

However, global measures using a somewhat different technique, involving assessments of relative concentration of oxygen isotopes, have indicated that annual gross primary production on land may be greater than 180 x 10^{15} g C, while that in the sea is around 140 x 10^{15} g C. This would give a global figure of over 320 x 10^{15} g C, implying global net primary production of more than 160 x 10^{15} g C. This global figure is 60 percent higher than those global estimates based on summation of individual ecosystem measurements.

Another approach[4] has used a comprehensive set of satellite indices of photosynthetic activity in the ocean and on land, combined with a model of primary production, to generate a more integrated global estimate of NPP. There is considerable spatial and temporal variability, but on average annual NPP on land amounts to around 56 x 10^{15} g C, while that in the sea is around 48 x 10^{15} g C (see Table 1.1).

The carbon cycle and global biomass estimates

Carbon fixation by photosynthesis forms one crucial step in the carbon cycle[5]. Once fixed, the carbon will remain for a greater or lesser period within living tissues, that is, form part of the planet's biomass. As all cells and individual organisms have a limited lifespan, eventually the carbon will rejoin the non-living carbon pool (see Table 1.2). It may, however, remain as organic carbon compounds for a far greater period than it remained part of the biomass; e.g. the woody tissues of Paleozoic forests were formed several hundred million years ago but remain, fossilized, as a source of coal and oil. Eventually all carbon will recycle through the inorganic pool, as carbon dioxide in the atmosphere, in the soil or dissolved in the sea, or as inorganic carbon compounds (carbonates) in rocks or dissolved in the sea.

The great majority of carbon at any one time lies within the lithosphere, around

80 percent as carbonate and the remainder as organic carbon compounds. A large proportion of this carbon is effectively inaccessible to the biosphere in the short term but itself participates in the overall carbon cycle, mainly through tectonic activity. As seafloor crust is gradually consumed along subduction plate margins, carbon sediments are taken into the Earth's mantle and later released as carbon dioxide by volcanic and hydrothermal activity. Although the loss of carbon to the mantle is extremely slow, without volcanic activity tectonic processes would eventually exhaust the available carbon pool.

Most of the carbon incorporated in living organisms is associated with green plants, and almost all of this is in the form of cellulose-rich woody tissues. The total terrestrial animal biomass appears to be insignificant in comparison, probably more than two orders of magnitude less. The cyanobacteria and algae that are the primary producers in the ocean are estimated[4] to amount to only 0.2 percent of the biomass of all primary producers globally, although they generate in the region of half the global NPP (they cycle organic material much more rapidly than land plants, which also sequester large amounts in woody tissues).

The role of diversity in the biosphere

The biota play the pivotal rile in the major biogeochemical cycles, with different groups of organisms (e.g. nitrogen-fixing bacteria and photosynthesizing plants) mediating different processes. Simplistically, therefore, at least some biological diversity is necessary to maintain the biosphere as it currently operates, and microbial diversity is particularly fundamental. However, just how much diversity is needed, and how much redundancy, if any, is built into the system, remains unclear. Indeed, the relationship between biological diversity and a whole suite of ecological measures including stability, resilience and productivity remains incompletely understood[6], although there is an increasing volume of theoretical and experimental work[7,8] indicating that diversity may play an important role in long-term ecosystem functioning.

Human influence on the biosphere

There are believed to be more than 6 billion humans on the planet at present. A significant proportion of global net primary production is diverted to support this population. Using a relatively conservative definition of appropriation that takes account of the global agricultural and natural production used by humans[9], that proportion has recently been estimated at around one third of the terrestrial global total[10]. This result is very similar to that obtained in an earlier study[11], but because of uncertainties in almost all the figures on which it and the earlier estimate were based, the margin for error is very high. Less conservative definitions also attempt to take into account the loss of overall net primary production that may result from human actions (e.g. through severe land degradation or accumulation of waste) but which is not directly used by humans. This is done by estimating the net primary production of what is thought would have been the prevailing biomes in the absence of humans. A detailed analysis in one European country using this approach[12,13] suggested that around 50 percent of NPP there was appropriated by humans.

Human efforts to appropriate the products of photosynthesis and other actions associated with the development of complex societies have had enormous impacts on natural biomes and biogeochemical cycles. Over large areas of the Earth's surface, humans have replaced complex and species-rich natural habitats with simplified modified habitats specialized for agricultural production.

Total carbon content on Earth	10^{23} g
Amount buried in sedimentary rocks: organic	1.6×10^{22} g
carbonate	6.5×10^{22} g
Active carbon pool near surface:	$40\,000 \times 10^{15}$ g
of which	
Dissolved inorganic carbon in sea	$38\,000 \times 10^{15}$ g
Atmospheric CO_2	750×10^{15} g
Organic carbon in soil	$1\,500 \times 10^{15}$ g
Biomass on land	560×10^{15} g
Biomass in the sea	$5\text{-}10 \times 10^{15}$ g

Table 1.2
Estimated global carbon budget and biomass totals

Note: Biomass figure on land refers to plants.

Source: Adapted from Schlesinger[5].

Clearance by fire, burning of fuelwood and charcoal, soil cultivation, and fossil fuel use all increase movement of organic carbon into the atmosphere. Global cycling of nitrogen, phosphorus and sulfur has also been perturbed. Application of industrially produced fertilizer has doubled the rate at which nitrogen in fixed form enters the terrestrial cycle, and industrial processes have doubled movement of sulfur from the lithosphere into the atmosphere. Increasing levels of nitrogen and phosphorus lead to shifts in nutrient availability which can cause radical change in natural communities, and sulfur is a major contributor to acidification phenomena.

That human activities may have profound local impacts on natural biota is indisputable. What is now becoming clear is that these activities may also have planet-wide impacts, particularly on climate. Analysis of atmospheric samples trapped in polar ice cores indicates that present-day concentrations of atmospheric carbon dioxide and methane (CH_4) are unprecedented in the past 420 000 years[14]. Although their absolute concentration in the atmosphere is low (CO_2 around 360 and CH_4 at 1.7 parts per million by volume) these two gases play an extremely important role in determining atmospheric temperature. It is indisputable that the rise in these gases is a result of human activities, so clearly these activities are having some impact on global climate. The extent of this impact, particularly when compared with natural climatic fluctuations, remains a subject of great controversy.

REFERENCES

1 Hutchinson, G.E. 1970. The biosphere. *Scientific American* 233(3): 45-53.
2 Lenton, T.M. 1998. Gaia and natural selection. *Nature* 394: 439-447.
3 Parkes, R.J. 1999. Oiling the wheels of controversy. *Nature* 401: 644.
4 Field, C.B. *et al.* 1998. Primary production of the biosphere: Integrating terrestrial and oceanic components. *Science* 282: 237-240.
5 Schlesinger, W.H. 1997. *Biogeochemistry: An analysis of global change.* 2nd edition. Academic Press, San Diego and London.
6 Schulze, E.D. and Mooney, H.A. (eds) 1993. *Biodiversity and ecosystem function.* Springer, Berlin.
7 Grime, J.P. 1997. Biodiversity and ecosystem function: The debate deepens. *Science* 277: 1260-1261.
8 McGrady-Steed, J., Harris, P.M. and Morin, P.J. 1997. Biodiversity regulates ecosystem predictability. *Nature* 390: 162-165.
9 Wright, D.H. 1990. Human impacts on energy flow through natural ecosystems, and implications for species endangerment. *Ambio* 19(4): 189-194.
10 Rojstaczer, S., Sterling, S.M. and Moore, N.J. 2001. Human appropriation of photosynthetic products. *Science* 294: 2549-2552.
11 Vitousek, P.M. *et al.* 1986. Human appropriation of the products of photosynthesis. *BioScience* 36: 368-373.
12 Haberl, H. *et al.* 1999. Colonizing landscapes: Human appropriation of net primary production and its influence on standing crop and biomass turnover in Austria. *IFF-Social Ecology Papers* No. 57. Institute for Interdisciplinary Research of Austrian Universities, Vienna. Also see: http://www.cloc.org/conference/presentations/in4/npp-abstract.htm
13 Schulz, N. 1999. Effects of human land-use on the amount of biologically available biomass-energy throughout the landscape. An empirical case study about Austria. Conference paper available online in pdf at: http://www.univie.ac.at/iffsocec/conference99/htmlfiles/blue.html (accessed January 2002).
14 Petit, J.R. *et al.* 1999. Climate and atmospheric history of the past 420,000 years from the Vostok ice core, Antarctica. *Nature* 399: 429-436.

2 The diversity of organisms

SYSTEMATICS AIMS TO DEFINE SPECIES and sort them into a hierarchy of named groups congruent with the branching pattern of evolution. There is no single operational definition of what a species is, and taxonomy at all levels is subject to change as a result of new methods and data, but species diversity of better known organisms can often be assessed with useful accuracy. Globally, about 1.75 million species have been described and named, but the total including undescribed species might be up to ten times greater.

All species known are assigned on the basis of shared patterns of form and function to one of about 100 major groups (phyla). There are marked differences between phyla in overall morphology, physiology and mode of life. These differences imply the existence of major genetic diversity, and contribute directly to structural, trophic and other dimensions of diversity within ecosystems.

The phyla of living organisms fall into three primary lineages: the true bacteria, the archeans and other organisms. The first two are prokaryotes, the remainder (protoctists, animals, fungi and plants) are eukaryotes.

EVOLUTION AND SYSTEMATICS

A basic principle in evolution is that just as new individuals arise from ancestral individuals, so new populations arise from existing populations, and ultimately new species arise from existing species. The chief mechanism by which this occurs is believed to be reproductive isolation. For example, physiographic or climatic change may divide an existing single population into two or more separate populations, or individuals may colonize a new and geographically separate habitat. The genetic makeup of these isolated populations will diverge, mainly through natural selection acting on them, but probably also through other mechanisms. This genetic divergence may be manifested in various ways, physically, physiologically and behaviorally. If the period of isolation continues for long enough, the populations will diverge enough that they can be regarded as separate species. Each one of these species may itself in turn give rise to other species in due course, although some will die out without giving rise to any progeny. The surviving descendant species may themselves give rise to new species and so on through the long march of evolutionary time.

The result of this process is a branching tree-like structure – a phylogenetic tree – rooted in the distant past. In archeans and bacteria, but probably rarely in other groups, elements of two separate branches may combine into one, giving a reticulate rather than exclusively tree-like pattern. If it is assumed that all life on Earth had a common origin in the distant past (see Chapter 3), then all existing organisms form the topmost extremities of a vast and unimaginably complex single phylogenetic tree.

Systematics has two roles[1-3]. The first is to name the immense variety of different sorts of organisms that exist. The second is to try to elucidate the relationships between all these different organisms, that is to develop hypotheses of where they are positioned in the phylogenetic tree. Systematics provides the basic framework for the whole of biology, and is a fundamental discipline for biodiversity studies. Taxonomy is the subset of systematics that deals in particular with the definition, naming and classification of species and, in some cases, subspecific populations. The traditional output consists of species descriptions or revisions, or lists of species in a given group (possibly with hypotheses of their evolutionary relationships), or checklists of all the species in some higher taxon in a site or region.

Because the actual evolutionary events that generated the overall phylogenetic tree are lost in history, the relationships between organisms have to be inferred from the evidence to hand. The most important forms of evidence are the characters of organisms, both living and fossil. These characters may be genetic, morphological, biochemical or behavioral. Methods to reconstruct phylogeny generally use two working assumptions: that species sharing a large number of characters are likely to be related, and that species sharing some uniquely complex and specialized feature are likely to be more closely related than species not possessing this feature.

Groups and names

In the current system for naming species (nomenclature), each has a two-part scientific name (binomial), based on Latin or latinized Greek, comprised of the genus name (e.g. *Vipera*) and specific epithet (e.g. *berus*). The author of the specific epithet may be given after the binomial. By convention, both parts of the binomial are italicized when printed, and the author name is shown in parentheses if the species was originally put into a different genus.

Similar species (e.g. the European adder *Vipera berus* and asp viper *Vipera aspis*) are grouped together in the same genus (*Vipera*), similar genera in families (Viperidae), families in orders (Serpentes), orders in classes (Reptilia) and classes in phyla (Craniata or Vertebrata) up to the highest level, the kingdom. An organism can only be assigned to a single species, genus, family, etc., and the taxonomic system forms a hierarchy with each lower taxonomic level being nested entirely within each increasingly inclusive higher level. Although the traditional Linnaean hierarchy includes only the seven obligatory categories above, intermediate categories are some-

times used, and a further more inclusive category – the domain.

Groups such as mammals or snakes, that because of shared unique characters are considered to contain all the living descendants of a common ancestor, are called monophyletic groups. Groups with no shared unique characters but only unspecialized or non-unique characters in common, are termed paraphyletic groups. These typically are groups of related species left over after one or more clearly monophyletic lineages that evolved within the group have been recognized and named. Examples are fishes (the craniate vertebrates without the unique features of tetrapods), reptiles (the amniote vertebrates without the unique features of birds or mammals), and the entire kingdom Protoctista. Groups defined on characters that appear to have evolved more than once are called polyphyletic groups. Although the goal of most systematists is to recognize only monophyletic groups in order to be able to retrieve evolutionary relationships from a classification, many paraphyletic groups, such as the three named above, continue to be very widely used in practice.

Given a classification congruent with phylogeny, the taxonomic hierarchy becomes a device to store information on hypotheses about evolutionary history. Because hypotheses about relationships are always subject to revision as new information becomes available, or existing data are reinterpreted, the taxonomy of species is not fixed. The fact that species names are liable to change can cause confusion, for example when, as is commonly the case, conservation legislation uses a name no longer current.

Species concepts and diversity assessment

Despite the importance of 'the species', there is no unequivocal and operational definition of what a species is or how species can be recognized[4, 5]. There are several definitions, differing in mainly theoretical and often subtle ways, but much of the existing body of systematic knowledge has been built up around elements of 'the biological species concept'. This defines a

Relationships between organisms have to be inferred from their genetic, morphological, biochemical or behavioral characteristics.

species as a population of organisms that actually or potentially interbreed in nature, and that are reproductively isolated by morphological, behavioral or genetic means from other such groups. It is, however, applicable only to organisms where sexual reproduction is the norm.

In most real cases, especially where all the systematist has to hand is a collection of preserved specimens, whether criteria concerning reproductive isolation are met or not cannot in fact be tested, but an experienced worker will come to hold some particular level of morphological or other difference as deserving of species status. Where there is good evidence from fieldwork and geographic data attached to specimens that two somewhat similar populations occur in the same locality (i.e. are sympatric) but maintain their differences, they may be presumed not to interbreed, and will be treated and named as species.

Different taxonomists will often use different criteria for the same group of organisms, so that one specialist may regard a group of fundamentally similar populations as a single species, whereas another will treat each smaller distinct population as a separate species. In the latter case, often associated with the 'phylogenetic species concept', the assumption is that each distinct lineage once established on its own evolutionary course is *de facto* a separate species. Many populations that were formerly described at subspecies level (i.e. somewhat distinct, but regarded as part of a single polytypic species complex) have subsequently been elevated to full species in this way, particularly if geographically isolated. However, subspecies have often been named on the basis of just a few superficial features, not representative of the overall pattern of variation within species, and the formal subspecies category is now much less used than in the past (particularly among vertebrates).

Different taxonomic characters and criteria are used to classify species in different groups of organisms. Those used, for example, to define species of fungi are very different from those used to define species of bird, and their application

Different taxonomic characters and criteria are used to classify species in different groups of organisms.

demands taxonomists with relevant specialized knowledge. Some organisms are difficult or logically impossible to accommodate in any species concept involving criteria that assume all individuals reproduce by outbreeding, that is by sexual reproduction with another individual. Many higher plants are self-fertile or make use of various forms of asexual reproduction; in the latter case all individuals in a lineage are identical clones (assuming a very low or non-existent mutation rate). Among the prokaryotes, bacteria can readily receive extraneous genes through direct entry of genetic material from the fluid environment, or from viruses or other bacteria[6], and this 'horizontal' transfer is independent of reproduction. Different kinds of bacteria have traditionally been defined by the cytological or biochemical properties of colonies in culture but, more recently, comparison of DNA and RNA sequences from sample collections has been used to distinguish one lineage from another. In either case, the biological species concept, developed with reference to sexually reproducing, outbreeding animals and plants, is not appropriate. Such factors mean that 'the species' cannot provide a standard unit in which to evaluate all biodiversity because it does not define a single level in the hierarchy and its significance is not equivalent across all groups of organisms; it can, however, serve this purpose for most plants and animals.

The use of different species concepts by different systematists can make a very large difference to the number of species recognized in a group and to complications in nomenclature (these will both affect the outcome of biodiversity inventory, an

important application of systematics). Usually, however, only a small or very small number of taxonomists is working on any one group of organisms at any time, so that while the species level taxonomy of organisms is in a continual state of flux, it is generally not subject to radical and wholesale change. Exceptions arise where new techniques, chiefly molecular information and DNA data in particular, are applied to previously neglected groups or ones that have not been revised for many years. Nevertheless, the key point appears

to be that units corresponding more or less closely to the biologists' model of 'the species' do indeed exist in nature and, in animals in particular, they define themselves to an extent through their reproductive behavior. It has thus been possible to reach some measure of consensus on species-level classification of well-studied groups of larger organisms such as terrestrial vertebrates, and to estimate and compare the number and kinds of species in different sites, areas or countries.

NUMBERS OF LIVING SPECIES

From a practical point of view it is more important to know how many species, and which ones, occur in some spatially restricted area, such as a protected area or a country, than in the world overall. However, proper evaluation of each local situation requires some knowledge of the wider context and, where the goal is maintenance of global biodiversity in the face of increased risk, it is clearly important to have a sound appreciation of the full baseline range of diversity. This requires both an estimate of the number of known valid species, and an estimate of the number of unknown species, neither of which is readily available.

The number of known species can be estimated by collating data from systematists and the taxonomic literature. Although many species names are synonyms (i.e. different names inadvertently applied to the same species), this can be done with reasonable precision for more familiar and well-reviewed groups of species. Recent calculations of this kind suggest that around 1.75 million of the probably far larger number that exist have been discovered, collected and later named by systematists[7,8].

Any estimate of how many undiscovered and hence undescribed species are likely to exist in any given group, and in the biosphere overall, involves substantial uncertainty[7,8]. In taxonomic groups where individuals are readily visible, popular or economically important, and subject to sustained systematic attention, e.g. mammals and birds, the number of known species is certainly very close to the total number of species in the

Box 2.1 New species discoveries

The discovery of entirely new species of mammals and birds is rare, and often involves small and obscure forms. Remarkably, two large mammals previously unknown to science were discovered in one small area, the Vu Quang Nature Reserve in Truong Son, Viet Nam (along with many new species in other groups). The Vu Quang ox or soala *Pseudoryx nghetinensis* was described in 1993, followed a couple of years later by a giant muntjac deer *Megamuntiacus vuquangensis* from the same area. The world's smallest muntjac deer, the Truong Son muntjac *Muntiacus truongsonensis*, was recently found in another part of the same region in Viet Nam. The soala is of particular interest because it does not appear to fit neatly in any of the main bovid groups as currently recognized. It is now known also to occur in adjacent parts of Laos. However, claims that another new bovid species existed in Southeast Asia, described as *Pseudonovibos spiralis*, were premature because the distinctive horns described as new were later shown to be domestic cattle horns, apparently reshaped artificially.

group that exist. On average around 25 new species of mammals and and five of birds have been described annually in recent years[8]. Changing systematic opinion on which populations should be regarded as separate species and which should not, rather than completely new discoveries, is the major source of change in the number of named species in such groups.

The converse applies to groups whose individuals are small, difficult to collect, obscure and of no popular interest, e.g. many groups of invertebrate animals. Frequently there are so few systematists actively working on a group that the number of named species appears to be limited mainly by the rate at which collected specimens waiting on museum shelves can be studied and described, and changing opinion on which populations are separate species is insignificant. In some cases, where new sampling and collection methods have been used, unexpectedly large numbers of new species have been found (e.g. tropical forest canopy insects and marine sediment nematodes). If findings from such local work are extrapolated to global level the total number of species calculated to exist is many orders of magnitude greater than the number actually known. Some estimates suggest that most undescribed terrestrial forms are likely to be tropical forest beetles, but new molecular techniques are revealing unsuspected diversity among microorganisms[9].

Although the goal of systematics is to recognize and name species, and to maintain an ordered body of information on names and associated biological data, there is no master catalog of all known species. Developing such a resource has only become feasible with advances in information technology during the past ten years. However, while many systematic data, in the form of checklists and museum catalogs, are now available in digital form over the Internet, and more will become so, a harmonized catalog in this format of all known species remains a distant prospect.

Recent estimates of the numbers of known and possibly existing species in the world biota are given in Table 2.1. These are mostly large numbers, and the fossil record suggests that overall diversity has been increasing for some 600 million years up to the very recent past, but the numbers themselves are of little significance except in a wider context. Currently, much concern is focused on species numbers in relation to anthropogenic environmental change.

Box 2.2 Improving taxonomic knowledge and capacity

The Convention on Biological Diversity decided in 1998 to establish a Global Taxonomic Initiative (GTI)[10], in recognition of a major 'taxonomic impediment' to effective biodiversity management. The objective is to improve decision-making for conservation, sustainable use and benefit-sharing by increasing taxonomic knowledge and the number of trained taxonomists and curators. Similarly, the Global Biodiversity Information Facility (GBIF)[11] aims to develop an interoperable network of biodiversity databases and information technology tools that will enable users to access the world's stores of biodiversity information. In the same field, Species 2000[12] (a global network based in the UK and Japan), and the Integrated Taxonomic Information System (ITIS)[13] in North America have joined forces to create a unified Catalogue of Life, planned to cover all known species of living organisms. Basic reference data on 250 000 species had been collated by mid-2001, and the plan is to reach 500 000 by 2003.

DIVERSITY AT HIGHER LEVELS

Until van Leeuwenhoek observed microorganisms through a primitive microscope in the late 17th century, humans had been aware only of organisms visible to the naked eye (macroscopic) and regarded all living things as either plants or animals. At the end of the 19th century, with improved cytological techniques and new views on evolution, a third kingdom of organisms (Protista) was recognized for bacteria and other unicellular organisms. Around this time it became widely accepted that the cell was the fundamental unit of organization of all living organisms. Subsequent work, in the mid-20th century, recognized a basic distinction between two kinds of cellular organization – prokaryotic and eukaryotic. In prokaryotic organisms, the genetic material is free within the cell. In eukaryotes, the genetic material is linked to proteins and organized into chromosomes that are packed within a membrane-bounded cell nucleus. There are several other profound differences. In eukaryotes the enzymes needed to extract energy from organic molecules are organized into discrete membrane-bounded organelles (mitochondria) within the cell, and in the eukaryotes that photosynthesize (plants, some protoctists) the pigments and enzymes needed to fix solar energy are also in discrete organelles (chloroplasts) within the cell. On the basis of this, it became accepted that the major taxonomic divide in organisms was between the prokaryotes, containing only the bacteria, and eukaryotes, which included all other organisms, including many unicellular forms included in the original kingdom Protista.

Systematics has traditionally generated information on similarity and hypotheses of relationship on the basis of characters restricted to particular sectors of the phylogenetic tree, but molecular sequencing has broadened the range of useful evidence, with the potential for radically dissimilar groups to be compared and their phylogeny estimated. Analysis of small subunit ribosomal RNA (SSU rRNA) – a molecule that is universal, functionally constant (central to protein manufacture in all cells) and very highly conserved over time – has proved especially informative. This work has revealed that there are two kinds of prokaryote: the true bacteria and the archeans. The archeans were first known only by 'extremophiles', i.e. forms living under exceptional conditions of high temperature or salt concentration, but representatives are now known to be widespread alongside bacteria in less extreme habitats. At biochemical level, the bacteria and archeans are as different from each other as from eukaryotes, leading to the conclusion that all organisms can be assigned to three basic

Table 2.1
Estimated numbers of described species, and possible global total

Domain	Eukaryote kingdoms	No. of described species	Estimated total
Archaea		175	?
Bacteria		10 000	?
Eukarya			
	Animalia		1 320 000
	Craniata (vertebrates), total	52 500	55 000
	Mammals[23]	4 630	
	Birds[24]	9 750	
	Reptiles[25]	8 002	
	Amphibians[26]	4 950	
	Fishes[27]	25 000	
	Mandibulata (insects and myriapods)	963 000	8 000 000
	Chelicerata (arachnids, etc.)	75 000	750 000
	Mollusca	70 000	200 000
	Crustacea	40 000	150 000
	Nematoda	25 000	400 000
	Fungi	72 000	1 500 000
	Plantae	270 000	320 000
	Protoctista	80 000	600 000
TOTAL		**1 750 000**	**14 000 000**

Notes: This table presents recent estimates of the number of species of living organisms in the high-level groups recognized, and in some selected groups within them. Vertebrate classes are distinguished because of the general interest in these groups. The described species column refers to species named by taxonomists. Most groups lack a formal list of species. All estimates are approximations. They are inevitably inaccurate because new species will have been described since publication of any checklist and more are continually being described, and other names turn out to be redundant synonyms. In general, the diversity of microorganisms, small-sized species, and those from habitats difficult to access, are likely to be seriously underestimated. Among Archaea and Bacteria the figures of 175 and 10 000 are very rough estimates of 'species' defined on features shown in culture[22]; there appears to be no sound estimate of the total amount of prokaryote diversity. The estimated total column includes provisional working estimates of the number of described species plus the number of unknown species; the total figure is highly imprecise. Only a small selection of animal phyla is shown, but the figure for Animalia applies to all. Figures in the total row are for all species in all domains.

Source: Data mainly from United Nations Environment Programme[7] and Hammond[8]; vertebrates from individual sources indicated.

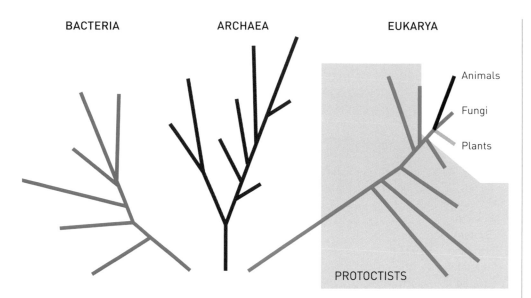

BACTERIA   ARCHAEA   EUKARYA

Animals

Fungi

Plants

PROTOCTISTS

**Figure 2.1
The phylogenetic tree**

Note: This diagram represents in highly simplified form the distance between the three domains of organisms and the general branching pattern within them; because of conflicting interpretations the root and branching sequence of the three domains are not represented.

Source: Adapted from Woese[15].

forms of life, or domains: Archaea, Bacteria and Eukarya, rather than two (prokaryotes and eukaryotes). The distance between lineages, in terms of amount of change in rRNA sequence, and their branching sequence in evolution, have been represented as a 'universal phylogenetic tree'[14-17].

Where the root of the tree is located, and its basal branching pattern, i.e. which two of the three domains are more closely related than either is to the third, remains open to discussion[18, 19]. It has been argued[20] that this deep branching took place in very early evolution, possibly before modern cell types had been established, when genes were widely spread by horizontal transfer (as persists to a lesser extent among extant prokaryotes) and in a sense, shared communally.

Reticulate evolution would have been common, with distinct lineages of organisms emerging only when the evolving cell became sufficiently integrated that horizontal gene transfer was reduced and vertical transfer down closed ancestor-descendent series became the norm. It has also been persuasively argued that the Eukarya arose through the permanent symbiotic fusion of a number of different prokaryotic organisms in one cell, with organelles such as mitochondria and chloroplasts representing the vestiges of different lineages of formerly independent prokaryotes[6, 18].

Within the Eukarya, the fungi, animals and plants form insignificant clusters overshadowed by the morphological and physiological diversity of the protoctists (protists). Protoctista is a name of convenience for the enormous and very diverse collection of all the small-to-microscopic eukaryotes that lack the distinguishing features of fungi, animals or plants. Further changes of perspective are expected as research on protoctist systematics continues. For example, a recent tendency[21] is to recognize the Chromista as another major group distinct from remaining eukaryotes. The chromists are aquatic species distinguished by structural and biochemical characters, and include organisms with the largest linear dimensions known (kelp), as well as microscopic but ecologically important organisms (e.g. diatoms, downy mildews).

The only known organisms that are not cells, or assemblages of cells, are viruses. They exist on the very boundary of most definitions of life. Consisting only of nucleic acids and protein, they are much smaller than the smallest bacteria, they can only replicate inside other living cells, and they are totally inert outside other cells, when they can survive for years in a crystallized state. Each type of virus may be more closely related to the organism in which it grows than to other viruses[6]. They are not discussed elsewhere in this book.

Some characters of the high level group-

**Table 2.2
Key features of the major
groups of living organisms**

Note: The four groups within
the Eukarya are commonly
regarded as kingdoms, the
highest formal category of
the Linnaean system of
taxonomy. The Archaea and
Bacteria have been treated as
a single group, or as separate
kingdoms, but a more recent
tendency is to treat Archaea,
Bacteria and Eukarya as
three separate domains in
recognition of the deep
phyletic divergence between
them, and fundamental
differences in RNA
composition.

Source: Principally Margulis and
Schwartz[6], also University of California[16]
and University of Arizona[17].

Archaea
Prokaryotic. Composition of the cell wall and of lipids in cell membranes differ from those in
Bacteria. Distinctive SSU rRNA, more similar in some respects to Eukarya than to Bacteria.
Reproduce asexually by cell splitting, or produce genetic recombinants without any fusion of cells by
accepting genes from other bacteria, or from the fluid medium, or through viruses, independent of
cell reproduction. Flourish in habitats with radical extremes of temperature or salinity that are
unavailable to other organisms (apart from bacteria in some cases), but also occur in other
environments.

Bacteria
Prokaryotic. Reproduce asexually by cell splitting, or produce genetic recombinants without any fusion
of cells by accepting genes from other bacteria, or from the fluid medium, or through viruses,
independent of cell reproduction. Metabolically uniquely versatile; key mediators of major
biogeochemical cycles. Permeate the entire biosphere, including other organisms, although dominant
only in exceptional habitats.

Eukarya

Animalia
Multicellular, mainly macroscopic, eukaryotes. Reproduce through fertilization of an egg by a
sperm, the fertilized egg (now diploid, i.e. a duplicate set of chromosomes) is called a zygote
and (except sponges) this forms a characteristic hollow multicelled blastula from which the
embryo develops. All heterotrophic.

Fungi
Mainly multicellular, micro- to macroscopic eukaryotes. Fungi develop directly without an
embryo stage from resistant non-motile haploid (one set of chromosomes) spores that can be
produced by a single parent. Sexual reproduction also results in haploid spores. Most consist
of a network of threadlike hyphae. Heterotrophs, vital to decomposition processes; form
mycorrhizal symbioses with plants, facilitating exchange of soil nutrients.

Plantae
Multicellular macroscopic eukaryotes. The fertilized egg develops into a multicelled embryo
different from blastula of animals. Alternate spore-producing generations and egg or sperm-
producing generations. Virtually all are terrestrial photosynthetic autotrophs.

Protoctista
Mainly microorganisms. Possess the features of eukaryotes, but lack the characteristics
of fungi, animals or plants. Extraordinary variation in life cycle and morphology. Early
evolution probably based on symbiotic relationships between different kinds of bacteria
forming lineages of composite organisms resulting in the protoctist grade of
organization. Include photosynthetic algae (formerly classed as plants) and heterotrophs
(formerly called 'protozoa').

ings of organisms (domains and kingdoms)
are outlined in Table 2.2. A synopsis of
information on each of the 96 phyla recognized
in the most recent comprehensive synthesis[6]
is presented in Appendix 1. The objective of
this material is to provide a convenient
overview of global organismal diversity, in
terms of higher taxon diversity.

REFERENCES

1 Minelli, A. 1993. *Biological systematics: The state of the art*. Chapman and Hall, London.

2 Vane-Wright, R.I. 1992. Systematics and diversity. In: World Conservation Monitoring Centre. Groombridge, B. (ed.) *Global biodiversity: Status of the Earth's living resources*, pp. 7-12. Chapman and Hall, London.

3 Mayr, E. and Ashlock, P.D. 1991. *Principles of systematic zoology*. McGraw-Hill Inc., New York.

4 Gaston, K.J. (ed.) 1996. *Biodiversity: A biology of numbers and difference*. Blackwell Science Ltd, Oxford.

5 Vane-Wright, R.I. 1992. Species concepts. In: World Conservation Monitoring Centre. Groombridge, B. (ed.) *Global biodiversity: Status of the Earth's living resources*, pp. 13-16. Chapman and Hall, London.

6 Margulis, L. and Schwartz, K.V. 1998. *Five kingdoms. An illustrated guide to the phyla of life on earth*. 3rd edition. W.H. Freeman and Company, New York.

7 United Nations Environment Programme 1995. Heywood, V. (ed.) *Global biodiversity assessment*. Cambridge University Press, Cambridge.

8 Hammond, P. 1992. Species inventory. In: World Conservation Monitoring Centre. Groombridge, B. (ed.) *Global biodiversity: Status of the Earth's living resources*, pp. 17-39. Chapman and Hall, London.

9 Williams, D.M. and Embley, T.M. 1996. Microbial diversity: Domains and kingdoms. *Annu. Rev. Ecol. Syst.* 27: 569-595.

10 GTI. http://www.biodiv.org/programmes/cross-cutting/taxonomy/ (accessed January 2002).

11 GBIF. http://www.gbif.org/ (accessed January 2002).

12 Species 2000. http://www.sp2000.org/ (accessed January 2002).

13 ITIS. http://www.itis.usda.gov/ (accessed January 2002).

14 Woese, C.R., Kandler, O. and Wheelis, M.L. 1990. Towards a natural system of organisms: Proposal for the domains Archaea, Bacteria and Eukarya. *Proceedings of the National Academy of Sciences* 87: 4576-4579.

15 Woese, C.R. 2000. Interpreting the universal phylogenetic tree. *Proceedings of the National Academy of Sciences* 97: 8392-8396.

16 University of California, Berkeley, Museum of paleontology. http://www.ucmp.berkeley.edu/alllife/threedomains.html (accessed January 2002).

17 University of Arizona, Tree of Life project. http://phylogeny.arizona.edu/tree/phylogeny.html (accessed January 2002).

18 Margulis, L. 1998. *The symbiotic planet: A new look at evolution*. Weidenfeld and Nicolson, London.

19 Philippe, H. and Forterre, P. 1999. The rooting of the universal tree of life is not reliable. *Journal of Molecular Evolution* 49: 509-523.

20 Woese, C.R. 1998. The universal ancestor. *Proceedings of the National Academy of Sciences* 95: 6854-6859.

21 http://www.ucmp.berkeley.edu/chromista/chromista.html (accessed January 2002).

22 Woese, C.R. 1998. Default taxonomy. *Proceedings of the National Academy of Sciences* 95: 11043-11046.

23 Wilson, D.E. and Reeder, D.M. (eds) 1993. *Mammal species of the world: A taxonomic and geographic reference*. 2nd edition. Smithsonian Institution Press, Washington DC and London. Available (accessed January 2002) in searchable format at http://nmnhwww.si.edu/msw/ Covers existing or recently extinct species, as of book publication year.

24 Clements, J.F. 2000. *Birds of the world: A checklist*. Pica Press, Sussex. Other lists are available online, e.g. a list based on Sibley, C.G. and Monroe, Jr., B.L. 1990. *Distribution and taxonomy of birds of the world*. Yale University Press, New Haven and London is available (accessed January 2002) at http://www.itc.nl/~deby/SM/SMorg/sm.html

25 http://www.embl-heidelberg.de/~uetz/LivingReptiles.html (accessed January 2002).

26 Estimate is sum of species numbers for Anura (4 701), Caudata (505) and Gymnophiona (caecilians) (159) retrieved from online resource Amphibia Web at http://elib.cs.berkeley.edu/aw/lists/index.shtml (accessed January 2002). Also see Duellman, W.E. 1993. *Amphibian species of the world: Additions and corrections*. Special Publication No. 21, University of Kansas Museum of Natural History, Lawrence, Kansas, and database with later revisions available online in searchable format at http://research.amnh.org/herpetology/amphibia/ (accessed January 2002).

27 Eschmeyer, W.N. *et al.* 1998. *A catalog of the species of fishes*. Vols 1-3. California Academy of Sciences, San Francisco. Available online in searchable format at http://www.calacademy.org/research/ichthyology/catalog/fishcatsearch.html (accessed January 2002).

3 Biodiversity through time

TWO FUNDAMENTAL PATTERNS CAN BE DISTINGUISHED in the fossil record. On one hand, new groups of organisms appear, diversify and generally persist for very long periods of time; on the other, most such groups and their included species eventually cease to exist. Most analyses of the fossil record show an erratic rise in overall biodiversity, increasing through the Mesozoic and Cenozoic and reaching a peak around the end of the Tertiary. However, diversity has been greatly reduced during several periods of radical environmental change during each of which more than half the multicellular species then living became extinct. Such mass extinction phases have provided important new opportunities for diversification in remaining lineages, and the spread of new communities. Evidently, the large number of species existing now on Earth is the result of a modest net excess of originations over extinctions during the 3 800 million years of evolution of life.

THE FOSSIL RECORD

Knowledge of the history of diversity through geological time is based on analysis of the fossil record. When fossils, as the inert mineralized parts or casts or imprints of dead organisms, are interpreted in a biological context they provide the only direct evidence of the history of life on the planet.

The fossils discovered and described by paleontologists represent more than a quarter of a million species, virtually all of them now extinct[1], but these are believed to make up only a very small fraction of all the species that have ever existed. For example, the fossil record of marine animals is far more comprehensive than that of terrestrial forms, but the marine sample is estimated to represent only about 2 percent of all the marine animals that have lived[2]. The fossil record overall may represent as little as 1 percent, or less, of all the species that have existed[1]. Clearly, statements about broad patterns in the evolution of life, and the ascendancy or extinction of groups of organisms, thus rest on a very narrow base of tangible evidence.

Macroscopic animals with hard skeletons that lived in shallow marine environments, where their remains could be buried by sediment, petrified, and later exposed in uplifted rock strata, are by far the most likely to be both preserved and found. The fossil record from the past 600 million years is thus dominated by mollusks, brachiopods and corals[1]. For many other kinds of organism, exceptional circumstances are required if a dead individual is to become preserved and found. With some exceptions, microscopic or soft-bodied organisms rarely leave discernable traces of their existence, and larger organisms are usually decomposed, disassembled and never discovered. Terrestrial vertebrate fossils are often of individuals that must have been preserved by the smallest of chances, perhaps sudden burial during a natural disaster, or as a result of the body falling into a rock crevice out of reach of scavengers.

All else being equal, the probability of preservation will rise greatly the more widespread and abundant the species is, and the longer it persists through time. Conversely, there is a very low probability that any individual of a numerically rare or restricted-range species with a short persistence time will die in circumstances conducive to fossilization and subsequently be found. Factors such as these mean that even the relatively better known groups are certain to be very incompletely known. The plant and animal species now living include a substantial number of rare or local species and it is difficult to imagine many of them being represented in the fossil record of our time as recovered in the future[3].

Because the fossil record gives an incomplete and biased view of the past history of life, the reconstruction of that history has been the subject of great debate. It has been generally accepted that the record can give a reasonable insight into past diversity in terms of taxonomic richness[2-4], particularly at higher taxonomic levels. On the other hand, recent analysis of a new database of fossil occurrence has indicated that sampling effects (reflecting variation in the nature of the fossil record and the way it is reported) can make a very significant contribution to the shape of global diversity curves. Preliminary results of this analysis have suggested, for example, that the widely accepted post-Paleozoic increase in marine diversity (see below) may to some extent be an artifact of the analytic methods used[5].

PATTERNS OF DIVERSIFICATION
The early history of life
In its earliest history, there was no life on the Earth. Now, there are about 1.75 million different species of all kinds known, with perhaps many times that number still unknown. Self-evidently, life has both arisen and diversified.

The planet Earth is between 4 500 million and 5 000 million years old. The oldest known rocks are about 4 000 million years (My) in age, and mineral grains (zircon) of apparently greater age have been reported. Carbon

'chemofossils' suggest that life had evolved by 3 800 My ago[6] but periodic intense ocean heating caused by extraterrestrial impacts is suspected to have restricted early organisms to thermophilic non-photosynthetic forms, possibly associated with hydrothermal vents[7,8]. Most of the principal biochemical pathways key to the modern biosphere were probably in place by 3 500 My ago[7]. The first tangible evidence of cellular organisms themselves in the fossil record consists of filaments and spheroids, believed to be the remains of prokaryote microorganisms, in rocks of about this age, including traces from an apparent Precambrian submarine thermal spring system of about 3 235 My in age[9].

Stromatolites (rock domes formed in shallow waters from multiple layers of sediment and bacteria) also first appear at around 3 500 My ago and are the most abundant fossils so far known during the 3 000 My up to the start of the Phanerozoic eon, marked by the start of the Cambrian period around 545 My ago. For much of this time, bacteria-like organisms were the only known life forms; gross morphological diversity was very low and many kinds apparently persisted for hundreds of millions of years, some outwardly indistinguishable from existing forms. Probable microbial mats, similar to stromatolites but developed on an exposed soil surface, have been interpreted as the first evidence of terrestrial ecosystems around 2 600 My ago[10].

The next major step in the evolution of life was the development of eukaryotic organisms. Biochemicals characteristic of eukaryotes have been found in shales 2 700 My old[11], far earlier than the first fossils, reported from

Figure 3.1
The four eons of the geological timescale

Arrows indicate approximate age of oldest confirmed fossils of the groups named.

Source: Adapted from Margulis and Schwartz[15].

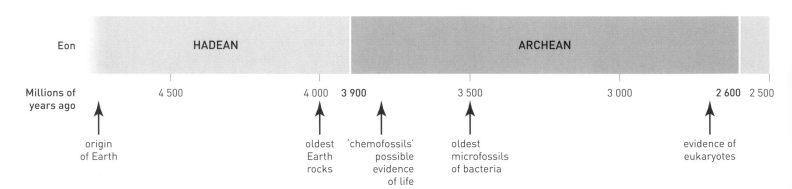

shales of around 1 500 My age[12]. These earliest eukaryote microfossils include a number of 'acritarchs' (widely interpreted as the cysts or resting stages of marine algae) some of which show cytological features characteristic of protoctists (see previous chapter)[12]. Diversity in acritarchs, and the rate at which different forms replaced one another in the record, were low until around 1 000 My ago when both species number and species turnover increased markedly[13].

Radiations around the early Phanerozoic boundary

For many years it was assumed that animals originated in the Cambrian period at the base of the Phanerozoic (the Phanerozoic is the eon of time characterized by presence of animal fossils; it includes the Paleozoic, Mesozoic and Cenozoic). This is now known not to be the case, as a wide range of fossil animals, including recognizable arthropods and possibly echinoderms, is now known from about 100 My before the Cambrian. Most fossils from this time, however, appear completely unrelated to extant forms, and consist mainly of enigmatic frond- and disc-shaped soft-bodied animals: the so-called Ediacaran fauna.

The lower Cambrian marks a dramatic change from this early fauna, with the sudden appearance in the fossil record (e.g. the Chengjiang and Burgess Shale faunas) of a wide range of animals, many with calcareous skeletons. It is generally accepted that this represents a genuine explosion of diversity that took place over only a few million years, and is not an artifact of the fossil record. The lower Cambrian thus represents the most important period of high-level diversification in the history of animal life on Earth. These archaic invertebrates had by the end of the Cambrian period, around 500 My ago, established all the basic body plans seen in extant animals, and many others besides. Each such basic lineage is recognized taxonomically at phylum level, and the range of morphological diversity was higher at 500 My ago than at any time before or since. As many as 100 different animal phyla may have existed during the Cambrian[14] including every well-skeletalized animal phylum living today (except perhaps the Bryozoa), whereas in the latest synthesis all extant animals are placed in 37 phyla[15].

Plants and animals began to extend into terrestrial habitats during the first half of the Paleozoic, with the first fossil material known from the late Silurian, around 400 My ago. At this point approximately 90 percent of the history of life to the present had already passed. Fossils suggest low diversity for the next 100 My until the later Devonian period. No new animal phyla appeared with the colonization of land, millions of years after the initial Cambrian radiation of animal phyla.

Diversity of marine animals in the Phanerozoic

The overall pattern of diversity (assessed as numbers of families) appears to show a possible early peak around the start of Phanerozoic time, followed by a plateau of somewhat higher diversity extending through most of the Paleozoic era, and then, after the end-Permian mass extinction (see below) a steady increase in diversity over remaining geological time[4]. Some recent methods of analysis[5] giving special attention to sampling

procedures suggest this broad view may be subject to significant revision.

Although the number of phyla has decreased markedly since the Cambrian, diversity at all lower taxonomic levels appears to have either increased overall or, in a few cases, remained more or less level. The number of orders of marine animals present in the fossil record climbed steadily through the Cambrian and Ordovician, slowing towards the end of the Ordovician to a figure of between 125 and 140, which has been maintained throughout the Phanerozoic.

The diversity of marine families represented in the fossil record shows a similar pattern of increase through the Cambrian (possibly falling during the latter half of the period) and Ordovician, leveling off at around 500, a figure which was maintained until the late Permian mass extinction. This extinction event resulted in the loss of around 200 families, but diversity increased subsequently to the modern level of around 1 100 families, with a number of temporary reversals during minor extinction events. The trend in number of species in the fossil record is even more extreme. From the early Cambrian until the mid-Cretaceous, the number of marine species remained low; since then, that is during the past 100 million years, it appears to have increased dramatically, perhaps by a factor of 10. (See Figure 3.3.)

Diversity of terrestrial organisms in the Phanerozoic

Although low to moderate peaks and troughs are evident in the record, the overall pattern of family diversity in terrestrial organisms shows a continuing rise from the Silurian to the present, thus differing somewhat from the pattern shown by many analyses of marine animals.

It is generally accepted that vascular terrestrial plants first arose in the Silurian, although some paleobotanists argue for a late Ordovician origin, a time when microfossils suggest that bryophyte-like (non-vascular) plants already existed[16]. Diversity increased during the Silurian, and then more rapidly during the Devonian, owing to the first appearance of seed-bearing plants, leading to a peak of more than 40 genera during the late Devonian. Diversity then declined slightly, but started to increase markedly during the Carboniferous, with 20 families and more than 250 species in the mid-Carboniferous record of the northern hemisphere. Following this, diversity increased only slowly until the end of the Permian. There was a marked decrease in diversity at the end of the Permian, coinciding with or preceding the mass extinction of animal species. Subsequent increase in diversity was slow, reaching around 400 species in the early Cretaceous, but apparently more rapid from the mid-Cretaceous.

This overall pattern masks important changes with time in the composition of the flora, most notably in the relative importance of the three main groups of vascular plant: the pteridophytes, gymnosperms and angiosperms (see Figure 3.4). The Silurian and early Devonian are marked by a radiation of primitive pteridophytes. During the Carboniferous, more advanced pteridophytes and gymnosperms developed and underwent extensive diversification. Following the late Permian extinction event, pteridophytes were largely replaced by gymnosperms (although ferns remained abundant) and these became the dominant group until the mid-Cretaceous. The dramatic

Figure 3.2
Periods and eras of the Phanerozoic

This is an expanded version of the most recent segment of the geological timescale shown in Figure 3.1.
The Phanerozoic is the eon of time extending from the base of the Cambrian, some 545 million years ago, to the present, and to which the entire fossil record was formerly thought to be restricted.

Source: Adapted from Margulis and Schwartz[15], Cambrian base date from International Subcommission on Cambrian Stratigraphy website[27].

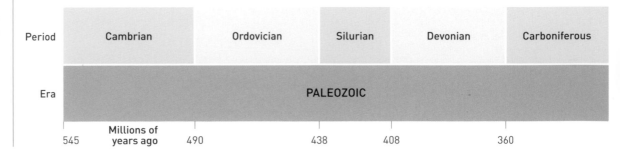

Period	Cambrian	Ordovician	Silurian	Devonian	Carboniferous
Era	PALEOZOIC				

Millions of years ago 545   490   438   408   360

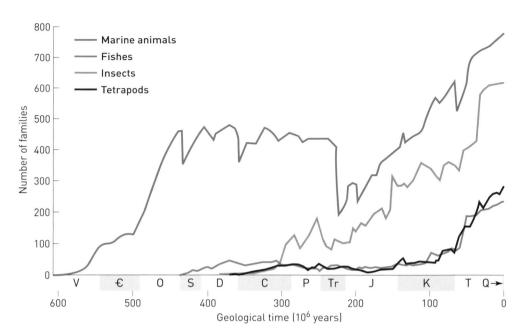

Figure 3.3
Animal family diversity through time

The lines plotted represent the number of families in the fossil record.

Notes: The blue line essentially represents marine invertebrate animals. Although a small number of vertebrate groups, notably fishes and a few tetrapod species, are included, these make up a tiny proportion of the total marine family diversity shown. The curve for fishes includes an increasing proportion of freshwater forms through the Cenozoic. Tetrapods are amphibians, reptiles, birds and mammals. The extent to which such results, based on sampling and interpretation of the fossil record, represent actual diversity has been subject to discussion.

Source: Marine animals, adapted from Sepkoski[2]; fishes and tetrapods, adapted after Benton[28]; insects, adapted from Labandeira and Sepkoski[18].

increase in plant diversity since then is entirely due to the radiation of the angiosperms, fossils of which first appear in the lower Cretaceous, although an early to mid-Jurassic origin, around 170 My ago, has been argued[17].

Colonization of land by animals has occurred many times; although the oldest body fossils of terrestrial animals date from the early Devonian, it is generally accepted that the primary period of land invasion by animals was the Silurian.

The overwhelming number of described extant species of terrestrial animals are insects. Their fossil record is more extensive than might be expected, but had been little studied until recently. Data on insect diversity at family level have been collated, based on nearly 1 300 families[18]. This analysis shows a very slow increase in families from the first appearance of insects in the Devonian, a rise

in numbers during the Carboniferous, and a steeply increasing rise throughout the Mesozoic to the Tertiary. The apparent explosion in insect diversity had previously been attributed to ecological opportunities provided by the expansion of flowering plants, but the insects are now known to have begun their diversification some 150 My before the flowering plants[18].

The fossil record of vertebrates includes around 1 400 families, with tetrapods (amphibians, reptiles, birds, mammals) somewhat outnumbering fishes. The bird record is much less substantial than that for other groups, probably because their light skeletons have been less frequently preserved.

Terrestrial vertebrates first appear in the fossil record in the late Devonian. Diversity remained relatively low during the Paleozoic, with around 50 families, and may have

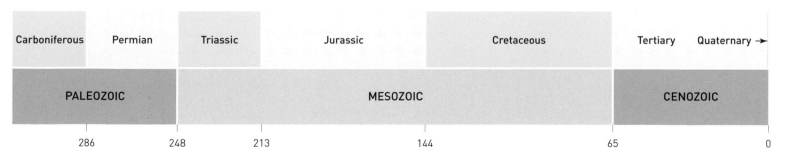

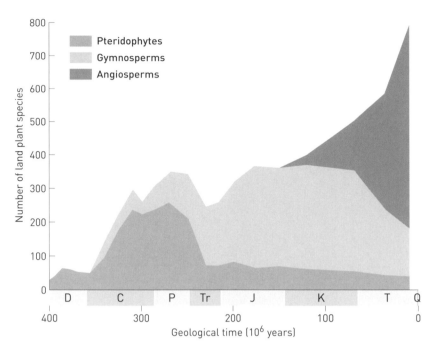

Figure 3.4
Plant diversity through time

Notes: Pteridophytes are ferns *Filicinophyta* and allies, gymnosperms are conifers *Coniferophyta* and allies, angiosperms are flowering plants *Anthophyta*. Note changing numerical dominance of each group over time.
This generalized diagram is to illustrate changing abundance of major groups over time; it does not represent short-term extinction events.

Source: Adapted from Kemp[24], after Niklas.

declined during the early Mesozoic. From the mid-Cretaceous the number of families started to increase rapidly, reaching a recent peak of around 340. Diversity of genera follows this overall pattern in a more exaggerated form. It appears that periodic increase in the number of tetrapod families is mainly a result of lineages becoming adapted to modes of life not already followed by other organisms, i.e. by adopting new diets or new habitats[19].

PATTERNS OF EXTINCTION

If living species represent between 2 and 4 percent of all species that have ever lived[20], almost all species that have lived are extinct, and extinction can be presumed to be the ultimate fate of all species.

Numerous estimates have been made of the lifespan of species in the fossil record; these range from 0.5 My to 13 My for groups as varied as mammals and microscopic protoctists. Analysis of 17 500 genera of extinct marine microorganisms, invertebrates and vertebrates, suggests an average lifespan of 4 My in these groups[1]. Given this average lifespan, at a very gross estimate, the mean extinction rate would be 2.5 species per year if there were around 10 million species in total. However, because of bias inherent in the fossil record, such lifespan estimates are likely to relate to widespread, abundant and geologically longer-lived species, in effect, the extinction-resistant forms; and so do not represent the biota as a whole[21]. If most species survived for less than 4 My, the overall extinction rate at any given time would have been correspondingly higher.

Major extinctions in animals

In general the Precambrian fossil record is too incomplete to allow detailed analysis of extinction rates. However, there is good evidence of a major loss of diversity during the Vendian period in latest Precambrian times, around 550 My ago, when the entire Ediacaran fauna disappeared (along with many acritarchs). Another wave of extinction affected archaeocyathid sponges, mollusks and trilobites during the lower Cambrian, some 530 My ago.

By far the most severe marine invertebrate mass extinction was in the late Permian (250 My ago). At that time, the number of families of marine animals recorded in the fossil record declined by 54 percent and the number of genera by 78-84 percent. Extrapolation from these figures indicates that species diversity may have dropped by as much as 95 percent. Other major extinctions in marine invertebrates occurred at the end of the Ordovician (440 My ago) (see Figure 3.3), when around 22 percent of families were lost, and during the late Devonian and late Triassic (21 percent and 20 percent, respectively). Around 15 percent of marine families disappeared at the end of the Cretaceous.

The vertebrate fossil record, especially for terrestrial tetrapods, is much less amenable to analysis of extinction rates than the invertebrate record, chiefly because it is less complete and less diverse. However, studies indicate that fishes have been subject to at least eight important extinction events since their recorded origin in the Silurian, while tetrapods have experienced at least six such events since their appearance in the late Devonian. Some of these events coincide with each other and with those recorded for marine invertebrates; in particular, the five major mass extinction events outlined above are paralleled by losses in vertebrate diversity. The most significant is the late Permian event, which, in terms of percentage loss, is the largest recorded extinction both for fishes (44 percent of families disappearing from the fossil record) and tetrapods (58 percent of families disappearing). The late Cretaceous event was more significant for tetrapods than for fishes, with at least 30 of the 80-90 families then in the fossil record disappearing at this time. These families were, however, virtually confined to three major groups which suffered complete extirpation – the dinosaurs, plesiosaurs and pterosaurs. Most other vertebrates were almost completely unaffected.

Major extinctions in vascular plants

Fewer major extinction events have to date been distinguished in the plant fossil record than in the animal record. Periods of elevated plant extinction appear in some cases more protracted than animal extinction events and not usually coincident with them[22]. However, there is good evidence for extensive reduction in woody vegetation at the end of the Permian, with widespread loss of peat forests in humid areas and of conifers in some semi-arid regions. A fourfold increase in atmospheric carbon dioxide around the Triassic-Jurassic boundary is correlated with a more than 95 percent turnover in the megaflora (i.e. leaf fossils, etc., as opposed to pollen or spores)[23]. The end-Cretaceous catastrophe appears to have had a major influence on the structure and composition of terrestrial vegetation and on the survival of species. Data from fossil leaves suggest that perhaps 75 percent of late

Cretaceous species became extinct, although data from fossil pollens indicate a lower though still significant level of extinction. During the Tertiary there are two other periods of widespread heightened extinction rates, during the late Eocene and from the late Miocene to the Quaternary, although in the latter extinction of taxa at generic level and above appears to have been mainly regional rather than global.

Mass extinctions

The very many species extinctions represented in the fossil record are not distributed evenly through time, nor do they occur randomly. In paleontology much attention has been devoted to mass extinction periods, during which some 75-95 percent of species then living became

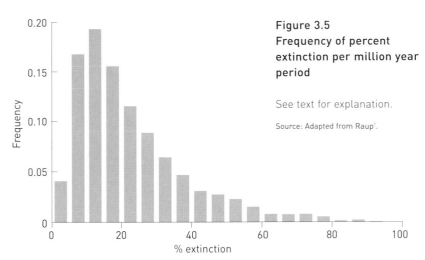

Figure 3.5
Frequency of percent extinction per million year period

See text for explanation.

Source: Adapted from Raup[1].

Figure 3.6
Number of family extinctions per geological interval through the Phanerozoic

Note: 76 geological intervals represented, average duration around 7 million years.

Source: Benton[29].

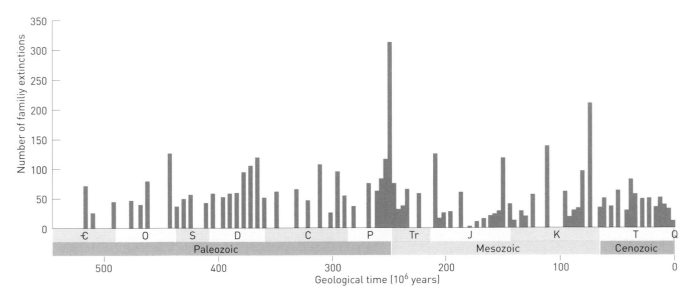

extinct during geologically very short periods of time, in some cases possibly as little as a few hundred thousand years or even less. Five such phases (based chiefly on extinction of marine species) are recognized during Phanerozoic time, late in each of the Ordovician, Devonian, Permian, Triassic and Cretaceous periods (Table 3.1).

Although each of the 'Big Five' mass extinctions had a very profound effect on then contemporary life, they are not isolated peaks standing out from a constant, very low, background rate. Rather, the extinction rate has varied continuously throughout the Phanerozoic with periods of more or less elevated rates of extinction, of which the most extreme can be characterized as mass extinction events. A frequency plot (Figure 3.5) of the percentage of species becoming extinct in each 1-My interval of the 600-My record of animal life provides an indication of this variation in extinction intensity[1]. The plot takes the form of

Period	Date (My ago)	Species loss (%)	Biotic change	Possible causes
Late Ordovician	440	85	The last but largest of several extinction events during the Ordovician. More than 25% of marine invertebrate families lost. Entire class Graptolithina reduced to a few species; acritarchs, brachiopods, conodonts, corals, echinoderms, trilobites, all much reduced.	Cooling; Warming Marine regression Marine transgression and anoxia
Late Devonian	365	80	Mass extinctions came at the end of prolonged period of diversity reduction. Rugose corals lost >95% of shallow water species; stromatoporoid corals reduced by half and reefs disappeared; brachiopods lost 33 families; ammonoids and trilobites severely affected. Fishes suffer only major mass extinction; all early jawed fishes (placoderms) disappear and most agnatha. First major crisis in plants; diversity greatly reduced, but spread of first tree, the gymnosperm *Archaeopteris*.	Marine transgression and anoxia
End Permian	250	95	The most severe extinction crisis: metazoan life came within a few percent of total extinction. Tabulate and rugose corals terminated, complex reefs disappear (return after 8-My gap); echinoderms almost wiped out; worst crisis in history of foraminifera; severe extinction in ammonites, brachiopods, bryozoa, mollusks. Some losses in early ray-fin fishes. Major loss of terrestrial vertebrates (75% of families) and insects (8 of 27 insect orders extinct). Mass extinction in plants: large plants including peat-forming trees lost (spread of small conifers, lycopods and quillworts); sudden unprecedented abundance of fungal spores at end of period.	Volcanism Warming Marine transgression
End Triassic	205	80	Mass extinction in marine invertebrates, especially brachiopods, cephalopods and mollusks; also mass disappearance of scleractinian corals and sponges. Several seed fern families lost; some land vertebrates lost, but evidence for mass extinction questionable.	Marine regression
End Cretaceous	66	75	Radical change in planktonic foraminifera; 85% of calcareous nanoplankton lost, also all ammonites, belemnites and many bivalves; losses in echinoids and corals. Many marine reptiles extinct (ichthyosaurs, plesiosaurs, mosasaurs); significant losses in freshwater and terrestrial vertebrates, including last dinosaurs (high turnover throughout dinosaur history – end Cretaceous unusual in that no replacements emerged). Mass extinctions in plants: highest (possibly 60% species loss) in angiosperms, lowest in ferns.	Impact of large meteor Volcanism Cooling Marine regression

an even left-skewed curve, with no substantial discontinuity between rare periods of high extinction rate (more than 60 percent of species extinct per one million years) and the most frequent rate (10-15 percent); the mean intensity is 25 percent of species extinct per 1-My interval. This suggests that mass extinctions may arise from causes not qualitatively different from those associated with extinctions at other times. It should also be remembered that because of the extremely long duration of the Phanerozoic during which species have been constantly becoming extinct, and the very short duration of the mass extinction events themselves, the latter account for only a small percentage (estimated at around 4 percent) of all extinctions.

The precise causes and timespans of each of the mass extinctions have been the subject of much debate and study[24,25]. It is now widely accepted that the late Permian mass extinction was a long-term event, lasting for 5-8 million years. It appears to have been associated with geologically rapid global physical changes (including the formation of the supercontinent Pangaea), climate change, and extensive, tectonically-induced marine transgression and increased volcanic activity. The late Cretaceous extinction is probably the best known, but in terms of overall loss of diversity is also the least important. There is strong evidence that this extinction event was associated with climate change following an extra-terrestrial impact, although this remains somewhat controversial. The late Ordovician event appears to be correlated with the global Hirnantian glaciation,

with three separate episodes of extinction spread over only 500 000 years. In all cases, however, the ability to determine accurately the timing and periodicity of extinction is heavily dependent on the completeness of the fossil record and the reliability and precision of stratigraphic analysis.

A mass extinction period is typically followed by a phase of 5-10 My of very low diversity, with a handful of species dominant in fossil faunas and floras. When diversity again increases the biota may be very different in composition from those preceding. In several instances, groups previously showing low diversity have radiated and spread widely following the demise of groups previously dominant; e.g. ray-finned fishes diversified following loss of placoderms; quillworts and seed ferns diversified for a time after loss of glossopterids; and the mammals radiated after loss of many terrestrial reptile groups. Reef organisms provide a classic example: reef communities have been lost several times during the Phanerozoic but with each reappearance the main reef-building forms have been different[25]. Important new evolutionary opportunities arise for the lineages that survive mass extinction events, which can radically redirect the course of evolution[25,26].

The present-day diversity of living organisms is the result of a net excess of originations over extinctions through geological time. Figure 3.6 shows the number of family extinctions in each interval of geological time, and the cumulative diversity of families, both marine and terrestrial.

REFERENCES

1 Raup, D.M. 1994. The role of extinction in evolution. *Proceedings of the National Academy of Sciences* 91:6758-6763.
2 Sepkoski, J.J. 1992. Phylogenetic and ecologic patterns in the Phanerozoic history of marine biodiversity. In: Eldredge, N. (ed.) *Systematics, ecology and the biodiversity crisis*, pp. 77-100. Columbia University Press, New York.
3 Jablonski, D. 1991. Extinctions: A paleontological perspective. *Science* 253: 754-757.
4 Benton, M.J. 1995. Diversification and extinction in the history of life. *Science* 268: 52-58.
5 Alroy, J. *et al.* 2001. Effects of sampling standardization on estimates of Phanerozoic marine diversification. *Proceedings of the National Academy of Sciences* 98: 6261-6266.
6 Mojzsis, S.J. *et al.* 1996. Evidence for life on Earth before 3,800 million years ago. *Nature* 384: 55-59.
7 Nisbet, E.G. and Sleep, N.H. 2001. The habitat and nature of early life. *Nature* 409: 1083-1091.
8 Farmer, J.D. 2000. Hydrothermal systems: Doorways to early biosphere evolution. *GSA*

Today 10. Available online (accessed January 2002) at http://www.geosociety.org

9   Rasmussen, B. 2000. Filamentous microfossils in a 3,235-million-year-old vulcanogenic massive sulphide deposit. *Nature* 405: 676-678.

10 Watanabe, Y., Martin, J.E.J. and Ohmoto, H. 2000. Geochemical evidence for terrestrial ecosystems 2.6 billion years ago. *Nature* 408: 574-578.

11 Brocks, J.J. *et al.* 1999. Archean molecular fossils and the early rise of eukaryotes. *Science* 285: 1033-1036.

12 Javaux, E.J., Knoll, A.H. and Walter, M.R. 2001. Morphological and ecological complexity in early eukaryotic ecosystems. *Nature* 412: 66-69.

13 Knoll, A.H. 1995. Proterozoic and early Cambrian protists: Evidence for accelerating evolutionary tempo. In: Fitch, W.M. and Ayala, F.J. (eds). *Tempo and mode in evolution: Genetics and paleontology 50 years after Simpson*, pp. 63-68. National Academy Press, Washington DC.

14 McMenamin, M.A.S. 1989. The origins and radiation of the early metazoa. In: Allen, K.C. and Briggs, D.E.G. (eds). *Evolution and the fossil record*, pp. 73-98. Belhaven Press, a division of Pinter Publishers Ltd, London.

15 Margulis, L. and Schwartz, K.V. 1998. *Five kingdoms. An illustrated guide to the phyla of life on earth.* 3rd edition. W.H. Freeman and Company, New York.

16 Wellman, C.H. and Gray, J. 2000. The microfossil record of early land plants. *Philosophical Transactions of the Royal Society of London. Series B. Biological Sciences* 355: 717-731.

17 Wikstrom, N., Savolainen, V. and Chase, M.W. 2001. Evolution of the angiosperms: Calibrating the family tree. *Proceedings of the Royal Society of London. Series B. Biological Sciences* 268: 2211-2220.

18 Labandeira, C.C. and Sepkoski, J.J. 1993. Insect diversity in the fossil record. *Science* 261: 310-315.

19 Benton, M.J. 1999. The history of life: Large databases in palaeontology. In: Harper, D.A.T. (ed.) *Numerical palaeontology: Computer-based modelling and analysis of fossils and their distributions.* John Wiley and Sons, Chichester and New York.

20 May, R.M., Lawton, J.H. and Stork, N.E. 1995. Assessing extinction rates. In: Lawton, J.H. and May, R.M. (eds). *Extinction rates*, pp. 1-24. Oxford University Press, Oxford.

21 Jablonski, D. 1995. Extinctions in the fossil record. In: Lawton, J.H. and May, R.M. (eds). *Extinction rates*, pp. 25-44. Oxford University Press, Oxford.

22 Crane, P.R. 1989. Patterns of evolution and extinction in vascular plants. In: Allen, K.C. and Briggs, D.E.G. (eds). *Evolution and the fossil record*, pp. 153-187. Belhaven Press, a division of Pinter Publishers Ltd, London.

23 McElwain, J.C., Beerling, D.J. and Woodward, F.I. 1999. Fossil plants and global warming at the Triassic-Jurassic boundary. *Science* 285: 1386-1390.

24 Kemp, T.S. 1999. *Fossils and evolution.* Oxford University Press, Oxford.

25 Hallam, A. and Wignall, P.B. 1997. *Mass extinctions and their aftermath.* Oxford University Press, Oxford.

26 Erwin, D.H. 2001. Lessons from the past: Biotic recoveries from mass extinctions. *Proceedings of the National Academy of Sciences* 98: 5399-5403.

27 International Subcommission on Cambrian Stratigraphy. http://www.uni-wuerzburg.de/palaeontologie/Stuff/casu6.htm (accessed January 2002).

28 Benton, M.J. 1989. Patterns of evolution and extinction in vertebrates. In: Allen, K.C. and Briggs, D.E.G. (eds). *Evolution and the fossil record*, pp. 218-241. Belhaven Press, a division of Pinter Publishers Ltd, London.

29 Benton, M.J. (ed.) 1993. *The fossil record 2.* Chapman and Hall, London. Data available online in interactive format at http://ibs.uel.ac.uk/ibs/palaeo/benton/ (accessed January 2002); also see http://palaeo.gly.bris.ac.uk/frwhole/fr2.families

4 Humans, food and biodiversity

THE LINEAGE LEADING TO THE HUMAN SPECIES EMERGED from an ancestry among the apes, almost certainly in Africa around 5 million years ago. Fossils usually attributed to the genus *Homo* itself date from the late Pliocene, perhaps 2 million years ago, and anatomically modern humans appeared some 100 000 years or more before the present.

Agriculture developed independently in several regions around 10 500 years ago, in association with increased population density. The global human population continued to grow only slowly until the 19th century, when revolutionary developments in agriculture, industry and public health triggered an exponential rise that has continued to the present day.

Agriculture is a means to channel the Earth's resources into production of human bodies. Humans have converted large areas of terrestrial habitat and use more than one third of net primary production on land. They are strongly implicated in the extinction of many large terrestrial mammal and bird species in prehistory, and are responsible for habitat change and exploitation that have caused the decline and extinction of many species in recent times.

HUMAN ORIGINS

The human species evolved as a natural element of diversity in the living world, and it is a simple ecological imperative that humans depend on other species and communities to supply the basic requirements of existence and to maintain biosphere function.

The creation of organic compounds by photosynthesizing organisms is the point at which the sun's energy enters the biosphere (Chapter 1); humans and other animals are unable to capture energy in this way and must consume and digest either primary producers or other organisms that are themselves dependent on primary producers, in order to obtain these energy-rich organic compounds for their own activities.

While humans are doing nothing fundamentally different from other animals, with the benefits of society and technology, which serve to increase the rates of resource extraction, they are uniquely successful at it. Self-evidently, humans have not arrived at their extraordinarily dominant position on the planet overnight. The growth of their influence can be traced back several million

years, to well before the Pleistocene when a stone-tool-wielding hominid first emerged somewhere in eastern Africa.

Climate during the past 2 million years

The Earth's climate appears always to have been in a state of flux. Generally the degree of accuracy in our understanding of global climate, and certainly the degree of resolution in the timescale of climate change, decreases the further back in time we go. It is therefore difficult to compare periods that in geological terms are recent with more distant times. However, it does seem that during the past 2 million years there have been numerous, intense climate changes that were at least as severe, or perhaps more severe, than any recorded earlier in the Earth's history[1]. It is during this period that hominids very similar to modern humans first appear in the fossil record. By the early Holocene, some 10 000 years ago, technologically sophisticated humans had spread to all the major land masses except Antarctica, and had evidently started to exert a major, and ever-growing, impact on the biosphere. Indeed there is

Map 4.1
Early human dispersal

A highly generalized view of the colonization of the world by advanced humans from the early Pleistocene onwards. Coastline is shown 150 m lower than at present and, with northern hemisphere ice cover, represents an approximation of that at periods of glacial maximum. Dates represent earliest well-established time of arrival; arrows indicate general net direction of dispersal, not actual routes. Presence and dispersal of early humans are not represented: the lineage is believed to have arisen in Africa and later fossil material dated to between 1.8 and 1.4 million years before the present is known from several sites in Eurasia.

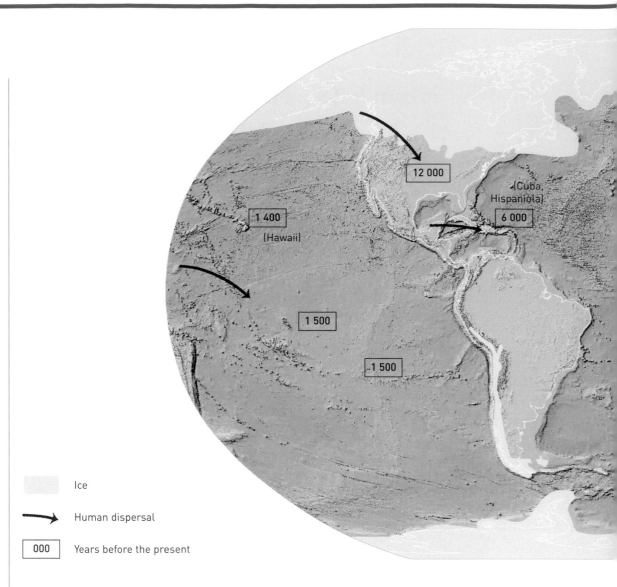

Ice

Human dispersal

| 000 | Years before the present

evidence that such impacts – for example in the extinction of large mammal and bird species – were already being felt considerably earlier than this. However, because the rise of humans coincides with a period of major climatic and ecological fluctuations, it is often difficult to disentangle the effects of the former from the latter, and so the precise nature of these impacts remains controversial.

The Tertiary period, which began some 65 million years ago and ended with the start of the Quaternary (the Pleistocene and Holocene) 1.8 million years ago, is characterized overall by a gradual decrease in global temperature and increase in aridity. Superimposed on this general pattern were many oscillations, occurring on a timescale of thousands of years[2]. These oscillations are

believed to be linked to cyclic variations in the Earth's position in its orbit around the sun, known as Milankovitch cycles. These cycles notwithstanding, the climate during virtually the whole of the Tertiary was notably warmer than at present.

Around 1.8 million years ago, at the very start of the Pleistocene[3], there was apparently rapid global cooling leading to the start of a period dominated by marked climate cycles of around 100 000 years' duration. For long periods of each cycle global temperatures were significantly lower than they are today, and extensive areas of the northern hemisphere land masses were covered in ice sheets. Detailed analysis of climate changes over the past 400 000 years or so, particularly through examination of Antarctic ice cores[4], indicate that each cycle

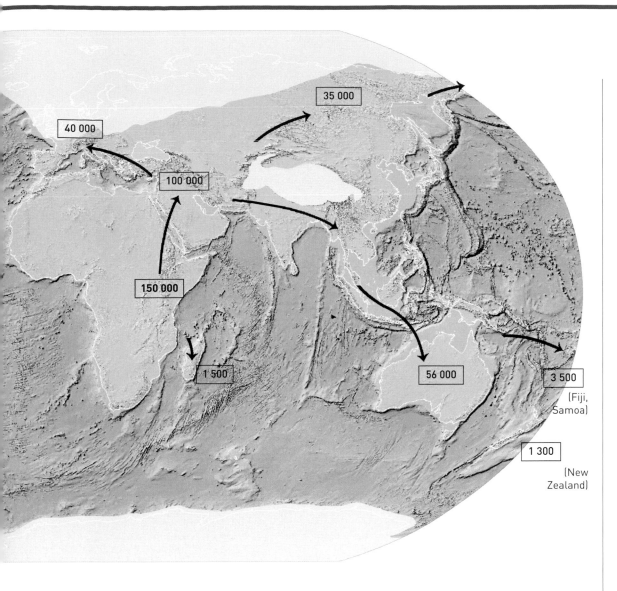

over this period has been broadly similar, with a gradual decline in average temperature, although with many minor and some large oscillations (relatively colder periods being referred to as stadials and warmer ones as interstadials), followed by a short period of intense warming in which temperatures rose from those of fully glacial conditions to those characterizing a warm interglacial state, perhaps sometimes over only a few decades.

Mean global temperature during glacial periods was around 6°C cooler than during interglacials, with cooling more pronounced at the poles than the equator. The mid-latitudes and equatorial regions were probably somewhat more arid than at present. The large amount of water locked up in the greatly expanded polar ice caps meant that mean sea

level during glacial periods was probably around 100-150 meters lower than at present.

Until around 11 000 years ago, each temperature peak was apparently followed by an almost immediate decline as the next glacial cycle began[4]. Overall, it appears that during the past half million years the Earth's climate has been as warm as, or warmer than, today's for only around 2 percent of the time[1]. It seems likely that this also holds true for the early Pleistocene. The end of the last glacial cycle, around 11 000-12 000 years ago, marked the start of the Holocene. Temperatures similar to today's have prevailed throughout the Holocene, making it by far the longest true interglacial period for at least the past half million years and probably for the last 1.8 million years.

Human origins and dispersal

The origins and early history of humans are among the most controversial subjects in paleontology. Remains are generally scarce and often open to varying interpretation. Nevertheless, consensus has emerged over the broad outlines. It is likely that the direct ancestors of humans – the hominid line – diverged from the apes in Africa during a cool, dry phase of the late Miocene, around 6 million to 5.5 million years ago. The primary evidence for this divergence is in the form of 'molecular clocks' calculated from comparison of human and ape genetic material[5]. Relevant fossil material from this time horizon is sparse and often of uncertain status, but material from Ethiopia dated to 5.8 million to 5.2 million years ago[6] is believed to represent *Ardipethecus*, generally accepted as the earliest known member of the hominid branch. Other species of this genus are known from before 4 million years ago, with various species of *Australopithecus* and *Paranthropus*, all from eastern Africa, somewhat younger in age.

Sometime during the middle to late Pliocene, 2.5 million to 1.8 million years ago, the genus *Homo* is thought to have evolved from *Australopithecus* stock[5,7]. Until comparatively recently it had been assumed that early man then remained confined to Africa until less than 1 million years ago. However, two well-preserved skulls recently found in the Caucasus (southern Georgia) have been dated to about 1.8 million years[8] (and variously attributed to *Homo ergaster*, otherwise known from Africa, or *H. erectus*). Elsewhere, the earliest stone tools from Turkey and southwest Asia have been dated to about 1.5 million years ago, and from northeast Asia to nearly 1.4 million years[9], while tools and fossils from East and Southeast Asia might be as old as 1.9 million years. These finds suggest that populations of early forms of *Homo* spread through the southeastern fringes of Europe as well as into Asia within at most a few hundred thousand years of the genus originating in Africa[5].

The earliest hominid remains elsewhere in Europe date from 780 000 years ago in Spain[10], followed by somewhat different material known from several sites widely distributed in northern Europe of about 500 000 years in age or less[11]. Fossil material from these two periods may be attributed to *H. antecessor* and *H. heidelbergensis*, respectively. The diversity of hominid material and the range of dates suggest that the spread of early hominids was complex and multidirectional, with possibly three lineages moving into Eurasia from Africa.

Modern humans, *Homo sapiens*, are believed to have appeared sometime after 200 000 years before the present. Anatomically modern human remains are known from around 100 000 in Africa and the Middle East, with more recent dates in East and Southeast Asia. The first appearance in Europe dates from about 40 000 years ago[12,13], with expansion apparently from the east toward the west, where in places premodern (Neanderthal) humans remained until about 28 000 years ago. One interpretation of the several lines of genetic evidence suggests, on the basis of mathematical models, that the modern human population is derived from a small founding population of perhaps 10 000 breeding individuals that existed sometime between 130 000 and 30 000 years before the present[12-14].

The earliest evidence of hominids outside Africa, Europe and Asia is much more recent. The age of the human skeleton found in 1974 at Lake Mungo in Australia, evidently the earliest human remains known on the continent, has been estimated at 62 000 years before the present[15], but a date close to 56 000 years is now widely accepted for human arrival on the continent[16].

In the Americas, the oldest good evidence of human presence is that of a coastal settlement in Chile dated around 14 000-15 000 years ago[17], although even this is far from universally accepted amongst archeologists. There are very controversial and now widely questioned claims for evidence of human settlement much earlier, most notably that dating from 32 000 years before the present from Pedra Furada in northeast Brazil[18]. The earliest unequivocal evidence of widespread occupation in the Americas comes from the so-called Clovis hunting culture whose oldest remains are generally dated at around 12 000 years before the present[19]. Settlement of the Caribbean islands including Cuba and Hispaniola appears to have taken place considerably later, around 6 000 years ago[20].

Archeological evidence indicates that colonization of the Pacific Islands east of New Guinea began around 4 000 years ago, when much of Melanesia and Micronesia was settled. Fiji and Samoa were probably colonized around 3 500 years ago, and the outliers of Hawaii, Easter Island and New Zealand within the last 1 500 years[21]. In the Indian Ocean, Madagascar was probably first settled around 1 500 years ago[22].

Technology

The earliest evidence of tool manufacture by hominids is from the Gona River drainage in northern Ethiopia where stone tools have been dated to at least 2.5 million years ago. These tools are small (generally less than 10 centimeters long) and simple, but are already of relatively sophisticated manufacture, suggesting that even older artifacts will eventually be found[23]. They are of essentially the same design as those dated from 2.3 million to 1.5 million years ago from other sites in East Africa (e.g. Lokalalei in the Lake Turkana basin in northern Kenya and the Olduvai gorge in Tanzania) indicating effective technological stasis for at least 1 million years (from 2.5 million to roughly 1.5 million years ago)[13]. Collectively these tools are referred to as products of the Oldowan stone tool industry[23].

Around 1.5 million years ago much more sophisticated and often larger tools including hand axes and cleavers suddenly appear in the archeological record in East Africa. These are referred to as Acheulian tools. The Acheulian tool industry spread into Europe, the Near East and India, and remained apparently relatively unchanged until around 200 000 years ago, showing similar temporal persistence to the Oldowan industry.

Evidence of tools and artifacts made from organic materials from the Paleolithic is understandably extremely scarce. In 1995, however, three large wooden implements very similar in design to modern-day javelins

and dated to around 400 000 years ago were discovered at Schöningen in Germany. These were found in association with a smaller wooden implement (probably a throwing stick), stone tools and the butchered remains of more than ten horses, and can be persuasively interpreted as throwing spears used in systematic, organized hunting[24].

There is also intriguing although indirect evidence from Southeast Asia dating to around 900 000 years ago that hominids at this time, at least in this region, were capable of repeated water crossings using watercraft. The evidence is in the form of stone tools dated to this age from the island of Flores in eastern Indonesia. Even at the time of the last glacial maximum (when global sea levels would have been at their lowest), reaching Flores from the Asian mainland would have required crossing three deep-water straits with a total distance of at least 19 kilometers. The impoverished – and typically island – nature of the Paleolithic fauna of Flores would appear to preclude the existence of any now submerged land bridge[25]. The next oldest, again indirect, evidence for the use of watercraft is the colonization of Sahul, probably sometime around 40 000 to 60 000 years ago. Even at times of lowest sea level, this would have necessitated crossing some 100 kilometers of open sea.

Fire

At some point in their evolutionary history, hominids clearly learnt to control, manipulate and, presumably later, to start fires. Determining even very approximately when this may have happened is difficult, and as

Watercraft may have played an important role in human dispersal for millennia.

Large-scale burning of vegetation is one of the major human impacts on the biosphere.

Large-scale burning of terrestrial vegetation is undoubtedly one of the major present-day impacts by humans on the biosphere, and is believed to constitute around one third of current annual anthropogenic carbon dioxide emissions. Evidence of earlier impact is invariably circumstantial. It is difficult to demonstrate widespread biomass burning in the fossil or subfossil record and even harder to demonstrate a link with human activities – natural fires may be caused by storms or volcanic activity and may be expected to vary in extent, frequency and intensity according to prevailing climatic and ecological conditions.

However, the abundance of elemental carbon in marine sediments off Sierra Leone in Africa can be persuasively interpreted as a measure of intensity of biomass burning in sub-Saharan Africa. Analysis of a core from these sediments covering the past million years or so indicates that inferred fire incidence in the region was low until around 400 000 years ago. Since then, five episodes of intense burning of vegetation can be inferred, all except the most recent coinciding with periods when the global climate was changing from interglacial to glacial. The current peak is unique in that it is occurring during an interglacial period. The change from interglacial to glacial climate is generally associated with increased aridity, so that vegetation may be expected to be more vulnerable to fire. It would also be expected that a substantial fuel base in the form of woody biomass would have accumulated during the warmer, wetter, interglacial. It may well be merely coincidental, but it is intriguing that the period of increased fire incidence in the Sierra Leone core – around 400 000 years ago – coincides with the timing of the first widespread evidence for hominid use of fire.

with all else to do with human evolution, controversial. This is chiefly because the existence of natural fires caused by lightning strikes and volcanic activity greatly complicates the interpretation of the archeological and geological record – an association between artifacts and evidence of burning does not necessarily indicate a direct link between the two[26].

The earliest dated associations between artifacts and burning that could be construed as deliberate use of fire are from Africa. Stone tools and splintered bones, which can be interpreted as evidence of butchery, have been found associated with clay baked at several hundred degrees for several hours at Chesowanja in Kenya, in deposits around 1.4 million years old[26, 27]. The characteristics of the clay are consistent with formation beneath a campfire, but are also consistent with formation around a slow-burning tree stump that could be associated with a natural bushfire[28]. Charred animal bones and other evidence of human occupation from just over 1 million years ago have been found at Swartkrans cave in South Africa, although similar problems of interpretation apply.

Numerous sites in Europe and Asia provide evidence of human occupation and associated fire from the mid-Pleistocene, some 400 000 years ago. The best known of these is at Zhoukoudien in China, although even here the evidence for deliberate use of fire is widely considered equivocal[29, 30]. Others are at Torralba-Ambrona (Spain), Terra Amata (France), Westbury-sub-Mendip (England) and Vertesszollos (Hungary)[31].

Food resources
The early australopithecine members of the hominid lineage appear, on craniodental evidence, to have been well adapted to consume hard but brittle items, such as nuts and seeds, and soft items, such as fruits[32]. Use and development of stone tools by later hominids is usually thought of as associated

with meat consumption, and cut and hammer marks on large mammal bones from hominid sites are evidence of this, but microscopic wear patterns found on some tools indicate they were used to scrape wood and cut coarse plants such as reeds or grass[33]. Oldowan and Acheulian stone tools are rarely found more than 10-20 kilometers from their rock sources, suggesting that their owners ranged over a relatively small area[33], so if hominid populations at the time were small in size, the environmental impacts of targeted resource extraction would have been minimal.

The ratio of different forms of carbon and nitrogen incorporated in human bone collagen can provide information on the relative importance of food from terrestrial, coastal or freshwater habitats. Such 'stable isotope analysis' of collagen extracted from archeological remains, coupled with other material evidence, suggests that pre-modern humans (Neanderthals) in Europe, although omnivorous and opportunistic, behaved as predatory carnivores[34]. Similar evidence for slightly later human remains suggests a significant broadening of the resource base, owing particularly to increased use of freshwater resources[35]. Remains found at Mediterranean sites suggest further diversification of human diet and the means of food gathering[36],

associated with an apparent rise in human numbers during the later Paleolithic and approaching the start of the Neolithic. Here, there was increasing use of agile small game, such as partridges and hares, and apparently decreasing reliance on slow-maturing, easily collected forms such as tortoises and shellfish, the average size of which decreased over time at a rate consistent with the effects of excess harvesting, possibly indicative of increasing human density[36].

Origins of agriculture

Until the end of the Pleistocene, humans evidently depended on hunting and gathering of wild resources for their sustenance. Around the end of this period a radical change was initiated – the emergence of crop-based agriculture and domestication of livestock – phenomena which appear to have arisen independently in Africa, Eurasia and the Americas.

The earliest direct evidence for animal domestication, of the dog *Canis*, dates from around 14 000 years ago in Oberkassel in Germany[37]. However, analysis of 'molecular clocks' appears to show that the dog diverged from its wild ancestor, the wolf *Canis lupus*, far earlier than this – perhaps more than 100 000 years ago[38]. It is quite possible that

The dog diverged from its wild ancestor, the wolf, more than 100 000 years ago.

domestication of the dog as a guard animal and aid in hunting predates domestication of other animals and plants as food sources by many tens of thousands of years, but in the absence of archeological evidence this remains speculative.

The study of plant and animal domestication for agricultural purposes is a rapidly expand-

age. Wheat *Triticum*, barley *Hordeum*, rye *Secale*, pea *Pisum* and lentil *Lens*, cattle *Bos*, sheep *Ovis*, goat *Capra* and pig *Sus* were all domesticated in this region, and formed the basis of the Neolithic peasant economy that spread steadily through much of western Eurasia into surrounding areas after about 10 000 years before the present. The system integrated the use of food plants, cereals especially, and domesticated animals for fertilizer and power as well as food.

Although early plant domesticates are known elsewhere in the world, integrated agricultural systems appear to have taken longer to develop, perhaps in part because of the absence of domesticated animals. For example, in Middle America seeds of domesticated squash *Cucurbita* are known from about 10 000 years ago, maize from 6 300 years and beans from 2 300, with village-based farming economies evidently taking several thousand years to develop[40]. Farming systems based on rice cultivation in Asia appear to date from 7 500 years ago.

Analysis of mitochondrial DNA has indicated that a multiple maternal origin is general among domestic livestock species, i.e. female animals from more than one wild stock have contributed, at different times and places, to the present genetic diversity among each breed[41]. For each of the four major breeds (cattle, sheep, goats, pigs), this evidence is consistent with archeological findings in suggesting a primary center of origin around the Fertile Crescent region, but also suggests additional centers in Asia.

Most of the major crops on which humans presently depend have been grown continuously since the early or middle Holocene. They have been constantly selected over this period and have developed large amounts of useful genetic variation. Indeed the success of individual crop species over wide geographical areas is partly determined by their flexibility in evolving and sustaining genotypes suitable for local environments. Conventional breeding involves the selection and crossing of desirable phenotypes within a crop in order to create more productive genotypes. The process of harvest, storage and sowing alone may have assisted in the

---

Box 4.1 Loss of diversity in agricultural genetic resources

With the rise of industrial-scale agriculture and commercial breeding, many local agricultural genetic resources, both crops and livestock, have been lost and replaced by modern, genetically uniform types specialized for superior production in higher input systems. There is much evidence that many local varieties possessed features of adaptive value in a particular environment and cultural context, and the precautionary principle argues that the diversity of domesticated forms and their wild relatives should be conserved where possible in order to maintain options for future breeding improvements. A complete picture of the global reduction in local genetic resources is not available, but there is abundant evidence at national level of the enormous scale of genetic erosion in crop plants.

China	Wheat varieties	About 1 000 (10 percent) of 10 000 varieties used in 1949 remained in 1970s
Korea (Rep.)	Garden landraces	About 25 percent of landraces of 14 crops grown in home gardens in 1985 remained in 1993
Mexico	Maize varieties	Only 20 percent of maize varieties planted in 1930s remain
USA	Varieties of apple, cabbage, field maize, pea, tomato	Only 15-20 percent of varieties grown in 1804-1904 available at present

A growing number of countries have documented and evaluated livestock resources, see Appendix 3 for summary of mammal breed status.

---

ing field, based on study of fossil pollen records, archeological remains, the genetics of present-day crops and their close relatives, and remaining indigenous agricultural systems[39]. The development of agriculture began independently in different continents and proceeded at different rates, while early cultivators undoubtedly continued to rely heavily on hunting and gathering from the wild.

The first development of agriculture as an integrated system for food production was based in the Fertile Crescent, composed of the uplands of Anatolia and western Iran and the arid lowlands to the south, with the first tangible evidence approaching 11 000 years in selection of traits such as non-shattering seed heads, uniform ripening of seeds, uniform germination, large fruits and seeds, and easy storage. Breeding methods have increased the rate of introgression of desired genetic traits into new cultivars; genetic modification is now possible at the level of incorporating individual genes directly into genomes.

Domesticated crops and livestock have been transported around the world probably since full-scale agriculture began, e.g. wheats are recorded in areas outside their presumed center of origin at least 8 000 years before the present. Some crops became increasingly widely distributed after the 1500s when European colonists moved out of their home continent.

CURRENT FOOD AND NUTRITION

Perhaps as many as 7 000 of the 270 000 described plant species have been collected or cultivated for consumption[42], but very few, some 200 or so, have been domesticated, and only a handful are crops of major economic importance at global level[43]. The variety of species used is limited more by production and cultural factors, such as tradition and palatability, than by nutritional value. Twelve crop plants together provide about 75 percent of the world's calorie intake. These comprise (in alphabetical order): bananas/plantains, beans, cassava, maize, millet, potatoes, rice, sorghum, soybean, sugar cane, sweet potatoes and wheat[42]. More than 40 mammal and bird species have been domesticated, around 12 of which are important to global agricultural production (see Appendix 3). Some, such as cattle, pigs, goats, sheep and chickens, are fundamental to many non-industrial agri-cultural systems, and provide a wide range of products in addition to foodstuffs.

Although the global consumption of wild terrestrial species cannot be assessed acc-urately, the amount is likely to be insig-nificant in comparison to products from just three domestic forms: pigs, cattle and chickens. On the other hand, fishes from wild sources are an important nutrient source, and the amount harvested and consumed is known to be greatly under-reported in official statistics.

At world level, cereals are the most important single class of food commodities, providing around 50 percent of daily calorie intake and 45 percent of protein. In contrast, meat provides around 15 percent of the protein, and fishery products only some 6 percent, or 15 percent of all animal protein. In each case, when comparing sources of calories, protein and fat in the global human diet, just two or three commodities stand out from a large number of commodities of lesser importance. Rice and wheat together provide around 40 percent of the world supply of both calories and protein, while milk and pigmeat are key sources of fat, calories and protein (see Table 4.1).

At country level, a much wider variety of plant species are important in that they make a significant contribution to human nutrition[44]. Around 22 species and groups of species provide more than 5 percent of the per capita supply of either calories, protein or fat in at least ten countries of the world. Notes on the origin, uses and genetic resources in these 22 crops are tabulated in Appendix 2, together with briefer information on the remaining 50 or so crops that are also nutritionally important but to a smaller number of countries.

Although the vast bulk of human food supply on a global level is derived from

Table 4.1
Top ten food commodities, ranked by percentage contribution to global food supply

Note: This information is partly determined by the way food commodities are aggregated for reporting purposes; for example, fishery products collectively provide about 6% of the hypothetical global protein supply, but no individual fishery commodity is important enough to appear in the table.

Source: FAO food balance sheets[46], 1997 data.

% calories/person/day		% protein/person/day		% fat/person/day	
Rice (milled equiv.)	21	Wheat	22	Pigmeat	14
Wheat	20	Rice (milled equiv.)	15	Soybean oil	10
Sugar (raw equiv.)	7	Milk (excl. butter)	9	Milk (excl. butter)	9
Maize	5	Pigmeat	5	Rape and mustard oil	6
Milk (excl. butter)	4	Bovine meat	5	Sunflowerseed oil	6
Pigmeat	4	Poultry meat	5	Palm oil	5
Soybean oil	2	Maize	5	Fats, animal, raw	5
Potatoes	2	Vegetables (other than		Bovine meat	4
Vegetables (other than		tomatoes, onions)	3	Poultry meat	4
tomatoes, onions)	2	Eggs	3	Butter, ghee	4
Cassava	2	Pulses (other than			
		peas, beans)	2		
Other	31	Other	25	Other	34
Global mean: 2 745 calories/person/day		Global mean: 73.4 grams/person/day		Global mean: 70.1 grams/person/day	

Map 4.2
Livestock breeds:
numbers and status

Color represents the number of mammal breeds in each country. Recording of breeds is incomplete globally. Pie charts represent the proportion of all mammal breeds associated with each FAO region assessed as threatened (gray) or extinct (black).

Source: Charts calculated from FAO *World Watch List for Domestic Animal Diversity* (3rd edition, 2000); country data derived from FAO Domestic Animal Diversity Information System (DAD-IS) database, available online at http://www.fao.org/dad-is/ (accessed February 2002).

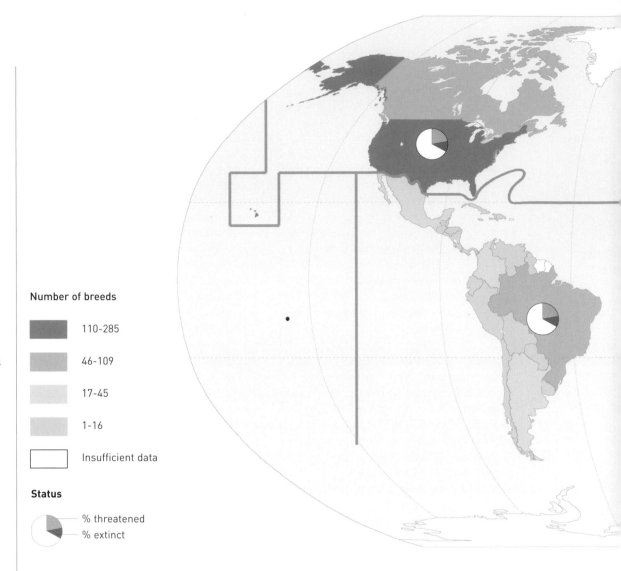

Number of breeds

110-285

46-109

17-45

1-16

Insufficient data

Status

% threatened
% extinct

farming, a large number of people worldwide, including many who are principally agriculturalists or pastoralists, make extensive use of wild resources. In parts of Africa, bushmeat (meat from wild animals) supplies most of the protein intake; similarly some 80 percent of people in sub-Saharan Africa are believed to rely largely or wholly on traditional medicines derived almost exclusively from wild sources. Indeed, traditional medicine continues to be the source of health care for the majority of people living in developing countries, and is widely incorporated in primary health care systems. Wild plants in many areas are extremely important as famine foods when crops fail, or may provide important dietary supplements, and use of fuelwood and charcoal from wild sources is almost universal in the developing world.

A small number of people continue to derive most of their requirements from wild sources by hunting, fishing and gathering. In a large sample (220) of such societies, nearly 40 percent have a high dependence on fishing, around one third are highly dependent on gathering and 28 percent on hunting of terrestrial resources[45]. Fishing and gathering tend to be alternative activities, both of them complementary to terrestrial hunting. In contemporary conditions, fishing tends to be more important where temperature is lower (especially in high northern latitudes) while the converse applies to gathering. People in such communities also derive shelter, medicine, fuelwood and esthetic or spiritual fulfillment from wild species, and the immense variety of plant and animal species used has been well documented.

Global food supply

National data on reported food commodity supplies are collated by the Food and Agriculture Organization of the United Nations (FAO)[46]. Given information on the food value of these commodities, and the size of the human population, it is possible to estimate the average national food supply per person. Dietary food value can be broadly assessed in terms of energy or materials. The former is conventionally measured in calories (calories/person/day), the latter in terms of weight of protein or fat (grams/person/day). These standard measures take no account of the vitamins and minerals (micronutrients) that are required for maintenance of full health. The nutritional value of the human diet varies geographically and over time, and to a great extent according to the state of

development and purchasing power of the people concerned.

The human population of the world doubled between 1950 and 1990, reached around 6 billion in the late 1990s, and will continue to grow for decades, albeit more slowly because of decreased reproduction rates. At global level, there has been enough food available in recent years to supply the entire human population with a very basic diet, largely vegetarian, providing 2 350 kcal per day. In 1992 the average food supply was estimated at 2 718 kcal daily (after losses during storage and cooking), made up of 2 290 kcal from plants and 428 kcal from livestock products[47]. Thus the global food supply was nominally sufficient for the calorific needs of a population 15 percent larger than the estimated population; a similar small annual

Diet class	Class 1 Rice	Class 2 Maize	Class 3 Wheat	Class 4 Milk, meat, wheat	Class 5 Millet, sorghum	Class 6 Cassava, plantain taro, yams
Main geographic distribution	Asia	Central and S. America (north)	N. Africa, West Asia, S. America (south)	Europe, N. America, Australia	Sahel, Namibia	Central Africa, Madagascar
Human population (thousands)	2 920 923	514 911	664 507	942 924	61 867	357 361
% of world population	54	9	12	17	1	7
Increase in supply needed by 2050	x 2.37	x 1.96	x 2.84	x 1.13	x 4.82	x 7.17

Table 4.2
World diet classes

Notes: Population calculation based on 1997 data, from FAO website. Excludes Japan and Malaysia, in anomalous position using six-part classification. Total population in 117 countries in sample: 5 462 493 000. The penultimate row in fact shows the percent each class forms of this total population in the sample countries; this is assumed to be an acceptable surrogate for the total world population.

Source: FAO[47].

surplus has existed since the 1970s[48]. Taking these aggregated global data at face value, it appears that a more than sufficient amount of food has been produced annually during the past two decades to maintain the world's human population. Global food supply has doubled since the 1940s; the increases in supply over the last two decades are attributed mainly to increased productivity (69 percent) and secondarily to an increase in production area (31 percent).

Each year during the past two decades, between 850 million and 900 million people have been undernourished[48]. Given that there has been sufficient food available in the world overall, undernourishment must be an effect of unequal access to the appropriate amount and type of food. Unequal distribution is evident at macro and micro scales: some countries are more productive and richer in resources than others and, whether at national or village level, high-status social groups secure better diets than others, compounded in some cases by gender differences. Poverty is a key cause of undernutrition, and often also an effect of it, forming a self-reinforcing cycle from which it is difficult to escape without appropriate outside intervention. Put in different terms 'food insecurity is a problem of lack of access resulting from either inadequate purchasing power or an inade-

quate endowment with the productive resources that are needed for subsistence'[47]. Although the 1996 World Food Summit in Rome called for a 50 percent reduction in the global number of chronically malnourished people by 2015, the absolute number of people affected remains high, and if trends continue this target will not be met[49].

Regional variation

The 'global average diet' is a simplifying abstraction that ignores an enormous amount of regional, national and local variation in food sources and in supply. Nor does it take account of micronutrient availability (vitamins, minerals, trace elements), i.e. substances that do not directly contribute to energy or protein intake but which are nevertheless essential elements of a healthy diet.

The human diet can be assessed in several ways. The FAO devised a simple classification to serve as the basis for an analysis of food requirements in relation to population growth[47]. In this scheme, diet is assessed in terms of calorie sources as recorded at national level in the FAO food balance database, and countries with similar diet structures are clustered together in one of six diet classes (Table 4.2, Map 4.3). Each class is named after the food product that best characterizes the diet (although this is not necessarily the principal calorie source).

The countries in each food class tend to share broad demographic features. Rice countries (class 1) have high population densities, higher than average mortality rates and little diet diversification. Maize countries (class 2) generally have population densities near the world average and low mortality, especially infant mortality. Wheat countries on average have low population density, but this masks serious land and water shortages in many. Class 4 countries, with a diet described as mixed 'milk-meat-wheat', include the world's most highly developed nations, with fertility, mortality and population growth rates well below the global average. Millet countries (class 5) in general tend to have high population growth, high fertility and low life expectancy; the diet provides only a marginal surplus of energy supplies over requirements. Diet class 6 countries contain the human populations most at risk from food insecurity. The diet is characterized by roots and tubers, and on average does not provide basic energy requirements. These populations show high fertility, high mortality and a rapid growth in numbers. Poverty is widespread and the infrastructure weak, but there are significant reserves of under-exploited arable land.

These major diet patterns have been used to help characterize the improvements to production and food supply that may be needed in order to meet the per capita energy needs of the population projected to exist in 2050. In terms of diet class, availability in class 4 could remain little changed, supply in classes 1, 2 and 3 would need to approximately double, and classes 5 and 6 would need at least a fourfold increase.

In some cases, food security may be better ensured by diversification rather than increasing yields. The limitations of agriculture based on uniform varieties and a high input of fertilizers and pesticides become more acute for farmers who rely on poor resources or marginal environments. By growing diverse and locally adapted crops farmers can bring about greater security in food production and more efficient use of limiting resources. Many traditional agricultural systems manage to varying degrees a high diversity of both cultivated and wild food species (Table 4.3), and in a sense reduce the distinction between wild plants and domesticated crops. Strong selection pressures exerted by natural processes and humans over several millennia and wide geographical areas have resulted in thousands of varieties within most crop species.

High agricultural diversity can not only provide insurance against crop failure in difficult agricultural environments, but tends to have nutritional benefits. Transition to more uniform diets, with high intake of fats and sugar, has resulted in declining nutritional status among numerous indigenous groups[50-52]. Low dietary diversity is associated with micronutrient deficiency, a problem far more common than general protein-energy malnutrition, and particularly prevalent in

children, pregnant women and breastfeeding mothers. Diverse cropping systems in backyard and home gardens, whether rural or urban, can lead to direct improvement in family nutrition and in some cases provide cash income. Even a small mixed vegetable garden is capable[53] of providing 10-20 percent of the recommended daily allowance of

Table 4.3
Examples of diversity in agricultural systems

Source: Multiple sources[74-86].

Country	People	Food production and diversity
Brazil	Kayapo	Over 45 tree species planted for food or to attract game; 86 varieties of food plants grown
Ecuador	Siona-Secoya	Major staples: 15 varieties of manioc, 15 of plantain, 9 of maize; pre-1978 traditional gardens yield 12 300 kilos of food or 8.8 million calories; 72% calories and 14.8% protein, 22.2% fats, 90.9% carbohydrates; post-1978 horticulture provides 67.8% calories, 10.2%, purchased
Indonesia	Java	500 species in home gardens in a single village; ability to support 1 000 people per km²
Indonesia	West Sumatra	6 main tree crops, many vegetables and fruit; 53 species of wild plants also protected and harvested
Kenya	Bungoma	100 species of fruit and vegetables; 47% households collect wild plants, 49% maintain wild plants in gardens
Kenya	Chagga	Over 100 species produced in gardens
Mexico	Huastec	Over 300 useful species found in managed forest plots called te'lom, 81 food species
Mexico	Migrant rural community in southeast	338 species of plants and animals in home gardens, including 62 species of wild plants
Papua New Guinea	Gidra	Approximately 54% calories and 82% of protein come from non-purchased sources (wild, sago, coconut gardens)
Peru	Bora	22 varieties of manioc and 37 tree species are planted; 118 useful species found in fallow fields
Peru	Santa Rosa	168 species identified in 21 home gardens
Philippines	Hanunoo	System of intercropping with 40 crops in a single field; over 1 500 plants considered useful, of which 430 may be grown in swiddens
Sierra Leone	Gola forest	Of food items 14% are hunted, 25% are from fallow land, 8% are from plantations, 19% are from farm, swamp or garden, 21% are from streams and rivers, 13% is bought or given
Thailand	Lua	110 varieties of food plants and 27 wild food plants are found in swidden fallows

Group	No. of species	No. of individuals		Biomass (kilos)	
		Per km² of habitat	Globally	Per km² of habitat	Globally
Bacteria[87]	400 000	7.84×10^{21} - 1.18×10^{22}	4×10^{30} - 6×10^{30}	1.39×10^{6} - 2.14×10^{6}	7.06×10^{14} - 1.09×10^{15}
Collembola[88]	6 500	2.4×10^{7}	2×10^{15}	64	5×10^{9}
Termites[87, 89]	2 760	2.3×10^{9}	2.4×10^{17}	1 400	1.44×10^{11}
Antarctic krill[90]	1	1.43×10^{7}	5×10^{14}	4.29×10^{3}	1.5×10^{11}
Birds[91]	9 946	1 600 - 3 200	2×10^{11} - 4×10^{11}		
Elephants[92-94]	2	0.07 - 0.1	4.26×10^{5} - 6.31×10^{5}	85 - 126	5.34×10^{8} - 7.29×10^{8}
Great whales[94-96]	10	< 0.01	2.83×10^{6} - 3.6×10^{6}	52 - 65	1.89×10^{10} - 2.33×10^{10}
Domestic livestock (excluding pets)[46]	ca 15				7.3×10^{11}
Humans	1		6×10^{9}		3.9×10^{11}
Humans plus livestock					**11.2×10^{11}**

Table 4.4
Number of individuals and biomass, selected organisms

Notes: Calculations based on data in sources cited after group names. Biomass is estimated dry mass standardized on 30% wet mass. Number of birds per km² estimated by dividing global bird population (2-4 x 10¹¹) by land area minus area of extreme desert, rock, sand and ice (approx. 125 million km²). Estimates of abundance for Collembola represent minimum global totals. Whale estimates presume a sex ratio of 1:1. Mean Asian elephant mass estimated at 3 500 kilos; range estimated at 500 000 km². African elephant mass estimated at 4 250 kilos.

protein, 20 percent of iron, 20 percent of calcium, 80 percent of vitamin A and 100 percent of vitamin C.

HUMAN NUMBERS AND IMPACTS
Human population size

Information on early human population numbers is based heavily on inference from circumstantial evidence, and remains on an uncertain footing even when written historical material becomes available in some abundance for the past few hundred years.

Highly speculative estimates based on extrapolations of population densities of great apes[54] and on studies of contemporary human hunter-gatherers[55] indicate that the global late Pleistocene human population may have been between 5 million and 10 million. It seems likely that any increase in human population up to then had been a result of increasing the total area occupied by the species, rather than by any major increase in population density in already occupied areas. Information on human population size in historic times is fragmentary, and populations lacking written records are likely to be inadequately represented.

One analysis[54] distinguishes three main phases of population change. First is a cycle of primary increase in Europe, Asia and the Mediterranean brought about by the spread and further development of Neolithic agriculture, which appears to have allowed a great increase in population density. At the start of the Iron Age in Europe and the Near East, some 3 000 years ago, the world population may have been doubling every 500 years, and the total probably reached 100 million around this time or soon after. Growth appears to have slowed to reach near zero by around year 400, possibly because the limits of then current technology had been reached. After the Dark Ages in Europe, a second growth cycle began around the 10th century in Europe and Asia, with numbers rising to a peak of around 360 million during the 13th century, followed by a slight fall. The global population then increased slowly until the 19th century, when an increasingly rapid rise began as a result of revolutionary developments in agriculture, industry and public health (Figure 4.1).

Crucially, the rate of global population growth peaked during the late 1960s: it was then at just over 2 percent per annum, but is now about 1.7 percent. The absolute increase per annum has also peaked; it was around 85 million more people per annum in the late 1980s and is now about 80 million.

| 12 000 | 11 000 | 10 000 | 9 000 | 8 000 | 7 000 | Years ago | 6 000 |

Such trends suggest that the present global total of some 6 billion may not itself double, as all previous totals have done. The medium variant of the current UN long-range forecast suggests the total in 2050 may be 9.3 billion[56]. Although several countries in Africa have yet to shift to lower reproduction rates, thus making a further doubling quite possible, it may be that 'children born today may be thinking about their retirement at a time when the global population count will have stabilized – or even begun to decline'[57].

The exponential rise in abundance of a single species, to a position of global ecological dominance, in the sense of using a disproportionate share of natural resources, is without known precedent in the history of the biosphere. This was not achieved without significant, often adverse, effects on the environment, many of which stem from the agricultural activities required to maintain human numbers. There is no single species in which so many individuals, of such large body size, are distributed so widely over the planet. There appear to be few macroscopic species, the Antarctic krill *Euphausia suberba* being one of these few, in which the number of individuals approaches the size of the present human population. No animal of comparable size has a population remotely similar in number to that of humans. Biomass provides a measure of the way in which global net primary production or NPP (see Chapter 1) is partitioned. If the standing crop of domestic livestock is added to the human biomass, amounting to around 11.2×10^{11} kilos in total, then only the global biomass of bacteria as a whole is higher (Table 4.4).

The current human population is unevenly spread across the land surface of the Earth. While some areas, such as most of Antarctica, the interior of Greenland, and hyperarid hot deserts, have no permanent human presence, in others human densities may locally reach

an extreme of 1 000 inhabitants per hectare (e.g. in Calcutta and Shanghai). Map 4.4 shows the current density of human populations over the Earth, based on census counts within administrative units of varying size. Among the most striking features of this map are the large areas of very high population density in parts of China and the Ganges-Brahmaputra lowlands, also on Java, and the large area of high population density extending over most of Europe.

Human activities have now made themselves felt throughout the biosphere, but it might be expected that the degree of transformation of terrestrial landscapes would be related to ease of access and proximity to population centers. Settlements, ranging in size from villages to cities of many million inhabitants, are connected by networks of paths, railways and waterways that allow the influence of humans to diffuse far beyond these settlements.

Map 4.5 shows the results of a GIS-based analysis of the relative distance of points on the Earth's surface from all such human constructions. Those points most remote can be given a high 'naturalness' value (i.e. ignoring other possible impacts, they are likely to be least disturbed) and, conversely, points surrounded by a high density of human structures can be given a low value. Although human population density was not part of the analysis, there are, unsurprisingly, strong similarities between Maps 4.4 and 4.5.

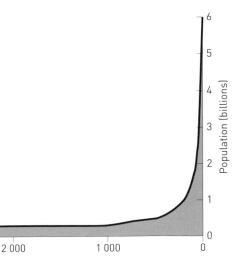

Figure 4.1
Human population

This graph shows the long period of many thousands of years during which the world human population remained small, followed by an exceptionally rapid rise to a total of 6 billion at present.

Source: Data from McEvedy and Jones[54]; FAOSTAT database[46].

Map 4.3
FAO world diet classes

A classification of country dietary patterns, based mainly on calorie sources. Each class is named after the foodstuff that best characterizes the diet. The classification of a few countries (e.g. Japan, Malaysia) appears anomalous, and some countries are not included in the classification.

Source: Analysis by FAO[47].

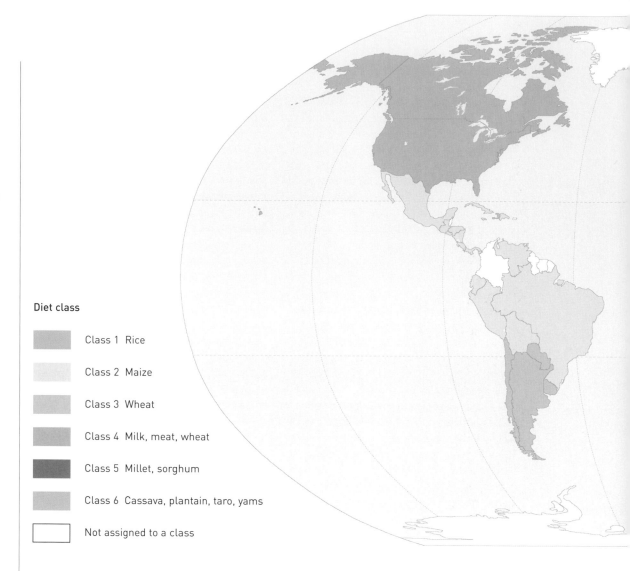

Diet class

	Class 1 Rice
	Class 2 Maize
	Class 3 Wheat
	Class 4 Milk, meat, wheat
	Class 5 Millet, sorghum
	Class 6 Cassava, plantain, taro, yams
	Not assigned to a class

The human share of global resources

Agriculture is a set of activities designed to secure a greater and more reliable share of the energy and materials in the biosphere for the benefit of the human species, i.e. to divert an increased amount of available energy toward production of human bodies. Naturally occurring plants and animals are replaced by specially cultivated or bred varieties that can produce nutrients efficiently from available resources, and in a form that humans can conveniently use. The growth and persistence of these selected species is subsidized by humans. In less-developed agricultural systems, this subsidy may be very small, perhaps just the removal of competition for light or grazing, i.e. plots are cleared or weeded and wild herbivores discouraged. Efficiency can be high but output is usually low. In western industrialized agriculture the subsidy is enormous. Competitors, pests and predators are removed from vast areas (through the use of herbicides and pesticides); fossil fuel is consumed to process, transport and apply any nutrients that limit production (nitrogenous fertilizer) and to store produce. Efficiency is high in some respects, e.g. use of space and labor, but much lower if all hidden costs of fossil fuel use and waste impact are considered. Output can be very high.

The area of land devoted to agricultural production is now a significant proportion of the global land surface. Five thousand years ago, the amount of agricultural land in the world was negligible. There is no direct evidence from the greater part of this period on the rate of expansion of agricultural land. Useful historical data relate to the past few

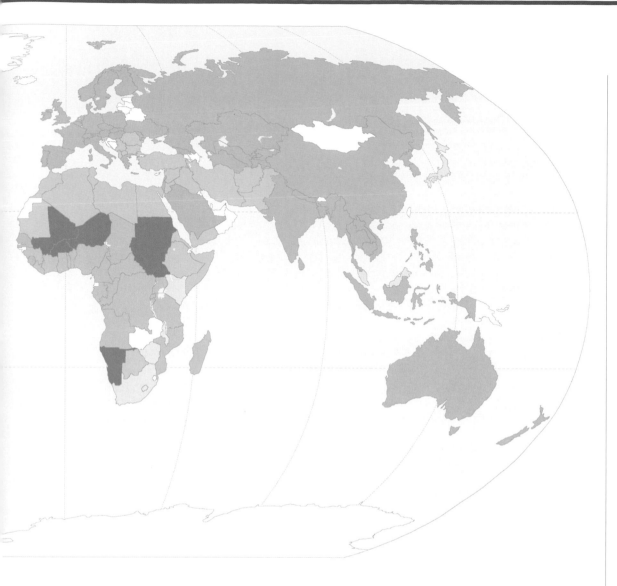

hundred years, and this evidence suggests that about 250 to 300 million hectares (ha) of land globally were devoted to crops in 1700. At present, arable and permanent cropland covers approximately 1 500 million ha of land, with some 3 400 million ha of additional land classed as permanent pasture (this figure includes rangeland and wooded land used for grazing). This represents a nearly sixfold increase in cropped land over the past three centuries (Table 4.5). The current extent of cropland is represented in Map 5.2. All the cropland is used to produce domestic plant material, and much of the land classed as pasture, together with large parts of the grassland and open shrubland landcover classes in Map 5.2 (of which 'permanent pasture' is a subset) is used to produce domestic herbivore biomass.

Most domestic herbivores are destined to become human biomass or to meet other human requirements, so can be counted as surrogate humans. The rise in livestock numbers has been accompanied by a decline in wild herbivores. For example, the extinct wild ox *Bos primigenius* of Eurasia and North Africa has been replaced in its former range by domestic cattle, of which approximately 1 360 million head exist in the world[46]. Similarly, the American bison *Bison bison* was reduced from perhaps 50 million before European arrival on the continent to

Table 4.5
Land converted to cropland

Source: 1700-1950 estimates from Richards[71]; 1980 and 1999 data from FAOSTAT database[46] (complete time series from 1961 available from this source).

Year	1700	1850	1920	1950	1980	1999
Cropland area (million ha)	265	537	913	1 170	1 431	1 501

under 1 000 at the end of the 19th century; although there has been significant recovery under management, it has been replaced over much of its former range by domestic cattle, of which 100 million head are present in the United States.

Data on the fauna of the Bialowieza forest in Poland (a forest remnant with populations of large wild herbivores and predators) have been used[58] to estimate the possible number and biomass of wild herbivores in heavily forested Mesolithic Britain. These estimates were compared with current herbivore populations in Britain. Indications are that there are now 40 times more domestic cattle than there were wild ox and 20 times more domestic sheep than there were wild deer; the overall large herbivore biomass has increased by a factor of ten (see Table 4.6). Similar values are likely to apply to other heavily populated industrialized countries. Not only are domesticated mammals far more abundant than their wild

relatives ever were, the latter are in many cases extinct or near extinction.

As mentioned in Chapter 1, calculations at global level and in a study of one country (Austria) estimate the proportion of terrestrial global production appropriated or diverted for human use at 30-50 percent (depending on how much change beyond direct harvesting is incorporated). One approach to assessing the effects of energy appropriation has used the empirical relationship between energy and species number[59]. There is much evidence suggesting that at a range of spatial scales and within different taxonomic groups, the diversity of species present in an ecosystem tends to be positively correlated with the amount of energy available (see Chapter 5). Accepting the empirical evidence for this relationship, it has been argued that the number of species present will decline if the amount of energy available for use declines. At global level, a conservative estimate using this relationship predicted that 3-9 percent of terrestrial species would be extinct or endangered by the year 2000[59]. The evidence from the Austrian study[60] is consistent with the species-energy relationship: the curve developed by Wright[59] predicts that with 41 percent of potential NPP at country level now being appropriated by humans, 5-13 percent of species should have been extirpated from the country; in fact, 8 percent of birds and 7-14 percent of reptiles have been lost.

The significance human resource appropriation has for other species, and for biosphere function, in part depends on the extent to which the resources are limited in availability. Space clearly is limited, and it may be that the main effects of appropriation will be exerted through direct changes to land-cover, and diminished availability of material resources for wild species, rather than through a diversion in energy flow.

SPECIES EXTINCTION AND HUMANS
Tracking extinction
Change in biological diversity has principally been assessed in terms of declining populations, species, communities and habitats. There has always been special concern about extinction, because this is a threshold from

Table 4.6
Estimated large herbivore numbers and biomass in Mesolithic and modern Britain

Source: After Yalden[58].

Mesolithic Britain			Modern Britain		
	No.	Kilos		No.	Kilos
			Sheep	20 364 600	916 407 000
Wild boar					
Sus scrofa	1 357 740	108 619 200	Pigs	853 000	127 950 000
Wild ox					
Bos primigenius	99 250	39 700 000	Cattle	3 908 900	2 149 895 000
Red deer					
Cervus elephus	1 472 870	147 287 000	Red deer	360 000	54 000 000
Roe deer					
Capreolus capreolus	1 083 810	21 676 200	Roe deer	500 000	10 500 000
Moose					
Alces alces	67 490	13 498 000	Introduced deer	111 500	5 017 500
			Feral sheep and goats	5 700	256 500
			Domestic herbivore total	25 126 500	3 194 252 000
			Wild herbivore total	977 200	69 774 000
Herbivore total	4 081 160	330 780 400	Herbivore total	26 103 700	3 264 026 000

which there is no turning back. As discussed in Chapter 3, the extinction of species is natural and expected, and self-evidently there have always been species at risk of extinction. It seems very likely, however, that recent and current extinction rates are considerably higher than would be expected without the influence of humans.

For several reasons it is difficult to record contemporary extinction events with precision. The species involved may well be unknown. Even if they have been discovered and named, they may be too small to be noticed without special sampling procedures. The entire process of decline and eventual extinction may take place over many years or even centuries in the case of particularly long-lived organisms such as many trees. The near-terminal stages in the process of species extinction are unlikely to be observed. Where such observations have been possible, it is because the species has been destroyed in part by unusually extreme hunting pressure (e.g. the passenger pigeon *Ectopistes migratorius*) or extreme ecological events (e.g. extinction of many native land snails in French Polynesia and Hawaii following introduction of the predatory snail *Euglandina rosea*), and has been the subject of sufficient interest to be closely monitored.

In other cases, positive evidence of extinction is lacking. Typically, many years elapse before sightings of a species become sparse enough to generate concern, and many more years are likely to pass before negative evidence (i.e. failure to find the species despite repeated searches) accumulates to the point where extinction is the most probable explanation. In other words, unless circumstances are exceptional, monitoring of recent extinction events has a resolution limit measured in years or decades. This is why it is not possible to state with precision how many species became extinct in a given month, year or even decade, nor to predict exactly how many species, let alone which ones, are going to become extinct this year or decade or century. The search effort and chance both play a part in determining whether a species not seen for decades is rediscovered, as shown by the occasional new encounter with

species once feared extinct (see Box 4.2). Hidden survivors are even more likely in plants, which may have propagules that can remain viable but unseen for extremely long periods.

Extinctions in the recent past are likely to be recorded with significant accuracy either where circumstances favor preservation of hard remains in good number (e.g. in caves, potholes or kitchen middens) or where early naturalists recorded the fauna or flora with sufficient care that they set a firm baseline against which the composition of the modern biota may be assessed. The detailed record of bird extinction in Hawaii is a result of the former circumstance, and the record of mollusk extinction on many islands a result of the latter.

Box 4.2 'Lazarus species'

Standing out as some good news against a background of widespread species depletion, a small but significant number of species have been considered extinct but rediscovered after a gap of several decades. In some cases, prolonged directed searches have been carried out, but others have emerged by chance. Sometimes a 'rediscovery' is in part a consequence of new systematic work confirming the species status of long-neglected populations, e.g. differentiation of the pygmy mouse lemur *Microcebus myoxinus* from other mouse lemurs on Madagascar. The coelacanth *Latimeria* is an extreme 'Lazarus taxon', being the living representative of an entire order (Coelacanthiformes) of early fishes thought to have become extinct some 80 million years ago. There are at least two population groups, one off southeast Africa (*L. chalumnae*) first discovered in 1938, and one in Indonesia (named *L. menadoensis*) discovered in 1998.

- Bavarian pine vole *Microtus bavaricus*: an alpine species believed lost when a hotel was built near its single known locality, but recently rediscovered nearby in the Austrian Alps.
- Fiji petrel *Pterodroma macgillivrayi*: known by one specimen collected in 1855, but regarded as extinct until a bird flew into a researcher's headlight one night in 1984.
- Jerdon's courser *Rhinoptilus bitorquatus*: known by just two museum skins and last recorded in 1900, until rediscovered in 1986 in a patch of scrub forest in southern India.
- Jamaican iguana *Cyclura collei*: believed to have gone extinct during the 1940s but rediscovered in 1990 in the remote Hellshire Hills.
- The Cranbrook pea *Gastrolobium lehmannii*: endemic to Western Australia where last collected from the wild in 1918 and since listed as extinct; rediscovered in 1995.

Table 4.7
Late Pleistocene extinct and living genera of large animals

Source: Adapted from Martin[73].

Area	Extinct within last 100 000 years	Still extant	Total genera	% extinct	Estimated timing of maximum extinction
Africa	7	42	49	14	No peak
North America	33	12	45	73	11 000-13 000 years ago
South America	46	12	58	79	11 000-13 000 years ago
Australia	19	3	22	86	ca 50 000 years ago

Map 4.4
Human population density

The relative density of human population based on census data relating to administrative units of various sizes.

Source: CIESIN, gridded population of the world, version 2, data available online at http://sedac.ciesin.columbia.edu/plue/gpw/ (accessed April 2002).

Population density

High

Low

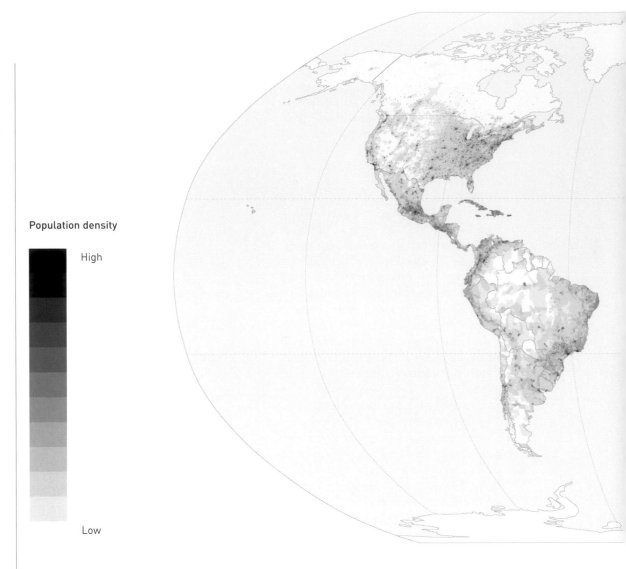

Early human impacts on biodiversity

A wide range of factors affects the frequency of occurrence of particular species in the fossil record, of which the abundance of that species in life is only one. There is not necessarily therefore any direct relationship between the former and the latter, so that deducing past changes in abundance of species from the fossil record is a problematic exercise. Cataloging, though not dating, extinctions is rather less contentious, although even this may be problematic as evinced by the existence of so-called 'Lazarus' taxa (those presumed extinct that are rediscovered alive.

One of the unusual features of the Quaternary period (the Pleistocene and Holocene) has been the disproportionately high extinction rates in the largest ter-restrial species, particularly mammals and birds (Table 4.7). These species are generally referred to as the 'megafauna', often defined as those with an adult mass of 44 kilos or more, although the term has not been used consistently.

The extinct American fauna include such well-known genera as the sabretooth cats *Smilodon*, giant ground sloths *Eremotherium*, glyptodonts *Glyptotherium* and mammoths *Mammuthus*, as well as a number of scavenging and raptorial birds, including the giant *Teratornis* and *Cathartornis*. Those in Australia include the marsupial equivalents of rhinoceroses (family Diprotodontidae) and lions (*Thylacoleo*), giant wombats (*Phascolonus*, *Ramsayia* and *Phascolomis*), the large emu-like *Genyornis* and the giant monitor lizard *Megalania*. In all, some 40

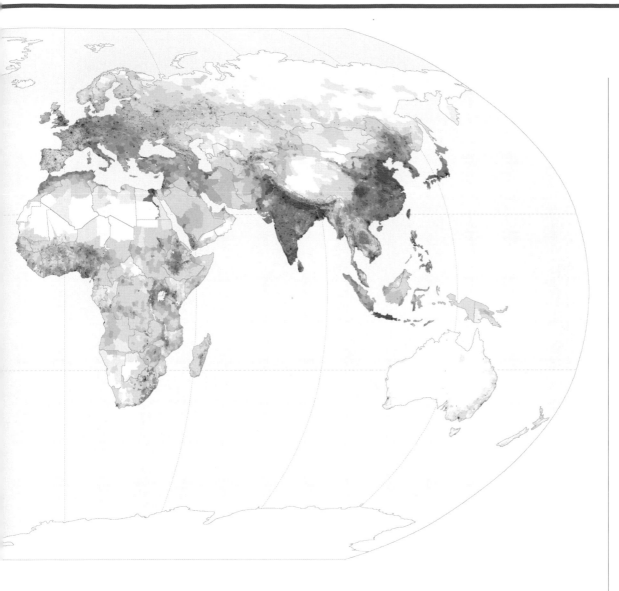

species of the larger Australian land mammals, reptiles and birds became extinct across the entire continent about 46 000 years ago[16], approximately 10 000 years after the first known human colonists.

These extinctions have been followed by a similar series on islands during the Holocene. On New Zealand, the moas (giant flightless ratite birds in the family Anomalopterygidae) became extinct after the first humans came to the islands. On Madagascar, two endemic hippopotamus species *Hippopotamus lemerlei* and *H. madagascariensis*, the elephantbird *Aepyornis maximus*, and a number of large to very large lemur species all appear to have died out 500 to 900 years ago. Similarly on the Caribbean islands, a number of large mammals, including several ground sloths (order Xenarthra, family Megalonychidae)

appeared to have survived until human occupation but to have died out at some point since then.

The precise causes of all these extinctions have been the subject of endless debate, which centers chiefly on the role of humans. At one extreme lies the 'blitzkrieg' hypothesis, applied particularly to the apparently sudden collapse (i.e. over a few hundred years) of the North American megafauna at the hand of humans. At the other are those who maintain that in most, if not all, cases the impact of early humans was negligible and climate change, particularly increasing aridity, was the cause.

Several features of the phenomenon seem to point persuasively to humans having played a pivotal role in most, if not all, of these extinctions. The most compelling is their

Map 4.5
Terrestrial wilderness

The wilderness value of any given point is essentially a measure of remoteness from human influence, assessed on the basis of distance from settlement, access routes and permanent manmade structures.

Source: GIS analysis by R. Lesslie (ANU), method developed for the Australian Heritage Commission.

Wilderness level

High

Low

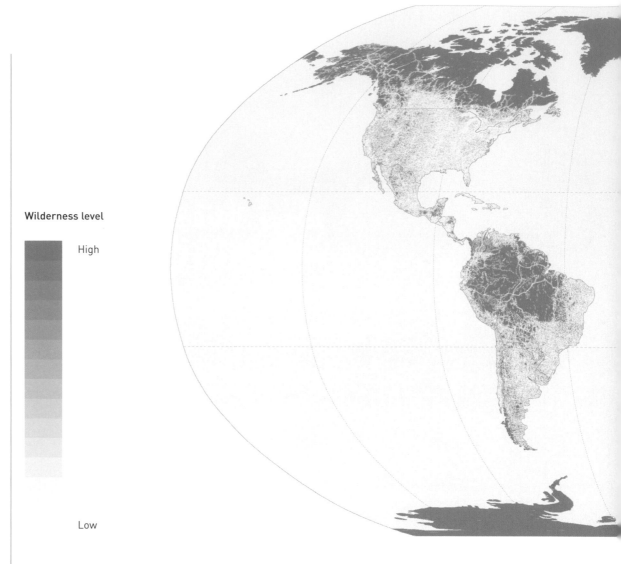

timing. In each case the arrival of humans seems to have preceded the major spate of extinctions (in as much as these can be dated), with no or very few such extinctions having been recorded prior to human arrival (compare Map 4.1 and Table 4.7). The cumulative weight of this coincidence is difficult to counter, and is supported by population modeling. For example, new archeological evidence suggests that the first Polynesian settlements in New Zealand date from the late 13th century, and moas were becoming scarce by the end of the 14th. This evidence, coupled with a mathematical model of population and predation, indicates that all 11 moa species were driven to extinction within 100 years of human arrival[61]. Similarly, a computer simulation[62] of population dynamics of humans and large herbivores in

North America accurately models megafaunal extinction in accord with archeological evidence for significant human arrival around 13 400 years ago and the first wave of extinction starting some 1 000 years later.

The situation is somewhat different in Africa and Eurasia, where megafaunal extinctions were relatively few, and were spread out over the whole of the Pleistocene. In Africa the peak was in the lower Pleistocene (21 genera extirpated from the region between 1.8 million and 700 000 years ago compared with nine between 700 000 and 130 000 years ago, and seven later than 130 000 years ago). It is noteworthy that this more gradual, earlier and less extreme pattern of extinctions has typified the region where humans evolved.

It is difficult to formulate an entirely climate-based model of extinction that can

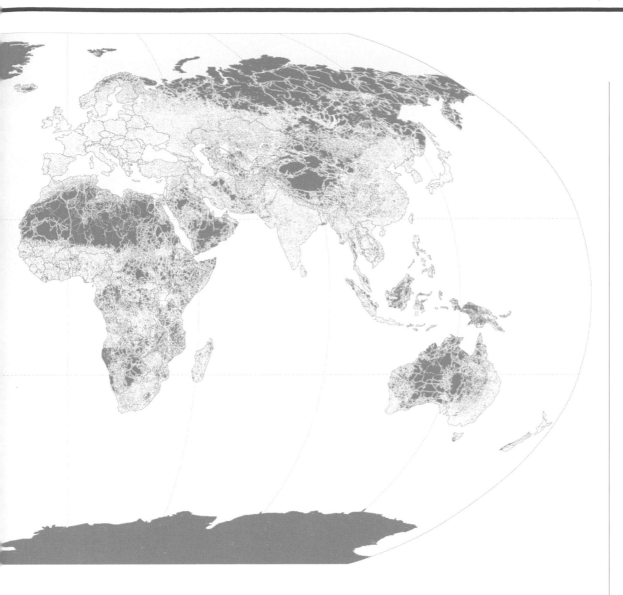

account for this asynchronicity – outside Africa and Eurasia, these species survived a series of climatic changes at least as extreme as those they faced at the start of the Holocene. The recent studies of *Genyornis newtoni* and other megafauna in Australia[16, 63] seem to indicate a widespread and largely synchronous disappearance from a wide range of habitat types during a period of relative climatic stability, strongly implying that some other agent was responsible. It is also noteworthy that the fossil and archeological evidence indicates that, on continents at least, these extinctions were not matched by parallel extinctions of smaller species (for example, as far as is known no insect species at all became extinct in Europe in the entire Pleistocene[1]). If climate change were the cause, it would be expected that these species would be at least as affected

as larger ones as in general their opportunities for long-range dispersal and migration are much more limited.

Even if humans are accepted as the major agents in these extinctions, evidence for the mechanisms involved in most cases remains elusive. It seems likely that extensive use of fire, and direct hunting of a fauna that had evolved in the absence of humans and was therefore unlikely to recognize them as potential predators, may have been sufficient cause to exterminate the large herbivores. The large carnivores and scavengers may then have suffered population collapses owing to the disappearance of their prey base.

Extinction in modern times
From the relatively sparse evidence that is available, it appears that amongst animals

Map 4.6
Vertebrate extinctions since AD1500

An indication of the number and former occurrence of recently extinct mammal, bird and freshwater fish species. Each of these three vertebrate classes is represented by a differently colored symbol, sized according to number of extinct species. Only extinctions that are fully resolved according to firm Committee on Recently Extinct Organisms (CREO) criteria are covered. The exception is Lake Victoria, indicated by a blue circle: some reports have suggested that up to a third of the 500 or so endemic fishes are extinct, but others note that available evidence is inconclusive. In many cases, including most islands and lakes, the position of the symbol indicates last record or core of former range. Where several species ranged more widely over a country, the symbol is positioned at the centre of that country.

Source: Based on several sources; see Appendix 4.

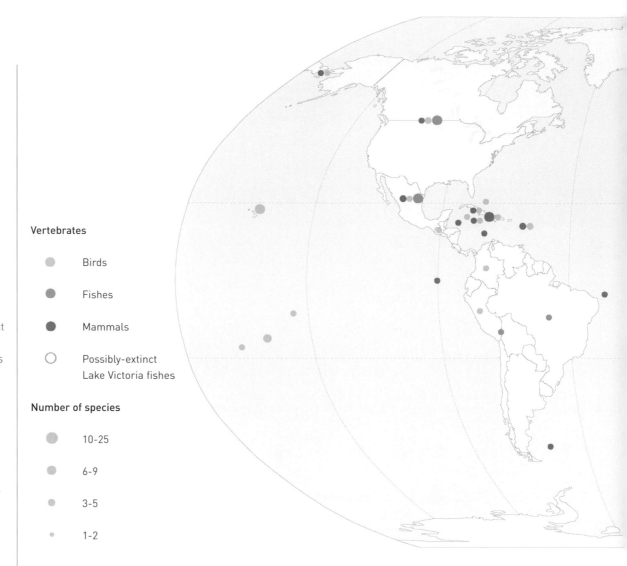

Vertebrates

- Birds
- Fishes
- Mammals
- Possibly-extinct Lake Victoria fishes

Number of species

- 10-25
- 6-9
- 3-5
- 1-2

more than 300 vertebrates, including at least 60 and possibly more than 80 mammals, more than 120 birds and around 375 invertebrates, have become extinct during the past 400 years (see numerical summary in Table 4.8 and list of extinct vertebrates in Appendix 4). Data for plants are much more equivocal, owing in part to the uncertain taxonomic status of many extinct plant populations. One source[64] lists some 380 extinct plant taxa and a further 370 or so classified as extinct or endangered (these include a number of infraspecific taxa and a number that although believed extinct in the wild are extant, and sometimes abundant, in cultivation).

Because mammals and birds tend to be relatively well recorded, and because they leave recognizable macroscopic skeletal remains, it is principally among these groups that known extinctions may be reasonably representative of actual extinctions. In these groups the known extinction rate over the past 400 years, based on data in Appendix 4, averages out at around 20-25 species per 100 years.

A crucial question then, is how this observed extinction rate compares with some hypothetical or expected background extinction rate. It is of course impossible to derive such a rate from observation of the modern world, as this has already been highly modified by human activity. The only reasonable comparison is thus with historical records, for which we must turn to the fossil record, discussed in Chapter 3. Although extinction rates have evidently been highly variable during the history of life on Earth, it seems that the average persistence time of species in the fossil record is around 4 million

years. If 10 million species exist at any one time, on this basis the extinction rate would amount to around 2.5 species annually. However, it is unclear how representative the fossil record is of species as a whole. It is likely that many rare species or those with very restricted distribution never appear in it at all. These rare species may almost by definition be expected to be inherently more prone to extinction than the species that are recorded and may therefore be expected to have a lower persistence time. This would mean that average species duration was less – perhaps much less – than 4 million years and the actual extinction rate in geological time considerably higher than the rate observed in the fossil record.

Applying a mean persistence time of 4 million years to birds and mammals (and assuming some 10 000 species of the former and around 5 000 of the latter), the background extinction rate would be around one species every 500 years and 1 000 years, respectively, so that current rates would be some 100 or 200 higher than background. Even if background rates in these groups were

Group	No. of extinct species
Mammals	83
Birds	128
Reptiles	21
Amphibians	5
Fishes	81
Insects	72
Mollusks	291
Other invertebrates	12

Table 4.8
Numbers of extinct animal species according to IUCN

Note: Alternative criteria used by the Committee on Recently Extinct Organisms (CREO) result in a lower number of mammals and fishes being regarded as certainly extinct than is given in this table (see full list in Appendix 4).

Source: Hilton-Taylor[67].

Map 4.7
Threatened mammal species

Color represents number of globally threatened mammal species in each country in 2000. Pie charts represent the proportion of the mammal fauna assessed as threatened at the national level in a small sample of countries. This is a highly generalized comparison because of differences in status assessment methods.

Note: To reduce ambiguity Alaska (United States) has for the purposes of this map been assigned to the same class as adjacent Canada rather than the conterminous United States.

Source: Global data from Hilton-Taylor[67], country data from selection of national Red Data books.

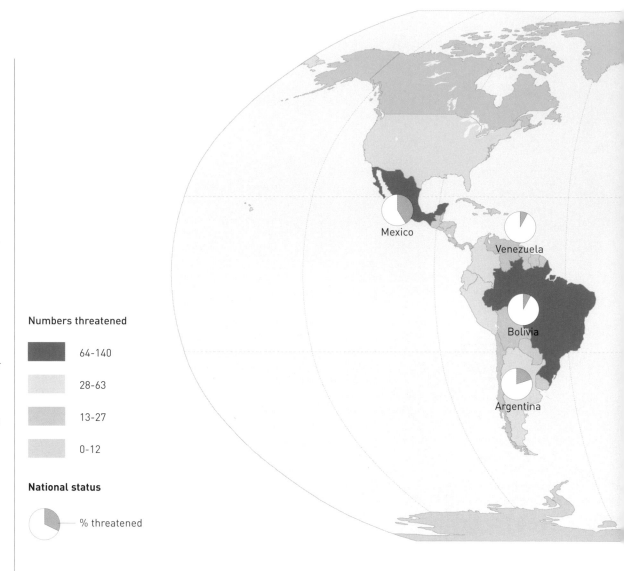

Numbers threatened

■	64-140
■	28-63
■	13-27
■	0-12

National status

○ % threatened

ten times this, the currently observed extinction rates would still be 10-20 times those expected. It seems therefore that, even if a high background extinction rate is postulated, recent extinctions are still much higher than might be expected (the alternative explanation is that the background rate was higher still). This elevated rate is particularly noteworthy as the Holocene appears to have been a period of relative climatic and geological stability, in which extinction rates might have been expected to be low.

Most known extinctions have occurred on islands, including 42 of the 61 resolved mammal extinctions (68 percent), and 105 of the 128 bird extinctions (82 percent). Reasons for the former are probably twofold. First, island species do appear to be particularly extinction-prone, by virtue of their limited ranges and usually small population sizes, and also because they have often evolved in the absence of certain pressures (e.g. terrestrial predators, grazing ungulates); if faced with these pressures (usually through human intervention) their populations may collapse completely. Second, it is much easier to arrive at some certainty that a given species is no longer present on an island of limited extent than that it has disappeared completely from a continental range, where the limits of its range were probably uncertain to start with.

Most known or probable continental extinctions have been among freshwater organisms, particularly fishes and mollusks. Many freshwater biota appear to have the characteristics of island organisms, in that they have limited and highly circumscribed ranges and they are similarly often sensitive to

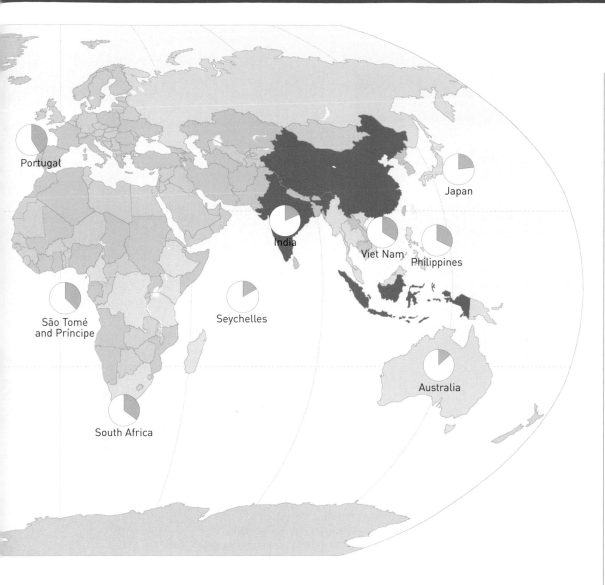

**Figure 4.2
Vertebrate extinctions by period since AD1500**

This graphic is derived from information in the 'Period' column of the list of extinct vertebrates given in Appendix 4. 'Early' refers to the first four decades of a century, 'mid' to the next three decades, 'late' to the final three decades. Only the extinction events regarded by Committee on Recently Extinct Organisms (CREO) criteria as 'resolved' are shown, and only the three groups recently assessed are represented.

Source: Derived from Appendix 4.

external pressures (e.g. introduction of predatory fish species, complete habitat destruction through drainage or dam construction).

Figure 4.2 represents the number of accepted extinction events among mammals, birds and fishes in each third of the five centuries since AD1500 (see also Map 4.6). The information in Appendix 4 allows some elements of this broad global picture to be disentangled:

- early extinctions among the remaining archaic fauna of large islands, notably mammals on Hispaniola, Cuba and Madagascar;
- somewhat later extinctions, especially among ground birds, on many isolated small islands, such as St Helena, the Mascarenes and others, during European colonial consolidation;

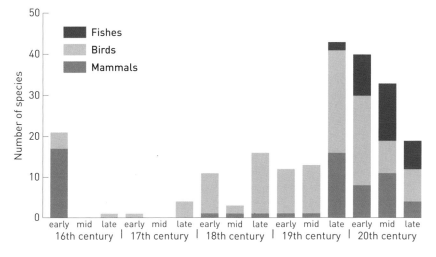

	No. of breeding bird species	No. globally threatened	% breeding species threatened
Continental countries			
Brazil	1 492	113	8
China	1 100	73	7
India	923	68	7
Colombia	1 695	77	5
Peru	1 538	73	5
Small islands			
French Polynesia	60	23	38
Solomon Islands	163	23	14
Mauritius (and Rodrigues)	27	9	33
São Tomé and Príncipe	63	9	14
Saint Helena and dependencies	53	13	25

Table 4.9
Island diversity at risk: birds

Note: The first five rows are the continental countries (Indonesia and Philippines excluded) with most threatened bird species; the other five rows are the small island groups with most threatened bird species.

Source: BirdLife International[70].

• increasing evidence of 20th century extinctions among fishes in many countries, and among marsupial mammals on the Australian mainland, with continuing losses of island species everywhere.

The apparent reduction in extinction rate from the late 19th century onwards may be in part attributed to management action designed to maintain highly threatened species, and there is indeed good evidence that populations of a small number of target species have recovered significantly, but it is sure also to reflect the difficulties of observing and documenting extinction events. As mentioned above, many years may pass before a possibly extinct species can be treated with certainty as extinct, and the declining extinction rate toward the end of the 20th century is most probably in part an artifact of this monitoring process.

Clearly, in view of our very incomplete knowledge of the world's species, and the fact that only a minute proportion of living species are being actively monitored at any one time, it is extremely likely that more extinctions are occurring than are currently known. Indeed, most predictions of present and near-future extinctions suggest extremely high rates. Most are based on combining estimates of species richness in tropical forest with estimates of rate of loss of these forests, and

predict species extinction on the basis of the general species-area relationship (which predicts a decline in species richness as area declines, see figure 5.1). It is widely believed that the great majority of all terrestrial species occur in tropical forests, and most of these species will be undescribed arthropods (notably beetles). At present rates of forest loss, it has been predicted that between 2 and 8 percent of forest species will become extinct, or committed to extinction, between 1990 and 2015[65]. Depending on whether higher or lower estimates of tropical species richness are used, extinction at this rate could entail loss of up to 100 000 species annually. As a cautionary note, it should be observed that few extinctions to date have actually been recorded in continental tropical moist forests, although monitoring species in these habitats presents great difficulty.

Threatened species assessment

Except for species lost as a result of random environmental factors, extinct species must largely be drawn from a pool of species that could be assessed as in decline or at risk, all of which face eventual extinction if negative trends or threats to their populations are not reversed. Various national and other programs have developed methods to assess the relative severity of risks faced by species, and to label species with an indicative category name. Conservation activities can then be prioritized on the basis of relative risk, taking account of other relevant factors, such as feasibility, cost and benefits, as appropriate.

The system[66] developed by the Species Survival Commission of IUCN-the World Conservation Union (IUCN/SSC) and collaborators in conjunction with its Red Data Book[67] and Red List Programme[68] has been designed to provide an explicit and objective framework for assessment of extinction risk, and to be applicable to any taxonomic unit at or below the species level, and within any specified geographical or political area. To be categorized as threatened, any species has to meet one of five sets of criteria formulated to permit evaluation of all kinds of species, with a wide range of biological characteristics. The criteria are defined on population reduction, population size, geographic area and pattern of occurrence, and quantitative population analysis.

The size and connectedness of different populations of a species influence the likelihood of its survival. In general, small isolated populations will be more sensitive than larger connected ones to demographic factors (e.g. random events affecting the survival and reproduction of individuals) or environmental factors (e.g. hurricanes, spread of disease, changes in the availability of food). Islands tend to have a much higher proportion of their biota at risk than continental countries because they start with many fewer species, all or many of which face the risks associated with a small range size (see Table 4.9 for an example using bird data). Biogeographic theory based around the species-area relationship, supported by much empirical evidence, predicts that each 'habitat island' created by fragmentation of a continuous habitat area will come at equilibrium to contain fewer species than previously. Human activities everywhere tend to promote fragmentation of natural and often species-rich habitats (e.g. primary tropical forest or temperate meadow grassland) and the spread of highly managed species-poor habitats (e.g. eucalypt plantations or cereal croplands). As a result, many species occur in just the kind of fragmented pattern that increases the risk of extinction.

Recent declines

Reduction in population numbers, or complete loss of a species from a site or an individual country (often termed extirpation) are far easier to observe than global species extinction, and appear liable to occur wherever humankind has modified the environment for its own ends. The conservation status of most species is not known in detail, and this certainly applies to the many million as yet undescribed species, but two large animal groups – the mammals and birds – have been comprehensively assessed and may be representative of the status of biodiversity in general. Approximately 24 percent (1 130) of the world's mammals and 12 percent (1 183) of the world's bird species are regarded on the basis of IUCN/SSC

criteria as threatened (see Table 4.10). Proportions are a great deal lower in other vertebrates, but none of these has been assessed fully. Empirical observations such as these give sufficient grounds for serious concern for biodiversity maintenance, regardless of any hypotheses that have been proposed regarding the future rate of extinction. Interestingly, the ratio of threatened mammals to threatened birds is near 1:1, as is the ratio of recorded, recently extinct mammals to birds, giving some indication that threatened species categories may be a reasonably reliable indicator of proneness to extinction, and also that mammals may as a group be somewhat more susceptible than birds to extinction.

Countless other species, although not yet globally threatened, now exist in reduced numbers and fragmented populations, and many of these are threatened with extinction at national level. The significance of loss of diversity at gene level implied by loss of local populations of species is not clear, although it has been argued that loss of resilience in response to environmental change is inevitable.

Table 4.10
Threatened species

Notes: Includes species assessed as globally threatened and assigned to categories 'critically endangered', 'endangered' or 'vulnerable' under IUCN criteria. Mosses not included here; most plant taxa listed were assessed for the *World list of threatened trees*[97] in 1998.

Source: Hilton-Taylor[67]; BirdLife International[70].

	No. of species in group	Approx. % of group assessed	Threatened species	% of total in group threatened
Vertebrates				
Mammals	4 630	100	1 130	24
Birds	9 750	100	1 186	12
Reptiles	8 002	<15	296	4
Amphibians	4 950	<15	146	3
Fishes	25 000	<10	752	3
Invertebrates				
Insects	950 000	< 0.01	555	0.06
Mollusks	70 000	<5	938	1
Crustaceans	40 000	<5	408	1
Others			27	
Plants				
Gymnosperms (Coniferophyta, Cycadophyta, Ginkgophyta)	876	72	141	16
Angiosperms (Anthophyta, flowering plants)	250 000	<5	5 390	2

Map 4.8
Critically endangered mammals and birds

The general distribution of almost all the 362 mammal and bird species categorized as 'critically endangered', the highest risk category, in 2000. Each circle represents one distribution record; some species are known from a single point locality, others are represented by a cluster of localities. On this map, a high density of symbols can represent many records of a single species, or single records of many separate species. Red and blue symbols represent mammals and birds, respectively.

Source: Mammals: species selection based on Hilton-Taylor[67], distribution research and mapping by UNEP-WCMC; birds: spatial data provided by BirdLife International (February 2002), further information in BirdLife International[70].

Critically endangered

○   Mammals

○   Birds

At global level, most of the species assessed as threatened are terrestrial forms (Table 4.11). The preponderance of terrestrial species is because the great majority of mammals and birds are terrestrial, and little or nothing is known of the population status of most aquatic species in most groups. Where significant numbers of aquatic species have been assessed, e.g. among crustaceans and mollusks, the proportion of threatened aquatic species rises markedly. Among fishes, the high number of freshwater species doubtless in part reflects the general lack of data on marine species, but to some extent indicates relative risk – many freshwater species being restricted to small and isolated habitat patches.

Evidence of the vulnerable nature of freshwater habitats and the risk faced by many aquatic groups is accumulating. For example, in the United States, freshwater groups are considerably more threatened than terrestrial groups (specifically, nearly 70 percent of the mussels, 50 percent of the crayfish and 37 percent of the fishes)[69].

Forest is an important habitat for a high proportion of the threatened terrestrial vertebrates. Among birds, for example, about 70 percent of the 1 186 species assessed as threatened in the year 2000 occur in forest, and 25 percent in grassland, savannah and scrub habitats[70]. Of the threatened forest birds, 41 percent occur in lowland moist forest and about 35 percent in montane moist forest[70]. Among the 515 mammals regarded as threatened in 2000 that were assigned to a habitat category, 33 percent occur in lowland moist forest and 22 percent in montane formations[67].

Table 4.11
Number of threatened animal species in major biomes

Notes: Counts include globally threatened species tabulated in each biome (IUCN categories 'critically endangered', 'endangered' or 'vulnerable'). Only mammals and birds have been comprehensively assessed. Some species, e.g. amphibians and migratory fishes, are counted in more than one biome row. Plants not tabulated: almost all assessed as globally threatened are terrestrial. Inland water includes saline wetlands, cave waters, etc., as well as freshwaters.

Source: Hilton-Taylor[67].

Numbers of globally threatened species can be mapped at country level (e.g. Map 4.7, mammals). Maps of this kind provide at best a broad overview of the occurrence of such species, shown as direct numbers or as a proportion of the total number in that group in the country. As with other country-level biodiversity analyses, this information could be used to help focus efforts to slow or reverse biodiversity decline. Data with improved spatial resolution (although plotted in a highly simplified way) are shown in Map 4.8, representing the individual ranges of all the mammals and birds assessed as 'critically endangered'. BirdLife International has generated individual distribution maps of all threatened birds. Among other applications, these are being used to develop the first-ever world map of all the threatened species in an entire major group of organisms. A preliminary version of this, as a global density surface, is shown in Map 4.9. Being geo-referenced rather than country based, and at relatively fine spatial resolution, this has considerable potential to focus bird conservation efforts effectively.

Biome type	Total	Mammals	Birds	Reptiles	Amphibians	Fishes	Crustaceans	Insects	Mollusks
Marine	315	25	105	9	0	163	0	0	13
Inland water	1 932	31	78	111	131	627	409	125	420
Terrestrial	3 627	1 111	1 144	283	143	0	0	438	508

Animal group

Map 4.9
Threatened bird species density

BirdLife International has developed digital range maps of all threatened bird species. This map is based on distribution data for the 1 186 species assessed as threatened in 2000. The data are plotted as a global density surface, representing the number of species potentially present in each location. The map is the first such treatment of any large group of threatened species.

Source: Density surface provided by BirdLife International, further information on threatened birds in BirdLife International[70].

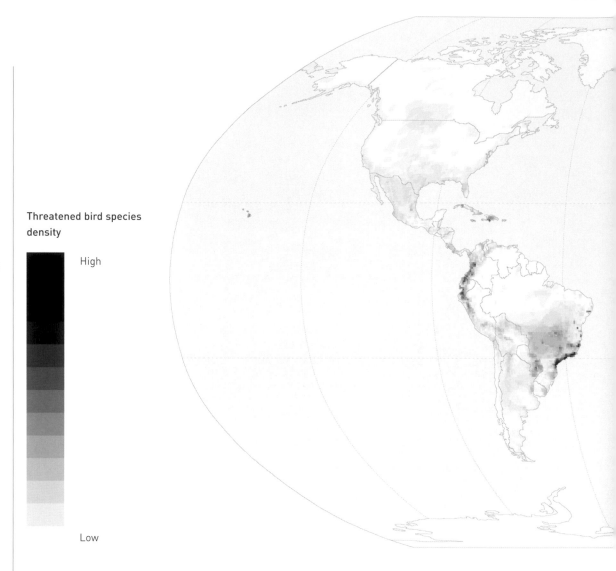

Threatened bird species density

High

Low

Proximate causes of recent declines

The continuing conversion of natural habitats to cropland[71] and other uses typically entails the replacement of systems rich in biodiversity with monocultures or systems poor in biodiversity. Habitat modification, from agricultural conversion and a variety of other causes, is in general the most important factor acting to increase species' risk of extinction. Among species assessed as globally threatened in 2000[67], habitat modification is the principal threat affecting more than 80 percent of the mammals, birds and plants; it is similarly predominant in several other major groups, notably in 95 percent of threatened bivalve mollusks, and is the main cause of loss in 75 percent of extinct freshwater fishes[98].

A second major source of biodiversity loss is the widespread introduction of species outside their natural range where they typically induce change at the community and ecosystem level. The effects of alien species are especially pronounced in closed systems such as lakes and islands. Introduced species, such as rats and cats, are cited as a cause of extinction in nearly 40 percent of the approximately 200 species where cause could be attributed; the majority were island forms[43]. Seven endemic snails in French Polynesia have been extirpated following the late 1970s introduction of a carnivorous snail species (*Euglandina rosea*) intended to control another introduced species (the giant African snail *Achatina fulica*), itself an agricultural pest. Accidental introduction of the brown tree snake *Boiga irregularis* to Guam in 1968 led to decline of the entire

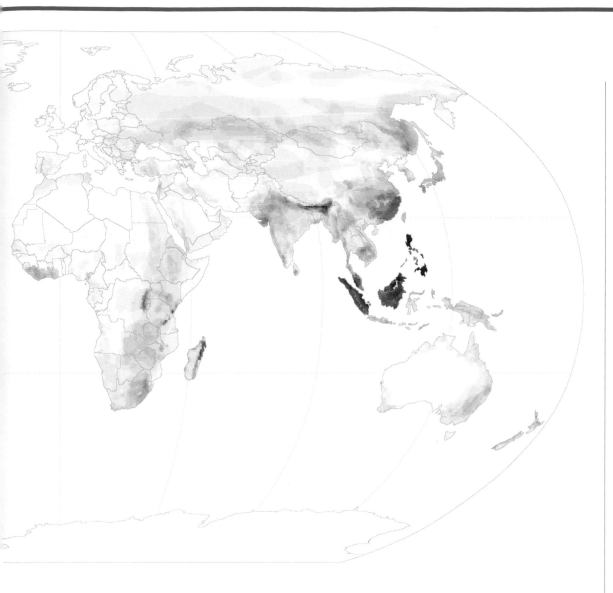

avifauna, of which one species is now thought extinct and one is extinct in the wild[70]. Introduction of the Nile perch (*Lates niloticus*) to Lake Victoria has contributed to the decline or extinction of nearly 200 native and endemic cichlid fishes. Overall, introduction events probably number in the thousands at the global level[72] and, although cases involving animals have been cited to illustrate the scope of the problem, introduced plants are in many places as pervasive and damaging.

High trade demand for certain species and products, whether for international markets consuming hardwoods, sea fish, live animals, and plants and derivatives, or local markets consuming commodities such as bushmeat or turtle eggs, can readily push exploitation beyond the production capacity of the resource. The direct impact of hunting, collecting and trade is the second most important threat category among globally threatened mammals and birds, with around 35 percent in each group affected[67].

Rapid environmental change, such as that associated with El Niño Southern Oscillation (ENSO) events, can have significant impacts on natural habitats. For example, in 1997-98 climate fluctuation associated with El Niño was implicated in the persistence and spread of fires in Brazil, Indonesia and elsewhere: an estimated 1 million hectares of savannah woodland burned in Brazil and a similar area of forests in Indonesia were affected by fire. The effect of events of this type will be multiplied many times wherever habitats are already fragmented and species are depleted.

REFERENCES

1   Coope, G.R. 1995. Insect faunas in ice age environments: Why so little extinction? In: Lawton, J.H. and May, R.M. (eds). *Extinction rates*. Oxford University Press, Oxford.

2   Adams, J. 1999. Sudden climate transitions during the Quarternary. *Progress in Physical Geography* 23(1): 1-36.

3   Van Couvering, J.A. (ed.) 1997. *The Pleistocene boundary and the beginning of the Quarternary*, Cambridge University Press, Cambridge.

4   Petit *et al*. 1999. Climate and atmospheric history of the past 420,000 years from the Vostok ice core, Antarctica. *Nature* 399: 429-436.

5   Larick, R. and Ciochon, R.L. 1996. The African emergence and early Asian dispersal of the genus *Homo. American Scientist* November-December.

6   Haile-Selassie, Y. 2001. Late Miocene hominids from the Middle Awash, Ethiopia. *Nature* 412: 178-181.

7   Wood, B. and Brooks, A. 1999. We are what we ate. *Nature* 400: 219-220.

8   Gabunia, L. *et al*. 2000. Earliest Pleistocene hominid cranial remains from Dmanisi, Republic of Georgia: Taxonomy, geological setting, and age. *Science* 288: 1019-1025.

9   Zhu, R.X. *et al*. 2001. Earliest presence of humans in northeast Asia. *Nature* 413: 413-417.

10 Bermúdez de Castro, J.M. *et al*. 1997. A hominid from the Lower Pleistocene of Atapuerca, Spain: Possible ancestor to Neandertals and modern humans. *Science* 276: 1392-1395.

11 Dennell, R. 1997. The world's oldest spears. *Nature* 385: 767-768.

12 Cann, R.L. 2001. Genetic clues to dispersal in human populations: Retracing the past from the present. *Science* 291: 1742-1748.

13 Rogers, A.R. 2001. Order emerging from chaos in human evolutionary genetics. *Proceedings of the National Academy of Sciences* 98: 779-780.

14 Barbujani, G. and Bertorelle, G. 2001. Genetics and the population history of Europe. *Proceedings of the National Academy of Sciences* 98: 22-25.

15 Thorn, A. *et al*. 1999. Australia's oldest human remains: Age of the Lake Mungo skeleton. *Journal of Human Evolution* 36: 591-612.

16 Roberts, R.G. *et al*. 2001. New ages for the last Australian megafauna: Continent-wide extinction about 46,000 years ago. *Science* 292: 1888-1892.

17 Dillehay, T.D. 1997. *Monte Verde: A Late Pleistocene settlement in Chile, Volume 1: The archaeological context and interpretation.* Smithsonian Institution Press, Washington DC.

18 Guidon, N. and Delibrias, G. 1986. Carbon-14 dates point to man in the Americas 32,000 years ago. *Nature* 321: 769-771.

19 Haynes, C.V. 1984. Stratigraphy and Late Pleistocene extinction in the United States. In: Martin, P.S. and Klein, R.G. (eds). *Quaternary extinctions*, pp. 345-353. University of Arizona Press, Tucson.

20 Harcourt, C.S. and Sayer, J.A. 1995. *The conservation atlas of tropical forests: The Americas.* IUCN-the World Conservation Union, Cambridge and Gland.

21 Pimm, L.S., Moulton, M.P and Justice, L.J. 1995. Bird extinctions in the central Pacific. In: Lawton, J.H. and May, R.M. (eds). *Extinction rates*, pp. 75-87. Oxford University Press, Oxford.

22 Dewar, R.E. 1984. Mammalian extinctions and stone age people in Africa. In: Partin, P.S. and Klein, R.G. (eds). *Quaternary extinctions*, pp. 553-573. University of Arizona Press, Tucson.

23 Semaw, S. *et al*. 1997. 2.5-million-year-old stone tools from Gona, Ethiopia. *Nature* 385: 333-336.

24 Thieme, H. 1997. Lower Palaeolithic hunting spears from Germany. *Nature* 385: 807-810.

25 Morwood, M.J. *et al*. 1998. Fission-track ages of stone tools and fossils on the east Indonesian island of Flores. *Nature* 392: 173-176.

26 Bird, M.I. and Cali, J.A. 1998. A million-year record of fire in sub-Saharan Africa. *Nature* 394: 767-769.

27 Gowlett, J.A.J. *et al*. 1981. Early archaeological sites, hominid remains and traces of fire from Chesowanja, Kenya. *Nature* 294: 125-129.

28 Isaac, G. 1981. Early hominids and fire at Chesowanja, Kenya. *Nature* 296: 870.

29 de Heinzelin, J. *et al*. 1999. Environment and behavior of 2.5 million year old Bouri hominids. *Science* 284: 625-629.

30 James, S.R. 1989. Hominid use of fire in the Lower and Middle Pleistocene: A review of the evidence. *Current Anthropology* 30: 1-26.

31 Hoopes, J.W. 1996. In search of nature. Imagining the precolombian landscapes of ancient Central America. Working paper for the Nature and Culture Colloquium, University of Kansas, Lawrence.

32 Teaford, M.F. and Ungar, P.S. 2000. Diet and the evolution of the earliest human ancestors. *Proceedings of the National Academy of Sciences* 97: 13506-13511.

33 Ambrose, S.H. 2001. Paleolithic technology and human evolution. *Science* 291: 1748-1753.

34 Richards, M.P. *et al*. 2000. Neanderthal diet at Vindija and Neanderthal predation: The evidence from stable isotopes. *Proceedings of the National Academy of Sciences* 97: 7663-7666.

35 Richards, M.P. *et al*. 2001. Stable isotope evidence for increasing dietary breadth in the European mid-Upper Paleolithic. *Proceedings of the National Academy of Sciences* 98: 6528-6532.

36 Stiner, M.C. 2001. Thirty years on the 'Broad Spectrum Revolution' and paleolithic demography. *Proceedings of the National Academy of Sciences* 98: 6993-6996.

37 Clutton-Brock, J. 1995. Origin of the dog: Domestication and early history. In: Serpell, J. (ed.) *The domestic dog: Its evolution, behaviour and interactions with people*, pp. 8-20. Cambridge University Press, Cambridge.

38 Vilà, C. *et al*. 1997. Multiple and ancient origins of the domestic dog. *Science* 276: 1687-1689.

39 Smith, B.D. 1998. *The emergence of agriculture*. Scientific American Library.

40 Smith, B.D. 2001. Documenting plant domestication: The consilience of biological and archaeological approaches. *Proceedings of the National Academy of Sciences* 98: 1324-1326.

41 Luikart, G. *et al*. 2001. Multiple maternal origins and weak phylogeographic structure in domestic goats. *Proceedings of the National Academy of Sciences* 98: 5927-5932.

42 FAO 1998. *The state of the world's plant genetic resources for food and agriculture*. Food and Agriculture Organization of the United Nations, Rome.

43 World Conservation Monitoring Centre 1992. *Global biodiversity: Status of the Earth's living resources*. Groombridge, B. (ed.) Chapman and Hall, London.

44 Prescott-Allen, R. and Prescott-Allen, C. 1990. How many plants feed the world? *Conservation Biology* 4(4): 365-374.

45 Pálsson, G. 1988. Hunters and gatherers of the sea. In: Ingold, T., Riches, D., Woodburn, J. *Hunters and gatherers 1: History, evolution and social change*, pp. 189-204. BERG, Oxford.

46 FAOSTAT database. Food and Agriculture Organization of the United Nations. Available online at http://apps.fao.org/ (accessed February 2002).

47 FAO 1996. Food requirements and population growth. World Food Summit technical background document TBD VI E4. Food and Agriculture Organization of the United Nations. Available online at http://www.fao.org/wfs/index_en.htm (accessed February 2002).

48 Uvin, P. *The state of world hunger*. Hunger report. Gordon and Breach. Available online at http://www.brown.edu/Departments/World_Hunger_Program/hungerweb/intro/StateofWorldHunger.pdf (accessed April 2002).

49 Committee on World Food Security 2001. Mobilizing resources to fight hunger. 2002 World Food Summit background document CFS:2001/Inf.7.

50 de Garine, I. and Harrison, G.A. (eds) 1988. *Coping with uncertainty in food supply.* Clarendon Press, Oxford.

51 Fernandes, E.C.M., Okingati, A. and Maghembe, J. 1984. The Chagga homegardens: A multi-storied agroforestry cropping system on Mount Kilimanjaro (northern Tanzania). *Agroforestry Systems* 2: 73-86

52 Juma, C. 1989. *The gene hunters: Biotechnology and the scramble for seeds.* Zed Books, London.

53 Marsh, R.R. and Talukeder, A. 1994. Effects of the integration of home gardening on the production and consumption of target interaction and control groups: A case study from Bangladesh. Unpublished paper.

54 McEvedy, C. and Jones, R. 1978. *Atlas of world population history.* Penguin Books, Harmondsworth.

55 Roberts, N. 1998. *The Holocene. An environmental history.* 2nd edition. Blackwell Publishers, Oxford.

56 UN Population Division 2001. *World population prospects. The 2000 revision.* Draft. ESA/P/WP.165. Available online at http://www.un.org/esa/population (accessed April 2002).

57 Smil, V. 1999. How many billions to go? *Nature* 401: 429.

58 Yalden, D.W. 1996. Historical dichotomies in the exploitation of mammals. In: Taylor, V.J. and Dunstone, N. *The exploitation of mammal populations*, pp. 16-27. Chapman and Hall, London.

59 Wright, D.H. 1990. Human impacts on energy flow through natural ecosystems, and implications for species endangerment. *Ambio* 19(4): 189-194.

60 Haberl, H. 1997. Human appropriation of net primary production as an environmental indicator: Implications for sustainable development. *Ambio* 26(3): 143-146.

61 Holdaway, R.N. and Jacomb, C. 2000. Rapid extinction of the moas (Aves: Dinornithiformes): Model, test, and implications. *Science* 287: 2250-2254.

62 Alroy, J. 2001. A multispecies overkill simulation of the end-Pleistocene megafaunal mass extinction. *Science* 292: 1893-1896.

63 Miller, G.H. *et al.* 1999. Pleistocene extinction of *Genyornis newtoni*: Human impact on Australian megafauna. *Science* 283: 205-208.

64 UNEP-WCMC species conservation database (plants). Available online at http://www.unep-wcmc.org/species.plants/plants-by-taxon.htm#search

65 Reid, W.V. 1992. How many species will there be? In: Whitmore, T.C. and Sayer, J.A. (eds). *Tropical deforestation and species extinction*, pp. 55-73. Chapman and Hall, London.

66 IUCN 1994. IUCN Red List categories. IUCN-the World Conservation Union, Gland.

67 Hilton-Taylor, C. (compiler) 2000. *2000 IUCN Red List of threatened species.* IUCN-the World Conservation Union, Gland and Cambridge. Also available online at http://www.redlist.org/ (accessed April 2002).

68 The Species Survival Commission of IUCN-the World Conservation Union. Red List material. Available online at http://iucn.org/themes/ssc/red-lists.htm (accessed February 2002).

69 Master, L.L., Flack, S.R. and Stein, B.A. (eds) 1998. *Rivers of life: Critical watersheds for protecting freshwater biodiversity.* Nature Conservancy, Arlington.

70 BirdLife International 2000. *Threatened birds of the world.* Lynx Edicions and BirdLife International, Barcelona and Cambridge.

71 Richards, J.F. 1990. Land transformation. In:Turner II, B.L. *et al.* (eds). *The Earth as transformed by human action*, pp. 163-178. Cambridge University Press, with Clark University.

72 United Nations Environment Programme 1995. Heywood, V. (ed.) *Global biodiversity assessment.* Cambridge University Press, Cambridge.

73 Martin, P.S. 1984. Prehistoric overkill: The global model. In: Martin, P. and Klein, R.G. (eds). *Quaternary extinctions*, pp. 354-403. University of Arizona Press, Tucson.

74 Juma, C. 1989. *Biological diversity and innovation: Conserving and utilizing genetic resources in Kenya*. African Centre for Technology Studies, Nairobi.

75 Alvarez-Buylla Roces, M.A., Lazos Chavero, E. and Garcia-Burrios, J.R. 1989. Homegardens of a humid tropical region in South East Mexico: An example of an agroforestry cropping system in a recently established community. *Agroforestry Systems* 8: 133-156.

76 Alcorn, J.B. 1984. Development policy, forests and peasant farms: Reflections on Haustec-managed forest contribution to commercial and resource conservation. *Economic Botany* 38(4): 389-406.

77 Davies, A.G. and Richards, P. 1991. Rain forest in Mende life: Resources and subsistence strategies in rural communities around the Gola North Forest Reserve (Sierra Leone). A report to the Economic and Social Committee on Overseas Research (ESCOR), UK Overseas Development Administration, London.

78 Michon, G., Mary, F. and Bompard, J. 1986. Multistoried agroforestry garden system in West Sumatra, Indonesia. *Agroforestry Systems* 4(4): 315-338.

79 Kunstadter, P., Chapman, E.C. and Sabhasri, S. 1978. *Farmers in the forest: Economic development and marginal agriculture in Northern Thailand*. An East-West Center Book, University Press of Hawaii, Honolulu.

80 Padoch, C. and de Jong, W. 1991. The house gardens of Santa Rosa: Diversity and variability in an Amazonian agricultural system. *Economic Botany* 45(2): 166-175.

81 Denevan, W.M. and Treacy, J.M. 1987. Young managed fallows at Brillo Neuvo. In: Denevan, W.M. and Padoch, C. (eds). *Swidden-fallow agroforestry in the Peruvian Amazon*. Advances in Economic Botany 5, pp. 8-46. New York Botanical Garden, New York.

82 Ohtsuka, R. 1993. Changing food and nutrition of the Gidra in lowland Papua New Guinea. In: Hladik, C.M. *et al.* (eds). *Tropical forests, people and food. Biocultural interactions and applications to development*. Man and the Biosphere Series. Volume 13. UNESCO and Parthenon Publishing, Paris and Carnforth.

83 Conklin, H.C. 1954. An ethnoecological approach to shifting agriculture. *Transactions of the New York Academy of Sciences* 17: 133-142.

84 Posey, D.A. 1984. A preliminary report on diversified management of tropical forest by the Kayapo Indians of the Brazilian Amazon. *Advances in Economic Botany* 1: 112-126.

85 Vickers, W.T. 1993. Changing tropical forest resource management strategies among the Siona and Secoya Indians. In: Hladik, C.M. *et al.* (eds). *Tropical forests, people and food. Biocultural interactions and applications to development*. Man and the Biosphere Series. Volume 13. UNESCO and Parthenon Publishing, Paris and Carnforth.

86 Michon, G. 1983. Village forest gardens in West Java. In: Huxley, P. (ed.) *Plant research and agroforestry*, pp. 13-24. International Centre for Research in Agroforestry, Nairobi.

87 Whitman, W.B., Coleman, D.C. and Wiebe, W.J. 1998. Prokaryotes: The unseen majority. *Proceedings of the National Academy of Sciences* 95: 6578-6583.

88 Rusek, J. 1998. Biodiversity of Collembola and their functional role in the ecosystem. *Biodiversity and Conservation* 7: 1207-1219.

89 Zimmerman, P.R. *et al.* 1982. Termites: A potentially large source of atmospheric methane, carbon dioxide, and molecular hydrogen. *Science* 218: 563-565.

90 Nicol, S. Time to krill? Antarctica Online: The website of the Australian Antarctic Program: http://www.antdiv.gov.au/science/bio/issues_krill/krill_features.html

91 Gaston, K.J. and Blackburn, T.M. 1997. How many birds are there? *Biodiversity and Conservation* 6: 615-625.

92 Said, M.Y. *et al.* 1995. African elephant database 1995. IUCN-the World Conservation Union, Gland.

93 Santiapillai, C. 1997. The Asian elephant conservation: A global strategy. Gajah. *Journal of the Asian Elephant Specialist Group* 18: 21-39.

94 Nowak, R.M. 1991. *Walker's mammals of the world*. 5th edition. Volume II. Johns Hopkins University Press, Baltimore.

95 Jefferson, T.A., Leatherwood, S. and Webber, M.A. 1993. *Marine mammals of the world*. FAO Species Identification Guide. United Nations Environment Programme, Rome.

96 Kemf, E. and Philips, C. 1995. *Whales in the wild*. WWF Species Status Report, WWF – Worldwide Fund for Nature, Gland.

97 Oldfield, S., Lusty, C. and MacKiven, A. 1998. *The world list of threatened trees*. World Conservation Press, Cambridge.

98 Harrison, I.J. and Stiassny, M.L.J. 1999. The quiet crisis: A preliminary listing of the freshwater fishes of the world that are extinct or 'missing in action'. In: MacPhee, R.D.E. (ed.) *Extinctions in near time: Causes, contexts, and consequences*, pp.271-332. Kluwer Academic/Plenum Publishers, New York.

5 Terrestrial biodiversity

TERRESTRIAL ECOSYSTEMS EXTEND OVER LITTLE MORE than one quarter of the Earth's surface but they are more accessible than aquatic habitats and so are far better known. The land supports fewer phyla than the oceans but most of the global diversity of species, and is characterized above all by an extensive cover of vascular plants, with associated animals and other groups of organisms.

Forest and woodland ecosystems form the predominant natural landcover over most of the Earth's surface. These systems generate around half the terrestrial net primary production, and forests in the tropics are believed to hold most of the world's species. Approximately half the area of forest developed in post-glacial times has since been cleared or degraded by humans, and the amount of old-growth forest continues to decline.

Grassland, shrubland and deserts collectively cover most of the unwooded land surface, with tundra on frozen subsoil at high northern latitudes. These areas tend to have lower species diversity than most forests, with the notable exception of Mediterranean-type shrublands, which support some of the most diverse floras on Earth.

Humans have extensively altered most grassland and shrubland areas, usually through conversion to agriculture, burning and introduction of domestic livestock. They have had less immediate impact on tundra and true desert regions, although these remain vulnerable to global climate change.

THE TERRESTRIAL BIOSPHERE
Land and water on the Earth's surface

Land extends over nearly 150 million square kilometers (km^2) or about 29 percent of the total surface of the planet, the remainder being covered by oceans. With the continents in their present position, more than two thirds of the land surface is in the northern hemisphere, and the area of land situated in the northern hemisphere above the Tropic of Cancer slightly exceeds that in the rest of the world put together (Table 5.1). A small proportion of the total land area is occupied by inland water ecosystems: lakes and rivers cover around 2 percent and swamps and marshland a similar amount. About half of the land surface, approximately 52 percent, is below 500 meters (m) in elevation, and the mean elevation is 840 m. A minor but significant proportion of the land surface is mountainous in nature, with alpine land-scapes tending to replace trees above 1 000 m at higher latitudes and above 3 500 m in the tropics.

The land as an environment for living organisms

The environmental conditions prevailing in any given place determine what kinds of organism can live there. Relevant environmental conditions include various aspects of the physical environment, and the other

Table 5.1
Global distribution of land area, by latitude bands

Region		Land area (million km²)
Northern hemisphere	North of Tropic of Cancer	74
	Equator north to Tropic of Cancer	26
Southern hemisphere	Equator south to Tropic of Capricorn	23.5
	South of Tropic of Capricorn (including Antarctica)	23.5

species that any given species directly or indirectly interacts with. These interactions may occur through a range of mechanisms, including competition, predation, symbiosis, mutualism and parasitism. For many living species, interactions with humans are now a significant feature of their environment.

The most fundamental distinction, at least for macroscopic organisms, is between terrestrial and aquatic environments. All living

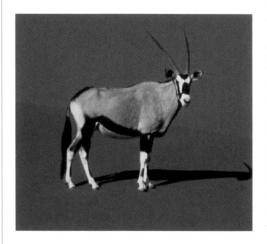

Most terrestrial organisms occupy an environment where water loss is a constant threat.

organisms require water because the basic processes of life take place in the aqueous medium found in cells. However, most terrestrial organisms, in contrast to aquatic ones, occupy an environment where water loss is a constant threat. The anatomical and physiological solutions to this problem are many and varied, including epidermal and/or cuticular layers with reduced permeability to water, water storage tissues, and metabolic or behavioral processes that conserve water. However, the greater the risk of dehydration, the more carbon and energy are needed for water conservation mechanisms, leaving less available for other adaptations. Thus, life tends to occur at low density and at low diversity in hyperarid environments, whether exceptionally hot or exceptionally cold.

There are a number of other extremely important differences between air and water as a medium for living things. It takes far less energy (about 500 times less) to raise the temperature of a given mass of air by 1 degree Celsius than to heat the same mass of water by the same amount, and water conducts heat far

more rapidly than air. As a result, while aquatic organisms are buffered against rapid fluctuation in their surroundings, terrestrial organisms can be subjected to wide extremes in temperature, corresponding to daily and seasonal variation in insolation.

The air surrounding terrestrial organisms is much less dense (about 800 times less) than water, and so land organisms must support themselves against the full effects of gravity, but are not subject to the large forces exerted on aquatic organisms by moving water.

Oxygen is freely and uniformly available in the atmosphere, but in water is far less concentrated and much more variable in time and space. On the other hand, mineral nutrients dissolve readily in water and it is possible for aquatic organisms to extract nutrients directly from their immediate surroundings (although in many aquatic habitats such nutrients are only present in low concentrations). On land, in contrast, mineral nutrients occur in soil where their distribution and concentration are spatially variable, and they are directly accessible primarily to microorganisms, plants, fungi and algae.

Global variations in terrestrial habitats

The wide variations in water availability and temperature regimes prevailing on the land surface interact in often complex ways with a number of other factors, including geology, soil, terrain, wind, fire regimes and human activities, to generate the immense range of environments apparent on the Earth's land surface. Classification of these into a manageable system is a major problem in biology, one not merely of theoretical interest but of considerable importance in the management and conservation of the biosphere. The problem arises largely from a need to divide the natural environment into a series of discrete bounded units for the purposes of mapping, measuring and monitoring habitats, whereas the world in reality appears to form a highly variable continuum.

Where gradients exist between different physical regimes (particularly of water and temperature), habitat types tend to intergrade imperceptibly, and it is impossible without arbitrary definitions to distinguish, for example, grassland with a few trees from open woodland with grass ground cover. Even broad categories, such as 'forest' or 'wetland', inevitably require arbitrary limits to be set, e.g. for the density of tree cover necessary before an area can be called a forest, or for the duration of flooding necessary before an area can be classified as a wetland rather than a terrestrial system. In such circumstances, it is important to keep in mind the inherent variability of ecosystems, rather than attach undue significance to their labels.

Vegetation and chlorophyll

The global distribution of actively growing vegetation can be visualized without classification or criteria relating to structure, to physiognomy or to species composition. Advanced very high resolution radiometer (AVHRR) satellite sensors measure the reflectance of vegetation, primarily of the green photosynthetic pigment chlorophyll, in the visible and the near infrared part of the spectrum. On land this can be interpreted as broadly equivalent to the density and vigor of green plant growth, represented as the normalized difference vegetation index (NDVI). This is plotted, aggregated over one year, in Map 5.1. This reveals a clear distinction between areas rich in standing growth of plants, whether cropland or natural vegetation, and areas where standing plant growth is sparse or absent – these are essentially the drylands and rangelands of conventional landuse classifications.

Spatial variation in chlorophyll density is only indirectly related to variation in primary production levels; net primary production depends further on soil and climate conditions and community dynamics within ecosystems (see Chapter 1). So, while Map 5.1 shows high chlorophyll in both tropical and high latitudes, net primary production is far higher in the former than the latter, where it is restricted mainly by seasonally unfavorable climatic conditions (compare with Map 1.2).

Landcover and ecosystems

Landcover classification is more directly concerned with differences in the physical aspects

of ground cover, mainly for landuse planning and management, than with biodiversity or the community aspects of vegetation cover. Many current landcover maps (Map 5.2) are based on interpretation of remote-sensing data that have the virtue of being quantitative in nature and available in time series suitable for monitoring applications. Because the source data are typically interpreted in terms of a classification system developed to take account of conditions on the ground, landcover maps are subject to some of the definition problems mentioned above, no matter how generalized the categories are at the highest levels (e.g. 'forest', 'cropland', 'urban').

One of the most useful ecological distinctions to be made is between areas with extensive or significant tree cover, and areas with few or no trees. Terrestrial plant growth is favored by high soil water availability and relatively elevated temperature during all or a major part of the year. Trees tend to be the main plant growth form in such conditions, and forest or woodland the main vegetation. Conversely, primary production is strongly limited by a shortage of soil water. Grasses and low shrubs tend to be the main plant growth forms in such dryland regions and, where vegetation exists, it consists mainly of grassland, savannah or shrubland.

Tree growth remains insignificant in those parts of the world where water is present in some form, but temperatures are too low for growth during all or part of the year.

Grasses and low shrubs are the main plant growth forms in dryland regions.

Map 5.1
Photosynthetic activity on land

A map illustrating the normalized difference vegetation index (NDVI) calculated from AVHRR satellite data and aggregated over the year 1998. NDVI values vary with absorption of red light by plant chlorophyll and the reflection of infrared radiation by water-filled leaf cells, and provide an indication of photosynthetic activity. The map reveals a clear distinction between areas rich in standing growth of plants, whether cropland or natural vegetation, and areas where standing plant growth is sparse or absent.

Source: Adapted from image provided by the SeaWiFS Project, NASA/Goddard Space Flight Center and ORBIMAGE.

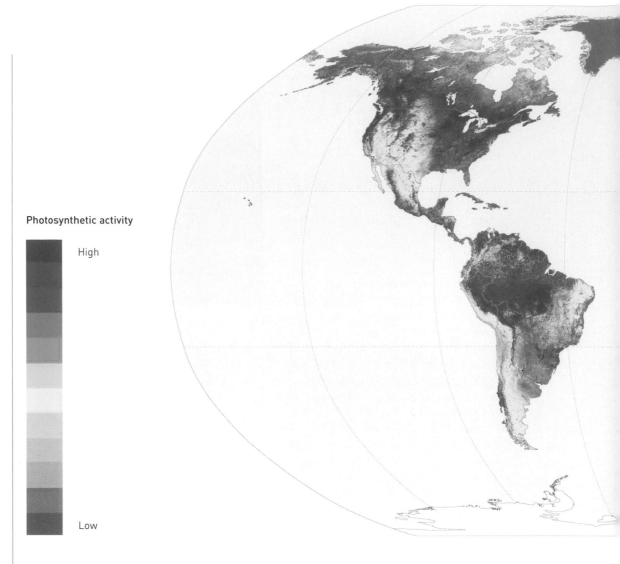

Photosynthetic activity

High

Low

For example, in polar regions there is an abundance of water, but because of permanent low temperature it is mainly in solid form and so unavailable to living organisms. In tundra areas at high latitudes the subsurface is permanently frozen and plant growth is restricted to the few summer months when thawing of the superficial layers makes liquid water available. In upper alpine regions, liquid water is often present in seasonal abundance, but temperatures are too low over the year as a whole to allow tree growth.

Many areas theoretically have sufficient rainfall and a temperature regime suitable for supporting significant tree cover but do not do so. The most important reason for this is undoubtedly human intervention but in some areas this may also be a result of natural causes. Nutrient availability may be limiting. Slopes may be too steep to allow formation and retention of soil, so that trees cannot anchor themselves. Minerals may be

Box 5.1 Defining ecosystems

The word 'ecosystem' was introduced by the plant ecologist Tansley in 1936, to refer to the communities of plants, animals, other organisms and the physical environment of any given place. The concept reflected a new and visionary approach to biological research which focused on the system-level flow of materials and energy between components at any given study site. The spatial boundaries of the system were originally of little or no significance. The term is now widely used, particularly in the context of environmental planning, to refer to broad biological communities of similar appearance, usually defined by physical, climatic, structural or phenological features. Ecosystem diversity is generally understood to refer to the range of different kinds of ecosystem, in this sense, within some defined area. In this usage, the 'ecosystem' is treated as a map unit and the spatial boundaries of the system assume major significance.

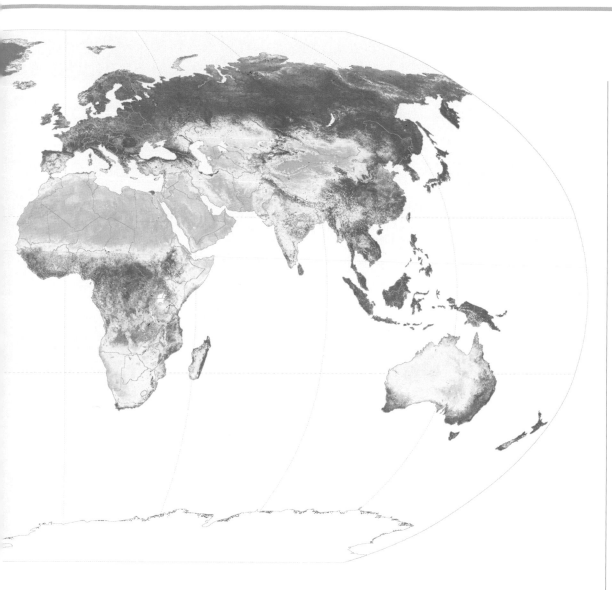

present in toxic concentrations, or the frequency and intensity of natural fires or floods may be too high. Large herbivores may also prevent establishment of significant tree cover, although as noted in Chapter 4 many of those large herbivores in the Americas and Australasia that might have exerted such an influence in the past are now extinct.

Landcover classifications can be used as the basis for habitat or ecosystem classifications (see Box 5.1), by incorporating elements of species composition and community structure. Ecosystem maps should have more direct application to biodiversity conservation and management than maps of landcover; however the development of such maps is constrained by the same problems of defining boundaries and developing a consistent approach to classification. Map 5.7 is a

representation of global forest cover, derived by applying a simple classification of five forest and woodland habitat types to the landcover data plotted in Map 5.2. Forest classification is discussed further below.

GLOBAL VARIATIONS IN TERRESTRIAL SPECIES DIVERSITY

Just as habitat types show great variation across the world's land surface, so does the number of species that may be found in any given place. The spatial heterogeneity, wide range of present physical conditions and complex history of the world's land surface all contribute to extreme variation in terrestrial biological diversity. Variation in species number is not strictly related to variation in habitat type because two areas that are structurally similar may have very different

Map 5.2
Global land cover

Map adapted from the global landcover classification developed by the University of Maryland. The Maryland classification includes 13 classes and was based on AVHRR remote-sensing data with a spatial resolution of 1 km. For presentation purposes the data have here been generalized to a 4-km grid. Also for clarity, the two needleleaf forest classes in the original classification have been combined, as well as the two broadleaf forest classes, and the urban and built-up category has been omitted.

Source: Data from University of Maryland Global Land Cover Facility. For full description see Hansen[98].

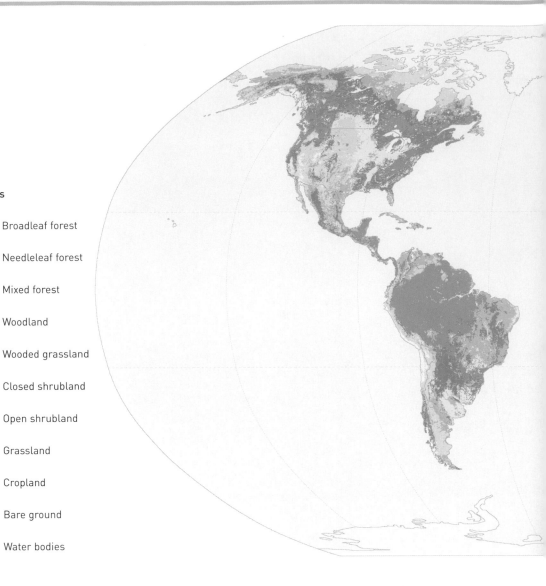

Cover class

Broadleaf forest

Needleleaf forest

Mixed forest

Woodland

Wooded grassland

Closed shrubland

Open shrubland

Grassland

Cropland

Bare ground

Water bodies

numbers of species present. The most important components of this variation can be expressed in three different, though not completely independent, ways:

- the kinds of organisms, particularly primary producers, that live in any one place – this influences the kinds of habitats found there and is a reflection of ecosystem diversity;
- the numbers of different kinds of organisms that live in any one place – this is usually assessed by measures of species diversity, although other classification systems (e.g. guilds, functional groups, higher taxa) can also be used;
- the individual taxonomic identity of the organisms that live in any one place – this is determined by biogeography and

influences, among other things, how a given area contributes to global biodiversity.

Measuring species diversity
Comparing the diversity of different parts of the world is complex because of the way diversity changes with scale. A wide range of observations has demonstrated that, as a general rule, the number of species recorded in an area increases with the size of the area, and that this increase tends to follow a predictable pattern known as the Arrhenius relationship, whereby: $\log S = c + z \log A$ where S = number of species, A = area and c and z are constants (see Figure 5.1). The slope of the relationship (z in the equation) varies considerably between surveys, although it is generally between 0.15 and 0.40, but some survey data do not fit the relationship at all. A

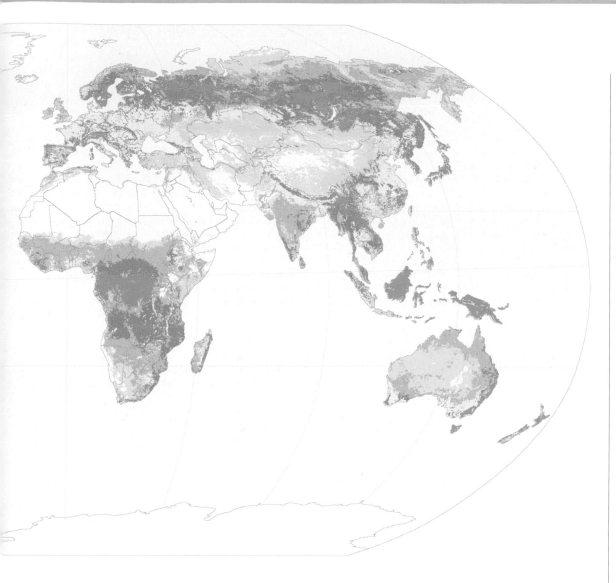

common generalization from this finding is that a tenfold reduction in an area (i.e. loss of 90 percent of habitat) will result in the loss of between 30 percent (with $z = 0.15$) and 60 percent (with $z = 0.40$) of the species originally present, or approximately half the species.

The relative diversity of different sites will often partly depend on the scale at which diversity is measured. Thus 1 m^2 of semi-natural European chalk grassland may contain many more plant species than 1 m^2 of lowland Amazonian rainforest, whereas for any area larger than a few square meters this could be reversed. In other words, when an area is sampled the number of species recorded increases with the size of the area, but this rate of increase varies from area to area, i.e. the slope of the Arrhenius relationship is not constant everywhere.

The reason for this increase may be quite straightforward. When small areas are sampled they are likely to be relatively homogeneous in terms of habitat type.

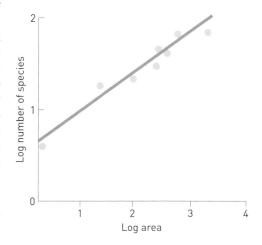

Figure 5.1
A typical species-area plot

Note: The data, consisting of species counts in a series of areas of different size, are plotted on logarithmic axes resulting in a straight-line graph, the slope of which (z) indicates the rate at which species number changes with changing area.

At a small scale, as sample area increases, so an increasing proportion of the species present in that habitat is likely to be included in the sample. Beyond a certain point, however, as larger areas are sampled so an increasing number of different habitats will be included in the sample area, each with new species that are likely to be included in the sample. The species/area relationship therefore increasingly reflects habitat hetero-geneity at larger scales. Ecologists attempt to take account of this by recognizing different kinds of biological diversity. The diversity within a site or habitat is often referred to as alpha (α) diversity while the differences between habitats are referred to as beta (β) diversity. Thus an area with a wide range of dissimilar habitats will have a high β-diversity, even if each of its constituent habitats may have low α-diversity. Differences in site diversity over large areas, such as continents, are sometimes referred to as gamma (γ) diversity.

Measures of diversity can refer simply to species richness but can also be more sophisticated statistical measures that take into account the relative abundance of different species in a given place. A variety of different measures of this kind has been developed (of which H', the Shannon-Wiener function, is a commonly used example). With many of these measures, an area in which all species are of similar abundance would generally be given a higher diversity measure than an area with the same number of species, a few of which were very abundant and the remainder rare. Deriving these statistical measures requires intensive sampling; for this reason, simpler measures of species richness tend to be more useful at larger scales.

Major patterns of variation in global species diversity

Despite the difficulties of establishing strictly comparable quantifiable measures, a wealth of empirical observations indicates that species richness in eukaryotes tends to vary geo-graphically according to a series of fairly well-defined rules. In terrestrial environments:

- warmer areas hold more species than colder ones;
- wetter areas hold more species than drier ones;
- areas with varied topography and climate hold more species than uniform ones;
- less seasonal areas hold more species than highly seasonal ones;
- areas at lower elevation hold more species than areas at high elevation.

The single most obvious pattern in the global distribution of species is that overall species richness increases as latitude decreases toward the equator (see Box 5.2). At its simplest this means that there are more species in total and per unit area in temperate regions than in polar regions, and far more again in the tropics than in temperate regions. This applies as an overall general rule, and within most individual higher taxa (at order level or higher), and within most equivalent habitats. The pattern can be seen in Maps 5.3, illustrating vascular plant species diversity, and 5.4, which represents country-level diversity in terrestrial vertebrates and vascular plants.

There is good evidence that moist tropical forests are the most species-rich environments on Earth. If current estimates of the

---

Box 5.2 Species and energy

Latitudinal variation in diversity on land is strongly correlated with, and may be largely explained by, variation in incident energy over the Earth's surface. The relationship between diversity and productivity, and related measures, has been the subject of long-standing debate in ecology, but recent studies have shown that at global or continental scale, organismal diversity, particularly as measured at higher taxonomic levels, is strongly correlated with available energy[1,2]. This kind of relationship has been demonstrated, for example, for flowering plants, for trees, lepidoptera, land birds and land mammals in a range of countries and continents[3-6], and for fishes in river basins at a global level[7]. One simplistic explanation for this may be that higher energy availability leads to increased net primary production (NPP), and this broader resource base allows more species to coexist. While the general relationship appears robust, the details are complex. Energy availability can be measured in several ways: as heat energy, as potential (PET) or actual evapotranspiration (AET), or as NPP, and which is the best predictor of diversity has yet to be determined. Some measure of the simultaneous availability of water and radiant energy may provide the best general predictor of potential macro-scale species richness[1,2]. More complete explanation for richness variation would need also to consider the roles of topography, history and edaphic factors.

number of as yet unknown species (see Chapter 2) in the tropical forest microfauna are accepted, these regions, extending over perhaps 7 percent of the world's surface, may conceivably hold more than 90 percent of the world's species. If tropical forest small insects are discounted, then coral reefs and, for flowering plants at least, areas of Mediterranean climate in South Africa and Western Australia may be similarly rich in species.

Topographic heterogeneity may be expected to play a significant part in determining species number for two reasons. First, such heterogeneity will increase habitat variability, thereby increasing the range of niches that can be occupied by different organisms. Second, depending on the size and mobility of organisms, the chances of geographic isolation and speciation increase in topographically diverse landscapes. The role of topography has been demonstrated statistically at continent scale for North American mammals[8], and at both landscape and patch scale for vascular plants[9, 10].

The available information on distribution of species is geographically very incomplete, and relates to only a small fraction of the 1.7 million known species. Geographically, western Europe has been more thoroughly sampled than elsewhere, while large areas in the tropics, particularly of South America and central Africa are very poorly known. Taxonomically, the larger mammals, birds, vascular plants and a few invertebrate groups, such as Odonata, are better known than other groups of species. Because of the uneven availability of information there has been considerable interest in identifying groups of individual species that may serve as surrogates for biodiversity more generally, or higher taxa, such as families, that might predict patterns of richness in their included species[11].

Maps 5.5 and 5.6 represent, respectively, the distributions of flowering plant families (phylum Anthophyta) and of terrestrial (non-aquatic) vertebrate families (phylum Chordata), plotted as a global family diversity surface for each group. The resulting maps share key features, notably a marked latitud-

inal gradient in family richness, but also have striking differences, particularly in areas of high diversity. For example, Africa appears to be very rich in vertebrate families, most notably in moist forest areas around the Gulf of Guinea and in the east, including less humid woodland and savannah habitats, but is relatively poor in plant families compared with other continents in the tropics. For flowering plants, this mirrors the sequence evident at species level, where there may be around 90 000 species in the neotropics, 40 000 in tropical Asia and 35 000 in tropical Africa[12].

Biogeography and endemism

While ecological factors influence which kinds of species, and how many of them, can persist in a given area, history has already determined which actual lineages are present. A complete explanation for global variation in biodiversity must therefore involve both historical events and current ecological processes. The former are implicit in any explanation of the origin of diversity, the latter in explanations of its maintenance; these being two separate, although intimately linked, problems. On land, continental drift resulting from plate tectonics, climate change, mountain building or sea-level change, and probably the evolutionary lability of different lineages, are among the important historical factors. Geographic features commonly restrict or prevent the further dispersal of species: for example, a large river can present a barrier to a terrestrial species, the sea is a barrier to non-flying island forms and land is a barrier to freshwater species.

Barriers to dispersal explain why the fauna and flora of ecologically similar areas in geographically separated parts of the world tend to be composed largely of different individual species. They also underlie the phenomenon of endemism. An endemic species is one restricted to some given area, which may be a continent or country, or more significantly a relatively small area, such as a mountain block, island or lake. Discrete areas of complex topography, particularly in the tropics, often have high endemism in a range of taxonomic groups, possibly because climate change has encouraged speciation by isolating different lineages at different times.

Communities and ecoregions

The most comprehensive attempts to describe and classify habitats try to combine elements of all three sources of variation outlined above. At fine scales these are generally based on community ecology. On land this is exemplified by the phyto-sociological approach developed principally in continental Europe during the 20th century.

UNESCO[15]	Closed forest	Trees ≥ 5 m tall with crowns interlocking
	Woodland	Trees ≥ 5 m tall with crowns not usually touching but with canopy cover ≥ 40%
US classification standards[95]	Closed tree canopy	Trees with crowns interlocking, with crowns forming 60–100% cover
	Open tree canopy	Trees with crowns not usually touching forming 10–60% or 25–60% cover
FAO[61]	Forest	Land with tree canopy cover > 10% and > 0.5 ha in area; trees should be able to reach a minimum height of 5 m
	Other wooded land	Land with either a crown cover of 5-10% of trees able to reach a height of 5 m at maturity; or crown cover of more than 10% of trees not able to reach a height of 5 m at maturity; or with shrub or bush cover > 10%

Table 5.2
Different definitions of forest cover

This was intended to describe and classify plant communities on the basis of dominant and other associated species, by inspection or quadrat analysis of vegetation patches, taking into account species identity, growth form and abundance. One major problem with this is that the more precisely a community is defined, the more site-specific it becomes, and hence the more limited its use in higher-level analysis and planning. A more recent related approach[13] attempts to delimit 'eco-regions', these being defined as relatively large units of land each of which contains a distinct assemblage of natural communities and species, with boundaries similar to the original extent prior to anthropogenic change. These are nested within a hierarchy having traditional biogeographic realms and biome systems (ecosystem types) as the first and second levels.

FORESTS

Forests and woodlands probably once covered about half of global land area and now cover about one quarter. They provide habitat for half or more of the world's species. They are responsible for just under half of the global terrestrial annual net primary production (Table 1.1), and they and their soils house about 50 percent of the world's terrestrial carbon stocks. In addition to carbon storage, forests perform many other important ecosystem services, such as regulating local hydrological and nutrient cycles, and stabilizing soils and watersheds.

Forests also provide a wide variety of products, including food and fuel, medicines, construction materials and paper, which are important both for human subsistence and for economic activity. Wood products are one of the most economically important natural resources. In the region of 3.3 billion cubic meters (m^3) of wood is extracted from forests and other habitats annually, the equivalent of several hundred million trees. Just over half of this volume is used as fuelwood and charcoal, of which developing countries consume 90 percent[14]. The remainder is industrial roundwood, which is processed into various wood products. Forests are frequently important culturally and play a significant role in the spiritual life of communities worldwide.

What is a forest?

Despite their importance in a number of different human contexts and the large amount of research that has focused on forest ecosystems, a precise definition of 'forest' remains elusive. Although it is generally accepted that the term indicates an eco-system in which trees are the predominant life form, the problem arises because of the broad range of systems in which trees occur and the difficulty, even, in deciding what constitutes a tree. For example, tree species may dominate at high altitude, but be barely recognizable as trees because of their spreading prostrate forms. Savannahs may possess large numbers of trees, but as they can occur at low density in association with other life forms it may be difficult to define precisely in which areas they are dominant. A variety of different definitions of forest have been proposed by organizations that evaluate and monitor natural resources (Table 5.2). Estimates of forest area may vary widely depending on the definition adopted. Of singular importance is the degree of canopy cover used as the threshold for dividing forests from non-forests (Table 5.2). As a consequence, the precise definitions employed should be borne in mind when comparing forest cover data provided by different institutions.

Forest types

There is great variation in the forms and types of forest distributed throughout the world. Information about this variation and the distribution of forest vegetation types is crucial to understanding the different roles of forests in supporting biodiversity, in carbon and hydrological cycles and other ecosystem processes, and in supplying wood and non-wood forest products. However, if deriving a satisfactory definition of forest is problematic, arriving at consensus on how to classify forests is even more difficult.

A number of global classification systems have been proposed, but as yet none has gained universal acceptance. The UNESCO (United Nations Educational, Scientific and Cultural Organization) system proposed by Ellenberg and Mueller-Dombois[15] is one such system. It includes nearly 100 forest and woodland 'subformations' and allows for yet finer subdivisions, but many of the characteristics that separate categories can only be determined in the field. Other classifications, such as the EROS Data Center seasonal landcover regions, with nearly a thousand classes, reflect more strongly the nature of landcover data obtained from Earth-orbiting satellites and the methods used in analyzing and classifying them[1]. This complex system has been translated into a much less complex one in the International Geosphere-Biosphere Programme (IGBP) classification, which includes seven forest and woodland types that reflect phenology and canopy closure world-

wide, but provides little other information on forest physiognomy, composition or environment within the class names[1]. Map 5.7 represents the global distribution of a range of forest types, based on forest physiognomy and phenology, here aggregated into five broad categories, discussed further below.

Temperate and boreal needleleaf forests
Distribution, types and characteristic taxa

Temperate and boreal needleleaf forests cover a larger area of the world than other forest types. They mostly occupy the higher latitude regions of the northern hemisphere, as well as high-altitude zones and some warm temperate areas, especially on nutrient-poor or otherwise unfavorable soils. These forests are composed entirely, or nearly so, of coniferous species (Coniferophyta). In the northern hemisphere, pines *Pinus*, spruces *Picea*, larches *Larix*, silver firs *Abies*, Douglas firs *Pseudotsuga* and hemlocks *Tsuga* dominate the canopy, but other taxa are also important. In the southern hemisphere coniferous trees, including members of the Araucariaceae, Cupressaceae and Podocarpaceae, often occur in mixtures with broadleaf species in systems that are classed as broadleaf and mixed forests.

Country	Forest definition	Forest area (km²)
Australia	Tree canopy cover > 20%	384 115
	Tree canopy cover > 70%	30 729
Senegal	Tree canopy cover ≥ 10% (includes dry woodland)	78 689
	'Closed' forest (canopy cover > 40%)	3 934

Table 5.3
Sample effects on forest area estimates of different forest definitions

Forest type	Area (million km²)
Temperate and boreal needleleaf	13.1
Temperate broadleaf and mixed	7.5
Tropical moist	11.7
Tropical dry	2.5
Sparse trees and parkland	6.9
Total	**41.7**

Table 5.4
Global area of five main forest types

Note: Based on forest cover as shown in Map 5.7. Estimates of this kind vary significantly with different source data and classifications.

Structure and ecology

The structure of temperate and boreal needleleaf forests is often comparatively simple, as conifer canopies are efficient light absorbers, reducing the possibilities for development of lower strata in the canopy. The tallest of these forests, the giant redwood forests of the west coast of the United States, may reach 100 m in height, but most are much shorter, and indeed some pine forests at high altitude or in arid environments are quite stunted.

The distribution of temperate and boreal needleleaf forest is limited at high altitudes and latitudes by lack of enough days with temperatures suitable for growth, and at lower altitudes and latitudes by competition with broadleaf species. In about a quarter of the area of these forests, deciduous conifers of the genus *Larix* replace the evergreen species. This is especially true in far northern continental areas with extremely low winter temperatures.

In many areas, wildfire is an important factor affecting the dynamics and maintenance of the forest ecosystem. Many coniferous species produce resins that increase flammability, and many are characterized by thick bark that increases the resistance of adult trees to fire-induced mortality. A number of tree species, such as the jack pine *Pinus banksiana*, have serotinous cones, which depend on the high temperatures of forest fires to open and release their seeds. Non-coniferous species are generally less resistant to fire than conifers, so periodic fires are an important factor in maintaining the composition and extent of these forests.

Biodiversity

Although tree species richness is low in most temperate and boreal needleleaf forests, many conifer species are of great conservation concern. A well-known example occurs in the giant redwood forests of northern California, where the redwood *Sequoiadendron giganteum* is considered vulnerable to extinction[4]. About 22 percent (140) of the world's 630 conifer species have been assessed as globally threatened[16]. Most of the threatened taxa are characteristic of mixed forests, particularly in the southern hemisphere. Old growth conifer stands, which may be many centuries in age, represent an irreplaceable gene pool and are an important habitat for many other organisms.

Species richness in these forests is commonly increased by a relatively high diversity of mosses and lichens, which grow both on the ground and on tree trunks and branches. For example, there are at least 100 species of moss growing in the coniferous forests between 1 300 and 2 000 m altitude on Baektu Mountain, on the Chinese-Korean border[5]. Mosses and lichens are important sources of food for many animals of coniferous forest.

Vertebrate richness is generally lower in boreal needleleaf forests than in broadleaf temperate and tropical forests. Many species are wide-ranging generalists, often with a holarctic distribution, e.g. wolf *Canis lupus*, brown bear *Ursus arctos*.

There are a number of animals of conservation concern that are dependent on temperate needleleaf forests. The northern spotted owl *Strix occidentalis caurina* requires large expanses of old-growth coniferous forest in the northwest United States to provide nesting habitat and adequate food resources; Kirtland's warbler *Dendroica kirtlandii* needs young regrowing jack pine as a nesting habitat.

Box 5.3 Fire in temperate and boreal forest

In general, as the frequency of fire increases, the intensity of individual fires tends to decrease, because of reductions in the standing amount of fuel. Increased fire frequency also tends to increase the diversity of the herb layer by severely affecting the shrub layer. Fires are caused by natural events such as lightning strikes and by human activities. Changing forest management has significantly altered fire regimes in coniferous forests. The 1988 fires that affected over 5 000 km^2 in and around Yellowstone National Park, United States, were attributed to the accumulation of fuel in the forests resulting from a long-term policy of fire suppression.

Climatic variation plays an important part in determining fire occurrence and severity. The Yellowstone fires, Canadian fires in 1989, and the 1987 Black Dragon fire in the boreal forest region of China and Siberia in 1987, were all ascribed to unusual drought conditions. The Black Dragon fire burned over 70 000 km^2, and qualified as the largest forest fire in recorded history[3]. There is concern that global climate change may increase the frequency and impact of fires in boreal coniferous forests.

Fire suppression programs have reduced the available habitat for this species to critical levels. While there is relatively little information available on the conservation status of invertebrates, many common old-growth species are known to become much rarer in modern managed forests, often through the loss of essential microhabitats[6].

Role in carbon cycle

Temperate and boreal needleleaf forests make a significant contribution to the global carbon balance, accounting for more than a third of the carbon stored in forest ecosystems (Table 5.6) and about 8 percent of global annual net primary production (Table 1.1). Furthermore, the soils under these forests store large amounts of carbon (up to 250 metric tons per hectare), some of which may be liberated by increasing decomposition rates related to climate change. Of particular note in this context are the giant conifer forests of the Pacific northwest of the United States. These forests may store more than twice as much carbon per hectare as tropical rainforests.

Use by humans

Global industrial roundwood production is dominated by coniferous species. Pines *Pinus*, spruces *Picea*, larches *Larix*, silver firs *Abies*, Douglas firs *Pseudotsuga* and hemlocks *Tsuga* from the needleleaf forests of the northern hemisphere are the major sources of softwood. Some conifer species from the southern hemisphere and the tropics also provide excellent timbers. Large-scale exploitation of natural coniferous and mixed forests is taking place around the Pacific Rim, notably in North America, Russia and Chile. Temperate and boreal needleleaf forests are also a principal source of pulpwood for paper production.

Other ecosystem services

These forests, like others, stabilize soils on sloping topography, especially in mountainous regions. Recognition of this function in early 20th century Switzerland was the basis for a new program of forest planting to control avalanches in the Alps. Coniferous forests have high recreational and cultural values, especially in regions such as northern Europe.

Temperate broadleaf and mixed forests
Distribution, types and characteristic taxa

Temperate broadleaf and mixed forests cover some 7.5 million km² of the Earth's surface (Table 5.4). They include such forest types as the mixed deciduous forests of the United States and their counterparts in China and

Table 5.5

Important families and genera, and numbers of species, in four areas of temperate broadleaf deciduous forest

Source: After Röhrig[8].

Family	Genus	Common name	Northeast America	Europe	East Asia	South America
Fagaceae	*Quercus*	Oak	37	18	66	
	Lithocarpus		1		47	
	Castanopsis				45	
	Cyclobalanopsis	Asian oak			30	
	Castanea	Chestnut	4	1	7	
	Fagus	Beech	1	2	7	
	Nothofagus	Southern beech				10
Aceraceae	*Acer*	Maple	10	9	66	
	Dipteronia				1	
Betualceae	*Betulus*	Birch	6	4	36	
	Alnus	Alder	5	4	14	
Salicaceae	*Salix*	Willow	13	35	97	
Juglandaceae	*Carya*	Hickory	11		4	
	Juglans	Walnut	5	1	4	
	Platycarya				2	
Leguminosae	*Cercis*	Redbud	1	1	2	
	Gleditsia	Honey locust	2		7	1
	Gymnocladus		1		3	
	Maackia				3	
	Robinia	Locust	1		1	
	Acacia	Acacia				1
	Albizia				10	
Magnoliaceae	*Liriodendron*	Tulip tree	1		1	
	Magnolia	Magnolia	8		50	
Oleaceae	*Fraxinus*	Ash	4	3	20	
	Osmanthus		2		10	
Rosaceae	*Malus*	Apple	1	1	8	
	Prunus	Cherry	3	4	59	
	Pyrus	Pear	1	1	5	
	Sorbus	Mountain ash	3	5	18	
	Kageneckia					2
	Quillaia					1
Tilliaceae	*Tilia*	Basswood, lime, linden	4	3	20	
Lauraceae	*Phoebe*				16	
	Sassafras	Sassafrass	1		2	
	Beilschmidia					1
	Cryptocarya					1
	Persea					1
Myrtaceae	*Amomyrtus*					1
	Myrceungenella					2
	Myrceugenia					8
	Nothomyrcia					1
Ulmaceae	*Celtis*	Hackberry	2	1	14	
	Ulmus	Elm	4	3	30	
	Zelkova				3	
Carpinaceae	*Carpinus*	Hornbeam	2	2	25	
	Ostrya		1	1	3	

Japan, the broadleaf evergreen rainforests of Japan, Chile, New Zealand and Tasmania, and the sclerophyllous forests of Australia, the Mediterranean and California. These last are characterized by a predominance of often evergreen trees with small, hard, leathery leaves.

Trees belonging to the Anthophyta and the Coniferophyta grow in mixtures in many of these forests, especially in the southern hemisphere. For example, the Valdivian and Magellanic rainforests of Chile include mixtures of *Nothofagus* (an anthophyte) with *Podocarpus* and members of the Cupressaceae (both conifers)[17]. Much of the forest of New Zealand was originally a mixture of the conifers *Podocarpus*, *Phyllocladus*, *Dacryocarpus* and *Dacrydium*, with such Anthophyta as *Metrosideros*, *Elaeocarpus* and *Weinmannia*. In North American mixed forests, pines *Pinus*, hemlocks *Tsuga* and cypress *Taxodium*, among other conifers, are mixed in various proportions with oaks *Quercus*, maples *Acer*, ashes *Fraxinus*, hickories *Carya*, beeches *Fagus* and other hardwoods.

The beech family (Fagaceae) is generally important in temperate broadleaf forests (Table 5.5), with such genera as *Castanopsis* and *Cyclobalanopsis* playing an important role in Japan and China, and many different *Quercus* species being important elements of hardwood forests in North America, Asia and Europe and of sclerophyllous forests in California and the Mediterranean. Sclerophyllous forests in Australia are largely made up of *Eucalyptus* species, as are the wet forests of Tasmania, which may also include *Nothofagus*[9].

Structure and ecology

Depending on the precise forest type, these forests tend to be structurally more complex than pure coniferous forests, having more layers in the canopy. It is not uncommon for an upper canopy layer to have as many as six distinct subcanopy and understorey layers below it. The tallest of these forests, the mixed forests of southern Chile and some *Eucalyptus* forests of Australia, can reach over 50 m in height. On the other hand, some sclerophyllous forests barely reach 5 m. Deciduous forests may support rich herb layers, which depend on the increased penetration of sunlight early in the growing season.

As in needleleaf forests, fire plays an important role in many types of temperate broadleaf and mixed forest. Both natural and anthropogenic fires are important ecological factors affecting the maintenance of forest structure and composition, especially in the *Eucalyptus* forests of Australia and the sclerophyllous forests of the Mediterranean and California. Spatial variation in forest structure and composition may be influenced by fire and other kinds of natural and anthropogenic disturbance. When canopy trees die, the resulting gaps in the canopy increase light availability locally, and such areas may be colonized by a different subset of the forest flora. These gap dynamic processes are important in maintaining stand diversity. Relatively few old-growth broadleaf and mixed forests remain in the temperate zones because of the historical exploitation of these forests by human populations.

Biodiversity

As might be expected from their structural diversity, temperate broadleaf and mixed forests tend to be richer in species than coniferous forests. Southern mixed hardwood forests in the United States are commonly composed of as many as 20 canopy and subcanopy tree species and may include as many as 30 overstorey species[10]. In comparison, European forests tend to be less species rich, while the deciduous forests of East Asia may be the richest of all[18] (Table 5.5).

In the late 1990s, some 370 temperate dicotyledonous tree taxa worldwide were

Table 5.6
Biomass and carbon storage in the world's major forest types

Note: The carbon storage figures are over- rather than underestimates as they incorporate no weighting for anthropogenic disturbance or for the variation in biomass and area among different forest classes within the broad types.

Source: After Adams[85] and Huston[28].

Forest type	Above-ground biomass per area (metric tons/ha)	Estimated total carbon storage including soil and roots (metric tons/ha)	Estimated global total carbon storage (petagrams = 10^{15} g)
Temperate needleleaf	200–1 500	300 (700 giant conifer)	394
Temperate broadleaf and mixed forest	150–300	350	262
Tropical moist forest	195–500	300	350
Tropical dry forest	98–320	250	62

considered to be of conservation concern[19]. Japan alone has 43 threatened endemic tree species, which are mostly characteristic of its temperate broadleaf forests[20]. *Fitzroya cupressoides*, an endangered large conifer once important for the local and international timber trade, is characteristic of mixed forests in southern Chile. It has been over-exploited and proves to be highly dependent for its regeneration on large-scale natural disturbances, such as landslides and lightning strikes. *F. cupressoides* is now listed in CITES (Convention on International Trade in Endangered Species of Wild Flora and Fauna) Appendix I, and commercial international trade is accordingly prohibited.

While temperate broadleaf forests generally support moderate animal diversity, species richness is often lower than in comparable tropical habitats. However, there is considerable geographical variation in richness in the northern hemisphere; the forests of East Asia are generally the most species rich[21]. As in the boreal needleleaf forests, many of the mammal and bird families of northern broadleaf forests are holarctic in distribution and can be found in other habitat types.

In contrast to most northern hemisphere forests, the temperate forests of southern South America, Australia and New Zealand contain several restricted-range mammals and birds. Analysis[22] of habitat requirements of Australian mammals indicates that the forests of southeast Australia and Tasmania are of particular importance for wildlife conservation.

Examples of temperate broadleaf forest species of special conservation concern include the huemul deer *Hippocamelus bisulcus* of the southern Andes, threatened by both habitat loss and hunting; a number of New Zealand forest birds including the kakapo *Strigops habroptilus* and some kiwi species *Apteryx* spp., which are threatened mainly by introduced predators; Leadbeater's possum *Gymnobelideus leadbeateri*, which is threatened by the loss of specific habitat within the montane ash forests of Victoria (Australia); the Amami rabbit *Pentalagus furnessi* of Amami Island (Japan), threatened

by habitat loss and introduced predators; and the European bison *Bison bonasus* of central and eastern Europe, with low genetic diversity and at risk from disease.

Box 5.4 Temperate forest bird trends

The figure shows the living planet index for a sample of 170 forest-occurring birds in North America and Europe. The prevailing trend during 1970-99 was for a small net increase over the period. This could be interpreted as reflecting a phase of relative stability in these forests during recent decades, following centuries of decline in area, but may also be correlated with local increase in forest area through the spread of plantations, and management in some forests may have exerted a positive effect. The separate North America and Europe samples (123 and 47 species) show extremely similar trends.

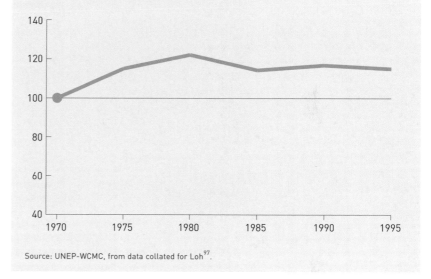

Source: UNEP-WCMC, from data collated for Loh[97].

Role in carbon cycle

Above-ground biomass of temperate deciduous forests in Europe, the United States and the former Soviet Union ranges from 140 to 500 metric tons per hectare depending on stand age and altitude, among other factors[23]. Soil carbon storage in these systems is lower than in the needleleaf forests, owing both to climatically favorable conditions for decomposition and to the inherent greater decomposability of leaf litter from broadleaf trees.

One survey[24] suggests that average soil carbon storage in these forests is somewhere between 135 and 160 metric tons per hectare. Thus, global carbon storage across the forest types included in this broad category may total as much as 231 petagrams (Table 5.6).

Use by humans

Temperate broadleaf and mixed forests have provided large amounts of timber over the centuries, but they have largely been replaced by tropical forests as the primary source of hardwood timber in global trade. Hardwood production continues in the temperate zones, for products such as wood chips, furniture and finishing wood.

Some of the non-timber products of temperate mixed forests include camphor from *Cinnamomum camphorum* in Japan

Tropical moist forests are often particulary rich in palms.

(though this has now largely been replaced by synthetics), and sweet chestnuts *Castanea sativa* from southern Europe. Mushroom production is also a major income source in some parts of Europe, North America and Asia.

Other ecosystem services

Services provided by temperate broadleaf and mixed forests include soil and watershed protection. This is especially important in the southern Andes[17] and other areas where steep topography is responsible for a high incidence of landslides, but it is also a benefit in Mediterranean regions where soils are prone to degradation. Natural temperate forests are important reservoirs of genetic material of trees such as eucalypts that are now commonly grown as plantation species. These forests, perhaps more than any other type, experience significant recreational use in many areas.

Tropical moist forests
Distribution, types and characteristic taxa

Tropical moist forests cover more than 11.5 million km^2 of the humid tropics (Table 5.4 and Map 5.7) and include many different forest types. The best known and most extensive are the lowland evergreen broadleaf rainforests, which make up over half the total area. These include, for example: the seasonally inundated varzea and igapó forests and the terra firme forests of the Amazon basin; the peat forests and moist dipterocarp forests of Southeast Asia; and the high forests of the Congo basin.

Mountain forests are generally divided into upper and lower montane formations on the basis of physiognomy. These include cloud forest – the middle- to high-altitude forests that derive a significant part of their water supply from cloud, and support a rich abundance of epiphytes. Mangrove forests (see Chapter 6) also fall within this broad category.

The high diversity of many tropical forests (see below) makes it difficult to characterize them taxonomically. However, some plant families are more prevalent than others. In neotropical moist forests, the legumes Leguminosae are particularly abundant and are often the most species-rich family. Other families that are generally among the richest in tree species in lowland neotropical moist forests are Moraceae, Lauraceae, Annonaceae, Sapotaceae, Myristicaceae, Meliaceae, Euphorbiaceae and Palmae[25]. In Southeast Asian lowland moist forests the dominant family is the Dipterocarpaceae; the Myrtaceae is also very speciose[26]. In African rainforests legumes are again important, and the ten richest families usually include the Olacaceae, Sterculiaceae, Dichapetalaceae, Apocynaceae, Sapindaceae and Ebenaceae. The other most abundant families in both Africa and Asia are often the same as those in the Americas[25].

Structure and ecology

Many tropical moist forests have canopies 40 to 50 m tall, and some have emergent trees that rise above the main canopy to heights of 60 m or more. Such large-stature forests are characteristic of lowland forests and some lower montane forests on relatively nutrient-rich soils. Another characteristic of these forests is a relatively high frequency of woody lianas and, especially in the neotropics, palms[24]. Moist tropical forests are also known for a high abundance and diversity of vascular epiphytes, which take advantage of the higher light availability found in the canopy and can survive because of abundant rainfall and high atmospheric moisture. On more nutrient-poor soils and at higher altitudes forest stature decreases substantially; communities such as those on white sands (bana and campina) and in upper montane environments (elfin forests) may be no more than a few meters tall. With increasing altitude, decreasing forest stature is accompanied by a reduction in the frequency of lianas and palms, and an increase in tree ferns (Cyatheaceae, Blechnaceae) and non-vascular as well as vascular epiphytes[27].

Unlike the other forest types discussed here, tropical moist forests have relatively little seasonal limitation to their growth, though seasonal drought may be a limiting factor, particularly in the semi-evergreen formations. However, the tropical moist forest environment is an intensely competitive one. Though solar energy inputs are high, canopy closure and complexity are also substantial, resulting in efficient capture of incident radiation and understorey light availability frequently much less than 2 percent of that above the canopy. This in turn limits the growth of understorey species and regenerating trees. Some species can tolerate low light availability, while others grow or regenerate only in gaps in the canopy. Such gaps are formed by the death of one or more canopy trees and represent a significant contribution to overall environmental heterogeneity, which is an important contributor to high diversity within tropical forests[28]. Infrequently, catastrophic disturbances such as blowdowns caused by hurricanes or convective storms may create large areas of regenerating forest[29, 30] and perhaps alter the long-term forest composition, as can logging and other forms of forest disturbance caused by human activities (see below).

Soil nutrients are another limiting resource in many forests. Soils in the humid regions of the tropics are notoriously poor in nutrients owing, among other factors, to loss through leaching by the high annual rainfall and to retention in the high-standing biomass. The formation of gaps in the canopy allowing regeneration of tree species is important for increasing the heterogeneity of available nutrients as well as the availability of solar energy. Forests such as the Amazonian várzea, which are seasonally inundated by sediment-bearing rivers, are an exception to this nutrient limitation.

Biodiversity

In numerical terms, global terrestrial species diversity is concentrated in tropical rainforests. Many theories have been proposed to explain this phenomenon[28]. Generally speaking, the wet tropical forests of Africa have a lower tree species richness than those of Asia and America (Table 5.7). However, there is great local variation in species richness. Within the Amazon basin, tree species richness ranges from 87 species per hectare in the east[31] to 285 species in central Amazonia[32] and nearly 300 species in the west[33].

The high diversity of tree species in lowland evergreen rainforests is mirrored in the diversity of epiphytes and lianas, which is also much higher in neotropical forests than in other regions[34]. Fifty-three families in the Anthophyta and at least nine pteridophyte (Filicinophyta and allies) families include epiphytes. Of nearly 25 000 species of vascular epiphytes, around 15 000 belong to the Orchidaceae. Nearly a thousand others are members of the pineapple family Bromeliaceae, which is primarily neotropical. Other groups having a high diversity of

Table 5.7
Tree species richness in tropical moist forests

Source: After Phillips *et al.*[94].

Region	No. of tree species (≥ 10 cm diameter) per hectare
Africa	56-92
Southeast Asia	108-240
Americas	56-285

Map 5.3
Diversity of vascular plant species

This map shows the species richness of vascular plants, plotted as a world density surface. It is based on some 1 400 literature records from different geographic units, with richness values as mapped calculated on a standard area of 10 000 km^2 using a single species-area curve. Value categories range between extremes of more than 5 000 species and fewer than 100 species per 10 000 km^2.

Source: Data and analysis © Wilhelm Barthlott (Botanic Institute and Botanic Gardens, University of Bonn). Reproduced by permission, with modification to colors. For further details see Barthlott[99] and website http://www.botanik.uni-bonn.de/system/biomaps.htm#worldmap (accessed March 2002).

Diversity

High

Low

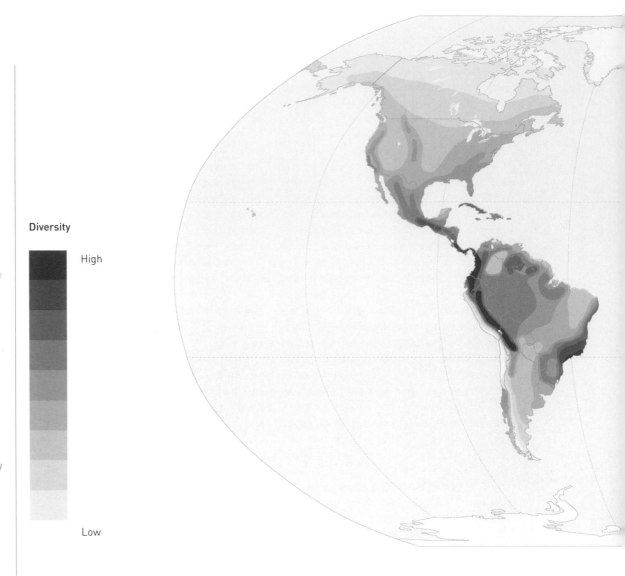

epiphytic species are the cactus family Cactaceae, the aroids Araceae, the pepper family Piperaceae and the African violet family Gesneriaceae.

Not all tropical moist forests have high species richness. Mangrove ecosystems have a low diversity of tree species despite their sometimes high productivity and high animal diversity (see Chapter 6). Extremely nutrient-poor soils, such as white sands, lead to the development of low-diversity forests including bana and campina[35]. As climate becomes more seasonal, tree species richness tends to decline (see dry forests, below). Increasing altitude also tends to reduce species richness, with montane forests typically having fewer tree species than lowland ones[36].

Regionally, the forests of Asia and South America are rich in animal species, generally more so than those of Africa[36]. Many forest animals are largely confined to moist forests, e.g. the okapi *Okapia johnstoni*, but some are widespread outside, such as leopard *Panthera pardus*[36]. In Africa, the Guineo-Congolean forest block contains more than 80 percent of African primate species, and nearly 70 percent of African passerine birds and butterflies[36]. About half of the 1 100 South American reptile species are found in moist forests, with around 300 of these endemic to the habitat[37]. Amphibian species are particularly diverse in tropical moist forests[38]; 90 percent of 225 species identified in the Amazon basin forests are endemic.

The importance and diversity of the fish communities in forest streams and rivers is often overlooked; the Amazon basin has the richest freshwater fish fauna known, with at

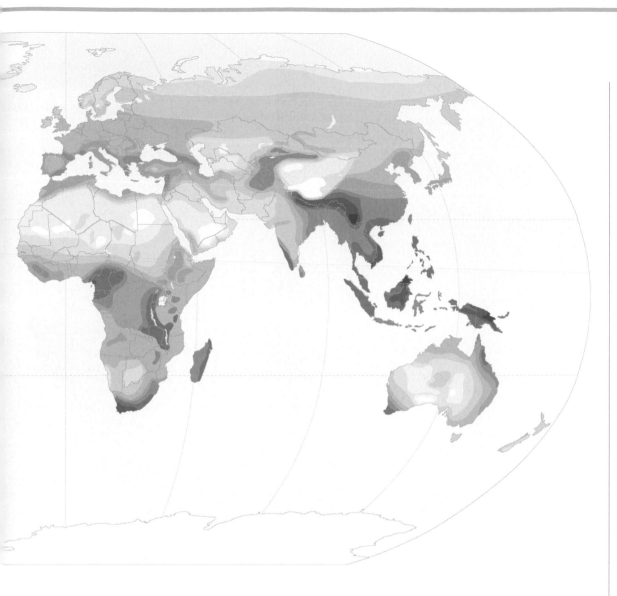

least 2 500 species, many important as major seed predators and dispersal agents[39]. Local species richness of insect and other arthropod groups in tropical forest canopies is much higher than in temperate forests[40]. Around one third of the animal biomass of the Amazon terra firme rainforest consists of ants and termites, and each hectare of soil is estimated to contain more than 8 million ants and 1 million termites[41].

Numerous tropical moist forest species are of conservation concern. Notable animals include the Sumatran rhino *Dicerorhinus sumatrensis* of Southeast Asia, endangered by habitat fragmentation and hunting; the bonobo *Pan paniscus* of the Democratic Republic of the Congo, threatened by habitat destruction and hunting for food; the Philippine eagle *Pithecophaga jefferyi* of the Philippines,

reduced to small fragmented populations through habitat loss and hunting; and the indri *Indri indri* of eastern moist forest on Madagascar, threatened by habitat destruction as is the recently rediscovered Edward's pheasant *Lophura edwardsi* of Viet Nam.

Role in carbon cycle

Lowland evergreen broadleaf rainforests can have high above-ground biomass (Table 5.6), though not as high as some giant conifer forests. Soil carbon, however, is relatively low in most tropical moist forests, with the exception of the peat forests of Southeast Asia and some swamp forests. On this basis it has been estimated that the remaining tropical moist forests store over 300 petagrams of organic carbon, or about one fifth of global terrestrial organic carbon. They account for

Map 5.4
Biodiversity at country level

Country-level biodiversity, represented by an index based on species diversity in the four terrestrial vertebrate classes and vascular plants, adjusted according to country area. Countries at the high end of the scale have a higher value of the index than would be expected on area alone. The index is unreliable for smaller countries (such as Togo and Luxembourg in this plot).

Note: To reduce ambiguity Alaska (United States) has for the purposes of this map been assigned to the same class as adjacent Canada rather than the conterminous United States.

Source: Based on national biodiversity indices developed by UNEP-WCMC; see Appendix 5.

Diversity

High

Low

nearly a third of global terrestrial annual net primary production (Chapter 1), and are therefore key to the global carbon cycle and in regulating global climate.

Use by humans
Tropical hardwood species contribute almost one fifth of world industrial roundwood production[2]. Of the several thousands of species (in more than 200 families in the phylum Anthophyta) that show commercial potential, a few hundred may be found in international trade. Important families include Dipterocarpaceae, with species of meranti and balau *Shorea*, and keruing *Dipterocarpus*; Meliaceae, with mahogany *Swietenia* and *Khaya*, and cedar *Cedrela* and *Toona*; and Leguminosae with rosewood *Dalbergia* and *Pterocarpus*.

The exploitation of tropical hardwood from moist forests has been the subject of much publicity in the past two decades, coinciding as it has done with increased rates of deforestation and forest degradation. Timber supplies from some countries are now widely exhausted, generating openings for other countries to take over as suppliers. A few major producers, however, continue to dominate supply. Indonesia, Malaysia, Brazil and India accounted for 80 percent of tropical log production in International Tropical Timber Organization (ITTO) countries in 1999-2000[2].

Important non-timber products from tropical moist forests include rattans, which are the second most important source of export earnings from tropical forests. A few other craft products and some medicinal products, such as

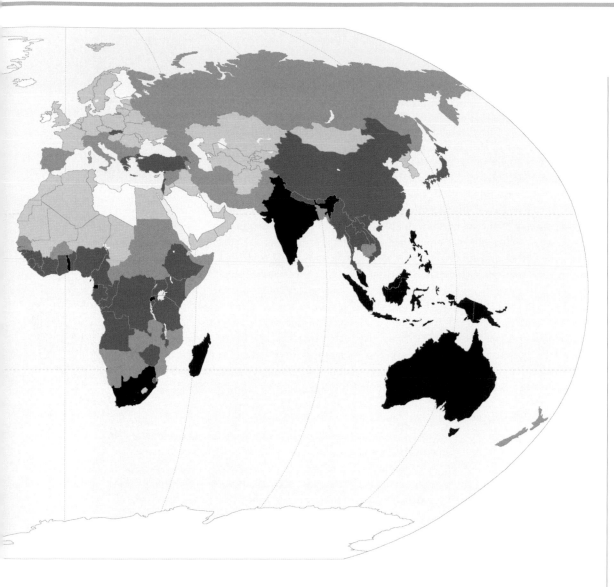

the bark of *Prunus africana*, are significant in international trade. Brazil nuts *Bertholettia excelsa* and native rubber *Hevea brasiliensis* are other extractive products from natural tropical moist forests that are important in international markets, and many tropical moist forests provide fruit, bushmeat and other products for local markets.

Other ecosystem services

Like other forest types, tropical moist forests often play an important role in soil and watershed protection. The high rainfall regimes of the humid tropics mean that exposed soil is particularly liable to leaching of mineral nutrients and to erosion. Montane forests, and especially cloud forests, serve to intercept and store water, thus regulating local and regional hydrological cycles[42]. Lowland forests

are also important in hydrological cycles; it has been estimated that about half the rainfall in the Amazon Basin is derived from water recycled by forest transpiration[43].

Regenerating forest, or forest fallow, is an important part of the cycle of shifting cultivation, which is vital for restoring fertility to areas that have been previously cultivated. This regeneration can only take place if nearby forest cover is adequate to provide a source of propagules.

Tropical dry forests
Distribution, types and characteristic taxa

Tropical dry forests are characteristic of areas in the tropics affected by seasonal drought. Such seasonal climates characterize much of the tropics, but less than 4 million km² of tropical dry forests remain. The seasonality of rainfall is usually reflected in the deciduous habit of the canopy trees, with most being leafless for several months of the year. However, under some conditions, e.g. less fertile soils or less predictable drought regimes, the proportion of evergreen species with leaves highly resistant to water loss increases ('sclerophyllous' forest). Thorn forest, a dense forest of low stature with a high frequency of thorny or spiny species, is found where drought is prolonged, and especially where grazing animals are plentiful. On very poor soils, and especially where fire is a recurrent phenomenon, woody savannahs develop[44, 45] (see 'sparse trees and parkland' below).

Perhaps the best-known tropical dry forest tree species is teak *Tectona grandis* (Verbenaceae), a deciduous hardwood characteristic of the seasonal forests of south and Southeast Asia, widely exploited for furniture and other uses, and now an important plantation species. In Southeast Asia the dipterocarps *Shorea* and *Dipterocarpus*, and *Lagerostroemia* (Lythraceae), and a number of species of legumes (Leguminosae) are also important components of seasonally dry forests[7]. In Africa, dry forests occur both north and south of the equatorial rainforests. In the north they are characterized by *Afraegele* (Rutaceae), *Diospyros* (Ebenaceae), *Kigelia* (Bignoniaceae) and *Monodora* (Annonaceae), among other taxa, while in the south the characteristic genera are *Entandophragma* (Meliaceae), *Brachystegia* (Leguminosae), *Diospyros*, *Parinari* (Chrysobalanaceae), *Syzigium* (Myrtaceae) and *Cryptosepalum* (Leguminosae)[46].

In the neotropics, tropical dry forests occur in Yucatan and Pacific slopes of Central America, on the leeward sides of Caribbean islands, in Venezuela, Caribbean Colombia, in northeast Brazil and in the Chaco region of Bolivia and Paraguay (see Map 5.7). The neotropical dry forests are quite rich in species and include Leguminosae, Bignoniaceae, Rubiaceae, Sapindaceae, Euphorbiaceae, Falcourtiaceae, and Capparidaceae as the families with the largest numbers of species. Important genera include *Tabebuia* (Bignoniaceae), *Trichilia* (Meliaceae), *Eryth-*

roxylum (Erythroxylaceae), *Randia* (Rubiaceae), *Capparis* (Capparidaceae), *Bursera* (Burseraceae), *Acacia* (Leguminosae) and *Coccoloba* (Polygonaceae)[47].

Structure and ecology

Tropical dry forests are generally of lower stature than moist forests, with canopy heights ranging from only a few meters to 30 m or occasionally 40 m[48]. The taller forests have multilayered canopies. Dry forests tend to have more small trees than moist forests and a lower above-ground biomass. The trees have a greater proportion of their total biomass below ground, as more extensive root systems help the trees to obtain water from the soil and avoid drought. Dry forests have a much lower incidence of epiphytes than wet forest, and tend to have both higher frequencies and higher diversity of vines and lianas[47].

Plants with specialized mechanisms for avoiding drought or conserving water are an important feature of these forests. The proportion of deciduous tree species is thought to increase steadily with decreasing annual rainfall, but factors such as the substrate and the between-year variation of seasonal rainfall patterns are also important. Many species have water storage tissues such as succulent stems or tubers, and specialized photosynthetic mechanisms that conserve water are especially common among the epiphytes. Most tropical dry forest trees tend to flower and sometimes to re-leaf before the end of the dry season, and stored water within the plant is essential to this pattern.

As in tropical moist forests, environmental heterogeneity is linked with increased species diversity. Gallery areas along water courses are one source of such variation, and they serve as a refuge for animals during the dry season[49]. Termite mounds provide an important source of environmental variation in African dry forests, adding to local topography and supplying high-nutrient microenvironments to the system, increasing tree species richness by 40-100 percent[46]. Fire is also a major factor determining the dynamics and extent of tropical dry forests.

Biodiversity

Though of lower species richness than tropical moist forests, dry forests still have appreciably more tree species than most temperate forests. The richest neotropical dry forests, which are not the wettest ones but those in western Mexico and in the Chaco of southeast Bolivia, have around 90 woody species per 0.1 hectare sample[47].

Although a comprehensive assessment has yet to be made, dry forests are thought to have high rates of plant species endemism relative to wet forests in the tropics[47]. Sixteen percent of the plant species of the Chamela dry forest in western Mexico are local endemics, and 20 percent of the flora of Capeira, Ecuador, are endemic to western Ecuador[50]. Many of the dipterocarps in Thailand's seasonal forests are national endemics and distinct from the species in the country's moist forests[7].

Vertebrate species diversity is lower in dry forests than in moist forests, but many dry forests have high rates of endemism among mammals, especially among groups such as insectivores and rodents, characterized by low body weights, low mobility and short generation times[49]. Among neotropical dry forests, those of Mexico and the Chaco have the highest numbers of mammal endemics (26 and 22 respectively[49]). Remaining areas of dry forest are often important refuges for once widespread species. The Gir forest of Gujarat (India) contains the only population of Asiatic lion *Panthera leo persica*, once found throughout much of southwest Asia; the dry forests of western Madagascar are inhabited by around 40 percent of the island-endemic lemurs; some, such as red-tailed sportive lemur *Lepilemur ruficaudatus*, are almost entirely confined to this habitat. Invertebrate species richness tends to be poorly known, but in groups such as lepidoptera and hymenoptera richness in some dry forest areas may be comparable to adjacent wet forest[51].

Because of their high degree of endemism and because degradation and conversion have progressed further than in wet forests, the biota of tropical dry forests are often highly threatened. Hunting, especially for the wildlife trade, and habitat conversion are

important pressures on dry forest animal species.

Threatened dry forest species include Spix's macaw *Cyanospitta spixii*, which is nearly extinct globally as a result of trapping and habitat loss in Brazil's northeastern caatinga region; the Chacoan peccary *Catagonus wagneri*, which was rediscovered in the Gran Chaco of central South America during the 1970s and is threatened by overhunting, habitat loss and disease; Verreaux's sifaka *Propithecus verreauxi* of western Madagascar, which is at risk from loss of spiny and gallery forest habitat; and the Madagascar flat-tailed tortoise *Pyxis planicauda*, which is restricted to the western *Andranomena* forest of Madagascar and is believed to be declining as a result of habitat destruction.

Many of the endemic lemurs of Madagascar are found in dry forest.

Map 5.5
Flowering plant family density

Global diversity of flowering plants represented as a density surface derived from distribution maps for 284 non-aquatic plant families. Scale indicates number of families present, up to a maximum of 182, divided into 15 classes.

Source: Compiled primarily from Heywood[100] by UNEP-WCMC.

Density

High

Low

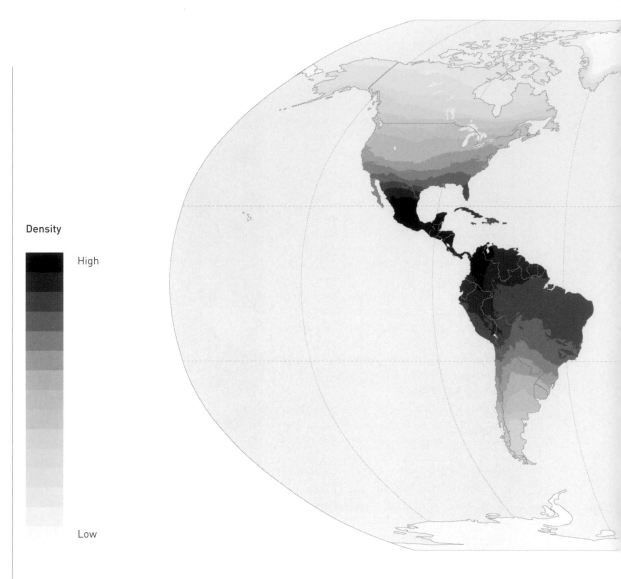

Role in carbon cycle

Their lower biomass means that tropical dry forests represent a smaller reservoir of stored carbon per unit area than the other forest types discussed so far. With a total biomass ranging from 98 to 320 metric tons per hectare[52] and soil carbon storage in the region of 100 metric tons per hectare[24], it is unlikely that relatively undisturbed tropical dry forests store more than 250 metric tons of carbon per hectare (Table 5.6). This, combined with the fact that little intact tropical dry forest remains worldwide, suggests that the total contribution of seasonally dry tropical forests to global carbon storage is far less than that of other forest types.

Use by humans

Notable among the economically important species of seasonally dry tropical forests is teak

Tectona grandis, which accounts for about 1 percent of reported global tropical timber exports. More than ten species of Thailand's seasonal forests are significant for the timber trade[7], with timber species such as mahogany *Swietenia* and several species of *Tabebuia* (Bignoniaceae) characteristic of neotropical dry forests. The southern dry forests of Africa also contain useful timber species such as *Entandrophragma* spp. Tropical dry forests also provide large quantities of fuelwood for local populations. A number of food plants are native to tropical dry forests and medicinal uses have been reported for many dry forest plant species[53]. Craft products are also important.

Other ecosystem services

Protection of relatively fragile soils is a vital ecosystem service provided by tropical dry

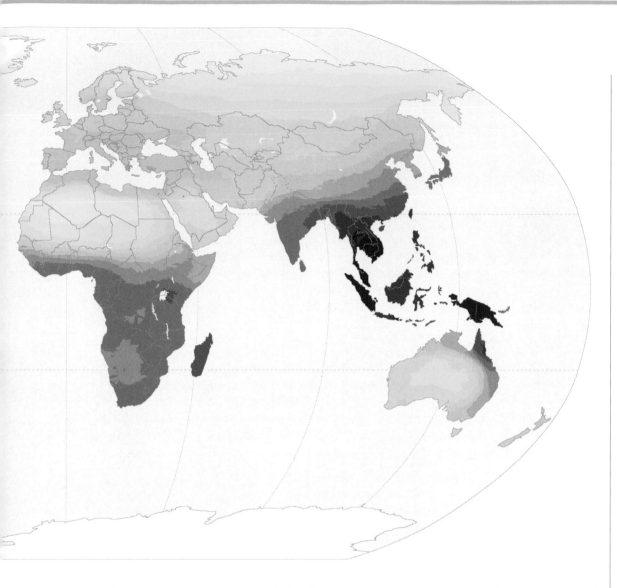

forests. Rains may be intense during the wet season, and erosion can be a severe problem in tropical dry forest areas, where soils are often thin and soil formation processes slow[54]. Tropical dry forests may also be important resources for native pollinators as well as nectar sources for domestic bees. Many dry forest trees produce conspicuous flowers with specialist pollination mechanisms. Their mass flowering provides a major nectar resource for pollinating insects at the end of the dry season when other such resources may well be limited[47, 55]. Honey production is one of the livelihoods being promoted for local communities in dry forest areas in Mexico and elsewhere.

Sparse trees and parkland
Sparse trees and parkland are forests with open canopies of more than 10 percent crown cover. They occur principally in areas of transition from forested to non-forested landscapes. The two major zones in which these ecosystems occur are the boreal region and the seasonally dry tropics.

At high latitudes, north of the main zone of boreal forest or taiga, growing conditions are not adequate to maintain a continuous closed forest cover, so tree cover is both sparse and discontinuous. This vegetation is variously called open taiga, open lichen woodland or forest tundra[56]. It is species poor, has high bryophyte cover, and is frequently affected by fire. It is important for the livelihoods of a number of groups of indigenous people, including the Saami and some groups of Inuit.

In the seasonally dry tropics, decreasing soil fertility and increasing fire frequency are related to the transition from closed dry forest through open woodland to savannah. The open woodland ecosystems include the more open *Brachystegia* and *Isoberlinia* woodlands of dry tropical Africa and parts of both the caatinga and cerrado vegetations of Brazil[46]. Open woodlands in Africa are more species rich than either closed dry forest or savannah. The cerrado supports a high diversity of woody plants, though many of them are of shrubby habit.

Animal diversity is generally low in forest tundra; few species are restricted to this habitat, many also occurring in boreal forest or tundra proper. The sparsely wooded tropical savannahs are generally more species rich than temperate forests or grasslands[57]. Wooded savannahs vary greatly. Those of America are relatively species poor, while African savannah sometimes attains a richness not far below rainforest in the same continent. Sparsely wooded areas in Australia are amongst the richest wildlife habitats on the continent, sometimes more so than adjacent wet forests[57, 58]. The large savannah vertebrates present in such high diversity in Africa are largely absent from other

The large savannah vertebrates of Africa are largely absent from other continents.

continents[59] (see Chapter 4). The density and biomass of tropical savannah soil invertebrates (mostly earthworms, ants and termites) is generally lower than temperate grasslands, but greater (at least in biomass) than tropical rainforests[60].

Species of conservation concern include the black rhinoceros *Diceros bicornis*, threatened primarily by hunting for their horn, and the golden-shouldered parrot *Psephotus chrysopterygius* of northern Queensland (Australia), threatened by the burning of seeding grasses during the breeding season and predation by feral cats.

Forest plantations

Forest plantations, generally intended for the production of timber and pulpwood, increase the total area of forest worldwide. FAO[61] estimates that forest plantations covered 187 million hectares in 2000, of which Asia accounted for 62 percent. This represents a significant increase from the 1995 estimate of 124 million hectares. Commonly monospecific and/or composed of introduced tree species, plantation forests tend to be less valuable as a habitat for native biodiversity than are natural forests. However, they can be managed in ways that enhance their habitat value. Plantations are also important providers of ecosystem services, such as maintaining nutrient capital and protecting watersheds and soil structure as well as storing carbon. They may also alleviate pressure on natural forests for timber and fuelwood production.

In some countries, wood from plantation sources makes up a significant portion of the industrial wood supply. For example, New Zealand is more than self-sufficient in wood production based on plantations[61]. New forest plantation areas are increasing globally at a rate of 4.5 million hectares per year, with particularly high rates of increase in Asia and South America[61]. In future, forest plantations are likely to play an increasingly important role in mitigation of greenhouse gases, as encouraged by the Kyoto Protocol.

Changes in forest cover

About half of the forest that was present under modern (i.e. post-Pleistocene) climatic conditions, and before the spread of human influence, has disappeared, largely through the impact of human activities. The spread of agriculture and animal husbandry, the harvesting of forests for timber and fuel, and the expansion of populated areas have all reduced forest cover. The causes and timing of forest loss differ among regions and forest types, as do the current trends in change in forest cover.

The temperate forests of western Europe have diminished by far more than the 50 percent estimated for forests globally. Much of this deforestation occurred between 7 000 and 5 000 years ago as Neolithic agriculture expanded[62]. The expansion of human populations and increasing demand for fuel during classical times and the Middle Ages put further pressure on European forests. Between Neolithic times and the late 11th century, forest cover in what is now the United Kingdom decreased by 80 percent. As European forests dwindled they became an increasingly valuable resource that was more carefully managed. Forest cover stabilized during the 19th century in much of western Europe in response to both improved management and reduced demand for forest products (owing to the increasing use of fossil fuels and changes in construction materials). Since the early 20th century, forest cover in Europe has expanded, often through the establishment of conifer plantations.

In eastern and central Europe and in Russia, forest clearance accelerated during the 16th and 17th centuries as sedentary agriculture expanded. One estimate[63] suggests that around 1 million km[2] of forest had been cleared in the former Soviet Union up to 1980. Timber exploitation continues to drive forest clearance in the coniferous forests of Siberia and parts of eastern Europe.

In North America, indigenous groups had impacts on the forests from at least 12 000 years ago, but most forest clearance took place after European settlement. Forest cover in eastern North America reached its minimum around 1860, but then increased with the westward movement of the agricultural frontier and subsequent urbanization and industrialization. Forests west of the Appalachians suffered the most severe impacts in the late 19th and early 20th centuries, but are still under pressure from demand for timber and pulp.

In Oceania, just as in North America, indigenous groups had significant impacts on the forests before the arrival of Europeans. This was especially true of the aboriginal use of fire in Australia, but European colonization greatly increased the rate of forest conversion. Over 230 000 km[2] of forest and 120 000 km[2] of woodland in Oceania are estimated to have been converted to cropland between 1860 and 1980[63].

In tropical Asia, Africa and Latin America, large-scale deforestation was precipitated by European colonial activities, including agriculture and timber exploitation. It is estimated that more than 1 million km[2] of forest and a similar amount of woodland in tropical developing countries were converted to cropland between 1860 and 1980[63]. The bulk of this conversion was in south and Southeast Asia, where forest area declined by 39 percent from 1880 to 1980.

Globally, tropical dry forest has lost the greatest proportion of its original area of the four major types of closed forest, nearly 70 percent. About 60 percent of the original area of temperate broadleaf and mixed forests has disappeared, and tropical moist and temperate needleleaf forests have lost about 45 percent and 30 percent of their original area, respectively.

Current trends in change in forest cover, which are shown in Table 5.8, reveal that the rates of deforestation continue to be high in

Region	Total forest area in 2000 (million ha)	Forest cover change 1990-2000	
		Annual change (thousand ha)	Annual rate of change (%)
Africa	650	−5 262	−0.78
Asia	548	−364	−0.07
Europe	1 039	881	0.08
North and Central America	549	−570	−0.10
Oceania	198	−365	−0.18
South America	886	−3 711	−0.41
World total	**3 869**	**−9 391**	**−0.22**

Table 5.8
Estimated annual change in forest cover 1990-2000

Note: Figures refer to natural forests and plantation forests combined.

Source: FAO[61].

Table 5.9
Global protection of forests within protected areas in IUCN categories I-VI

Note: Forest areas and protection as assessed in 1999.

Source: UNEP-WCMC[96].

Forest type	Global protection	
	(km[2])	(%)
Temperate and boreal needleleaf	675 470	5.4
Temperate broadleaf and mixed	457 535	7.0
Tropical moist	1 382 004	12.2
Tropical dry	413 524	11.2
Sparse trees and parkland	274 401	5.8
Total	**3 202 934**	**8.3**

the developing countries of the tropics, in both absolute and proportional terms. In contrast, temperate countries are losing forests at lower rates, or indeed showing an increase in forest area[64].

Pressures on forest biodiversity

The principal pressures on forests and their biodiversity are conversion to other landuses, principally forms of agriculture, and logging. Conversion of forest to agriculture is the main cause of tropical moist forest loss. This is largely due to expanding populations and the use of shifting cultivation at an intensity that does not permit adequate fallow periods. Government resettlement programs that have moved large numbers of poor farmers have increased the rate of land colonization and clearance in parts of Southeast Asia and Latin America. In some areas, land has been

Fuelwood and charcoal consumption more than doubled between 1961 and 1991.

converted to ranching principally as a means of gaining title in order to permit speculation in land values. Thus, population growth, poverty and inequitable land tenure are among the causes underlying deforestation by conversion to agriculture.

Timber extraction puts great pressure on biodiversity in both tropical and temperate forests. Global consumption of industrial roundwood was more than 1 521 million m^3 in 1998[14] and, on current trends, is projected to continue to rise. Although some timber species are naturally abundant, a factor that can help ensure their survival from commercial exploitation, many have suffered

extensive and irreversible population and genetic losses. Furthermore, rare species that are indistinguishable in the field from their commercially important relatives are in danger of extinction through overexploitation. This particular problem exists, for example, among the dipterocarp groups meranti, balau and keruing. A few hundred species may be traded under these names, and a significant proportion of these are geographically and ecologically restricted and so at high risk of extinction.

Furthermore, logging operations create access to forest areas that may otherwise have remained isolated. This improved access facilitates hunting and other activities that exert pressure on forest biodiversity, and may ultimately lead to colonization and conversion of the land to agricultural use. There is also strong evidence that logging can increase the probability of wildfire in temperate forests and even in tropical moist forests not usually subject to burning[65].

Particularly in the drier areas of the tropics, fuelwood extraction can have serious impacts on forests and open woodlands. Fuelwood and charcoal consumption more than doubled between 1961 and 1991, and is projected to rise by another 30 percent to 2 395 million m^3 by 2010[64]. About 90 percent of the consumption is in developing countries, but wood fuel may play an increasing role in some developed countries, increasing demand for wood still further[61].

In addition to loss of area, forest conversion and logging lead to changes in the condition or quality of the remaining forest. These can include fragmentation of large areas of continuous forest. Tropical forest fragments are distinct from continuous forests in both ecology and composition[66]. There are physical and biotic gradients associated with fragment edges, and forest structure undergoes radical change near the edges as a result of the impacts of wind and increased tree mortality. Some animal species are 'edge-avoiders' and decline in abundance in forest fragments, while others become more abundant. Some non-forest and even non-native species of plants and animals successfully invade forest fragments but not continuous forest. In addition to directly affecting canopy composition, removal of large timber trees may also affect the availability of seed for regeneration and may affect animal species that depend on the timber species.

Other factors that affect forests and their biodiversity include acid rain and global climate change. So far, most of the effects of acid precipitation, which is caused by industrial air pollutants, have been documented in temperate needleleaf forests and associated waterways of Europe and North America. The likely impacts of global climate change on forests are still being debated, but there seems to be general consensus that the boreal coniferous forests are particularly vulnerable to both range restrictions and increasing fire frequency[67].

Another forest type that has been shown to be vulnerable to climate change is tropical montane cloud forest, which depends upon clouds to supply it with atmospheric moisture. Research has shown that the mean cloud base is moving upwards on tropical mountains as a result of climatic shifts. The forest species are not able to migrate at a comparable rate and, in any case, range shifts will be limited by the land area existing at higher elevations. Local extinctions in cloud forest amphibians, including the golden toad *Bufo periglenes* assessed as critically endangered, have been attributed to climatic fluctuations that may be linked to long-term global climate change[68].

NON-FOREST ECOSYSTEMS

The parts of the Earth that are too cold, too dry or too severely affected by fire and/or grazing do not support forest or woodland ecosystems. However, as can be seen from Map 5.1, many of them do support active plant growth. Natural non-forest ecosystems include tundra (both arctic and montane), grasslands and savannahs, and shrublands. Less productive, but with unique elements of biodiversity, are the deserts and semi-deserts (Map 5.4).

Tundra

Tundra is the vegetation found at high latitudes beyond the limits of forest growth; the same term is sometimes used for similar vegetation at high elevation at lower latitudes, but these may be distinguished as 'polar tundra' and 'alpine tundra', respectively. In the Arctic, polar tundra occurs north of the northern tree line, which is determined by a number of climatic factors including the summer position of arctic air masses[69] and the depth of permafrost (permanently frozen

In polar tundra systems, temperatures are low for much of the year.

subsurface soil). Similarly, alpine tundra occurs above the climatic tree line on mountains, and its elevation varies in a complex fashion with latitude, continental or oceanic climate, and the maximum elevation and overall size of the massif[70].

The characteristics of polar and alpine tundra environments differ in many respects. In high-latitude polar tundra systems, temperatures are low for much of the year, while permafrost limits both drainage and root extension, and the growing season may last for as little as six to ten weeks. Rainfall is low, usually less than 200 mm per year, and at extreme latitudes may be so low that the environment is described as polar desert. At high elevation in temperate regions temperatures may be similarly low, although permafrost is rare. However, at high altitudes in the tropics, although low temperatures occur every night, high insolation causes warming during the day so that adequate temperatures for active plant growth occur

Map 5.6
Terrestrial vertebrate family density

Global diversity of terrestrial vertebrates represented as a density surface derived from distribution maps for all 350 non-aquatic families of mammals, birds, reptiles and amphibians. Scale indicates number of families present, up to a maximum of 126, divided into 15 classes.

Source: Compiled from multiple sources by UNEP-WCMC.

Density

High

Low

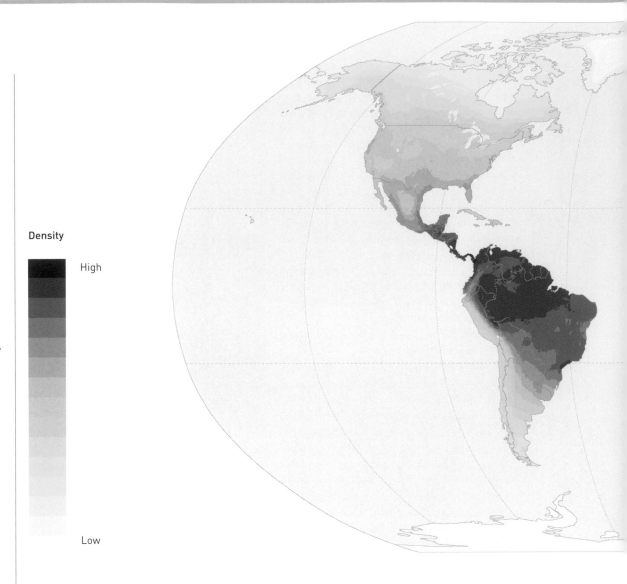

throughout the year and the diurnal temperature range is large.

Despite these environmental differences, polar and alpine tundra vegetation have some features in common. Both lack trees, but contain woody species growing in dwarf or prostrate forms, especially in locations with less extreme climates. As latitude or altitude increases, grasses, sedges, bryophytes and lichens increase in importance while shrubs decrease. Many plants have tussock or cushion growth forms. At extremes of latitude or elevation a high proportion of bare ground is characteristic.

In the arctic tundra, plants cover 80-100 percent of the ground[69], and cover decreases along the climatic gradient to polar desert. The important woody plants are birches *Betula*, willows *Salix* and alders *Alnus*. These

are all genera that occur as trees in temperate regions but as spreading, prostrate or dwarf forms, sometimes less than 20 cm tall, in tundra. Other shrub species are also important, including *Dryas, Vaccinium* and *Empetrum*. Interspersed with the shrubs and of increasing importance in more extreme sites are the sedges *Carex* and the cotton grasses *Eriophorum*, among other graminoids, and a high diversity of mosses and lichens. Biomass is often much lower above than below ground, and annual production is low. This combination means that the potential of these ecosystems for recovery following disturbance is limited[71].

Temperate alpine systems are characterized by many of the same taxa as the arctic tundra, but at lower latitudes other groups become important. At high altitude in the

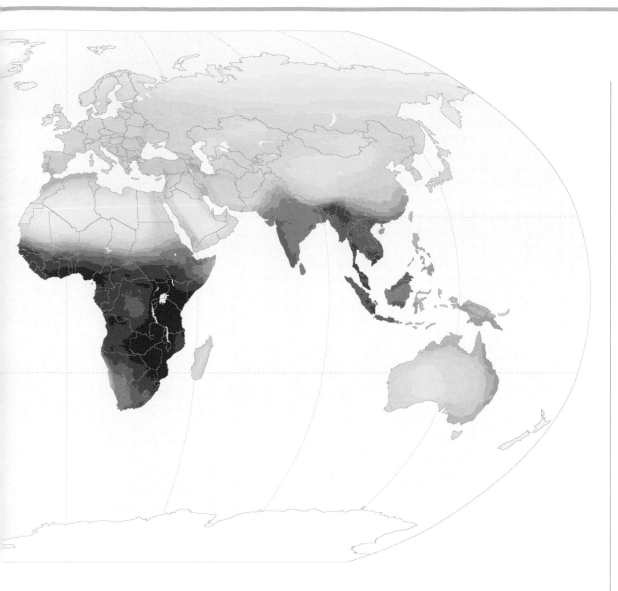

tropics, giant rosette plants are a notable feature of alpine communities. These distinctive plants, which have a number of morphological and physiological adaptations to the high insolation, large temperature fluctuations and desiccating conditions on tropical mountains, include *Espeletia* and *Puya* in the páramos of the Andes, and *Senecio* and *Lobelia* in the high mountains of Africa[72].

In comparison with forested ecosystems, both polar and alpine tundra are relatively species poor. It is estimated that species richness declines by a factor of between three and four between the boreal and arctic zones[71], and species richness in the polar desert is one fifth that of the tundra. The entire North American Arctic has a vascular flora of about 600 species[69]. Bryophytes and

lichens may add more than 300 additional species to the circumpolar flora. The tropical alpine systems are richer; Venezuelan páramos include more than 400 angiosperm species[72], but this is still much lower than in surrounding forests.

Because most arctic plant species have wide geographic distributions, few are of significant conservation concern, but many alpine areas are isolated by lowlands with contrasting climates so increasing endemism in many alpine floras. The floras of isolated mountains in North America have significant rates of endemism[73]; about 15 percent of the flora above the timber line in the Alps are locally endemic species[74] and about 80 percent of the flora at high altitude in East Africa and Ethiopia are endemic[75]. Restricted distributions tend to make species more vulnerable to extinction, while the harsh environment of alpine regions adds to the risk.

Animal species richness tends to be low; groups represented by several species in boreal forest are often reduced in diversity by up to one third in tundra habitat[71]. In contrast, a few groups, particularly water birds and waders, are able to exploit the large numbers of invertebrates found in tundra soil and can be both diverse and abundant[71]. Although the species richness of the most common in-vertebrate groups (Collembola and oribatid mites) decreases with increasing latitude, their total abundance may increase.

There are relatively few globally threaten-ed species that are completely dependent on tundra. An exception is one of the world's most severely endangered species, the once abundant Eskimo curlew *Numenius borealis* of the Americas, which nests almost ex-clusively within this habitat. Two globally threatened bird species, Steller's (*Polysticta stelleri*) and spectacled eider (*Somateria fischeri*) remain within the Arctic throughout the year. Although low in number of species, the Arctic is home to most of the world's geese and calidrid sandpipers[76].

Although the tundra accounts for less than 2 percent of global annual net primary production (Table 1.1), the high below-ground biomass and soil carbon in arctic tundra means that it makes an important contribution to global carbon stocks. Total biomass in tundra communities of the Russian Arctic falls in the range 7-30 metric tons per hectare, of which some 60-70 percent is below ground[77]. Sedge-moss communities in the North American Arctic may have 15-30 times as much biomass below ground as above ground[69], and three to eight times the amount of dead as live material may accumulate. Tundra soils may store around 200 metric tons of carbon per hectare[24].

Because of its inhospitable climate, tundra is not subject to severe pressure for conversion for other landuses. However, its lack of ecological resilience means that disturbances, e.g. those associated with settlements or long-distance pipelines, tend to have long-lasting effects. It is anticipated that the effects of global warming on the arctic tundra will be significant, as relatively large temperature increases are predicted for this zone[69]. These will cause changes in the permafrost regime and decomposition of accumulated soil organic matter, which in turn will release additional carbon dioxide into the atmosphere. Evidence suggests that the period of active plant growth has recently lengthened in parts of North America. There is also evidence that species are already migrating northwards in response to climate change, so that arctic tundra is likely to be compressed into a much smaller area of remaining appropriate climate.

Grasslands and savannahs

Grassland ecosystems may be loosely defined as areas dominated by grasses (members of the family Gramineae excluding bamboos) or other herbaceous plants, with few woody plants[40]. Grassland ecosystems are typically maintained by drought, fire, grazing and/or freezing temperatures[78]. In addition, they are often associated with soils of low fertility[40]. Savannahs are tropical ecosystems charac-terized by dominance at the ground layer of grasses and grass-like plants. They form a continuum from treeless plains through open woodlands to closed-canopy woodland with a grassy understorey. Some savannah areas therefore meet general definitions of grass-land (fewer than 10-15 woody plants per hectare[79]) while others meet the definition of woodland. Some polar and alpine tundra communities also meet the definition of grassland.

Around 20 percent of the Earth's land surface (excluding Antarctica) supports grassland, with these regions differing greatly in naturalness[40]. Temperate grasslands make up approximately one quarter of this area, and savannahs the remainder (Map 5.8). The most extensive areas of temperate grasslands are the prairies of North America, the pampas and campos of southern South America and the steppes of central Europe, southwest and central Asia and Russia. Temperate grass-lands are sometimes divided into formations of tall grass, mixed grass and short grass, which differ both floristically and ecologically; short-grass communities are usually associated with drier climatic regimes[80]. Tropical grasslands and savannahs include the llanos of the Orinoco basin in Venezuela and Colombia, the cerrado of central Brazil, and the savannahs of tropical and subtropical Africa. In addition to many species of grasses, the sedges (Cyperaceae) and many different groups of dicotyledonous herbs are also important.

Key ecological factors in grasslands are grazing pressure and the effects of fire. Most natural grasslands had, at one time, large populations of native grazing mammals. These have been replaced to a great extent by

environments, are often rich in organic matter and are therefore particularly vulnerable to conversion to cropland, with replacement of native grasses by their domestic derivatives (cereals) and other plants[79].

At very fine spatial scales, natural grasslands can be extremely species rich. For example, a square meter of 'meadow steppe' in the former Soviet Union may have 40-50 plant species[82], a tall grass prairie remnant of less than 2 hectares may contain 100 species, and 250 hectares may contain 250-300 plant species[79]. However, grassland communities tend to be similar over large areas, and

The world's grasslands and savannahs support distinctive plant and animal communities.

domesticated ungulates, which also exert a significant degree of grazing pressure (the magnitude depending on stocking densities). Grazing tends to increase abundance of less palatable species and to increase species richness in productive areas, or decrease it in less productive areas. At intermediate frequencies, fire tends to increase diversity and suppress invasion by woody species. Frequent fires favor grasses, which usually recover easily, whereas low fire frequency may allow the density of woody species to increase.

These factors have important consequences for vegetation structure in grassland and savannah systems. A high proportion of plant biomass, in the form of roots and rhizomes, is located underground; there is a high turnover of those parts of the plant above ground[81]. One important consequence of this is that grassland soils, especially in more humid

structurally simple, so that at the landscape scale diversity is relatively low compared with tropical moist forest or Mediterranean-type ecosystems. Grasslands tend to have low rates of endemism; however, the climatic and soil gradients within them have led to substantial ecotypic variation and high genetic diversity[79].

The world's grasslands and savannahs support distinctive plant and animal communities. Although species diversity tends to increase towards the tropics, it tends to be moderate or low at the landscape scale and above. Little more than 5 percent of the world's mammal and bird species are primarily dependent on grasslands habitats[40]. All these ecosystems hold, or formerly held, an array of native herbivores, and these in turn support a number of high-profile mammalian and avian predators.

Box 5.5 Grassland bird trends

The figure illustrates the living planet index in a sample of 29 grassland birds from North America (25 species) and Europe (4). The clear trend over four decades is downward, and as with forest birds the separate continent samples show a similar overall pattern.

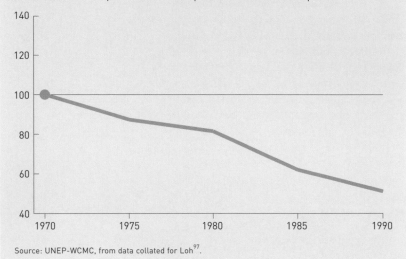

Source: UNEP-WCMC, from data collated for Loh[97].

The savannah communities of East Africa are typified by large herds of ungulate herbivores, including a remarkable diversity – more than 70 species – of antelope and other medium- to large-sized bovids. The biomass of ungulate herbivores here may rise to 30 metric tons per km[2], which is the highest recorded in any terrestrial environment[28]. Most grassland invertebrate biomass is found within the soil (most commonly nematodes, enchytraeid worms and mites) and may be in the order of 100 to 1 000 times as great as vertebrate biomass; soil invertebrate biomass above the soil is often dominated by Orthoptera[81].

Many grassland birds require large areas of habitat to take full advantage of sparsely distributed food resources, and fragmentation of natural grasslands has made conservation of these wide-ranging species difficult[40]. An analysis of the location of bird species in the neotropics indicated that nearly 12 percent of threatened birds (38 species) are confined to grasslands. In this region, the grasslands of southern Brazil and northern Argentina are particularly threatened as a result of agricultural improvement[83]. In North America, grassland species have experienced the most consistent declines of any group of birds monitored in a national survey of breeding birds; available data suggest that there has been a constant decline in these species over the past 30 years[84].

Among notable threatened mammal species in grasslands are the greater one-horned rhinoceros *Rhinoceros unicornis*, associated with tall riverine grassland in northern India and southern Nepal, where it is threatened by poaching and further loss of its restricted habitat. Other large mammals are the saiga antelope *Saiga tatarica* of central Asian grasslands, and the vicuña *Vicugna vicugna* in arid grasslands and plains in the Andes; both of these are threatened in parts of their range. Among birds, the short grassland habitat of the plains wanderer *Pedionomus torquatus* of southern Australia continues to be lost to cultivation; Rudd's lark *Heteromirafra ruddi* inhabits the montane grassland plateaus of eastern South Africa and is threatened by habitat degradation; while the lesser florican *Sypheotides indica*, of western India and Nepal, is critically endangered through loss of suitable grassland habitat.

Grassland biomass ranges from around 2 to over 80 metric tons per hectare[81]. As a rule, more than 50 percent of this is below the soil surface, and the ratio of root to shoot biomass ranges from below five in warm humid grasslands to as much as 30 in the desert grassland of Mongolia[81]. Usually more than half of the below-ground biomass is in the upper 10 cm of the soil, and soil carbon stocks may be as high as 250 metric tons per hectare[85]. Annual production reaches 30-50 metric tons per hectare per year in some warm humid grasslands. It is estimated that grasslands and savannahs together account for about a third of global terrestrial net primary production (Table 1.1), so despite their relatively low biomass grasslands play an important role in the global carbon balance.

Because the high below-ground biomass tends to increase the fertility of their soils, grasslands have been subject to high rates of conversion to agriculture. Less than half of North American grasslands remain in natural or semi-natural states[40], the steppes of the former Soviet Union have been extensively irrigated and converted to agriculture and much of the pampas has been converted to agriculture or grazing land. Some anthropogenic grassland consists of short-term monospecific sown pasture, while some areas support species-rich semi-natural grassland created over centuries by pastoralists in conjunction with livestock grazing.

Domestic livestock grazing is the most extensive human use of unconverted or anthropogenic grassland ecosystems, and of most arid or semi-arid ecosystems. Livestock have an impact on ecosystems through trampling, removal of plant biomass, alteration of plant species composition through selective grazing, and competition with native species. The impact of this on the biological diversity of these ecosystems has been variable. In some areas where the native vegetation is well adapted, the impact on plant species diversity has been relatively small. In other areas, where vegetation has not evolved in the presence of hooved herbivores, the changes have been great. Sometimes, particularly in tropical and semi-tropical grasslands, the dominant component of vegetation has shifted from grass to woody plants.

Overgrazing can lead to reduction in plant cover, loss and degradation of soil, and invasion by non-native plant species. In almost all cases, wild animal diversity has been greatly affected (mostly through competition and hunting, but also through spread of pathogens), so that the biomass of domestic livestock greatly exceeds that of native wild herbivores. In some areas, feral species (e.g. rabbits, camels, donkeys, horses, goats) may also have a marked impact on natural or semi-natural ecosystems.

Shrublands

Shrub communities, where woody plants, usually adapted to fire, form a continuous cover, occur in all parts of the world with 200-1 000 mm of rainfall[28]. In more arid areas including some semi-desert systems, shrubs are the dominant life form, but cover is discontinuous. The most distinctive and best-known shrublands are those of Mediterranean climate regions.

Mediterranean climates are typified by cool, wet winters and warm, or hot, dry summers. However, no single climatic or bioclimatic definition of a Mediterranean ecosystem has yet been established, so that these areas remain rather loosely defined. Mediterranean ecosystems encompass a wide range of habitat types including forest, woodland and grassland, but are typified by a low, woody, fire-adapted sclerophyllous shrubland (maquis, chaparral, fynbos, mallee) on relatively nutrient-poor soils. These systems occur in five distinct parts of the world: the Mediterranean basin; California (United States); central Chile; Cape Province (South Africa); and southwestern and south Australia. Each of these regions occurs on the west side of a continent and to the east of a cold ocean current that generates winter rainfall. They cover around 2.5 million km^2 in total, or between 1 percent and 2 percent of the Earth's surface (according to definition). More than two thirds of the total Mediterranean-type ecosystem area is found within the Mediterranean basin.

Region	Approximate area (million km^2)	Total no. of plant species
Cape Province, South Africa	0.09	8 550
Southwestern Australia	0.31	ca 8 000
California	0.32	5 050
Chile	0.14	ca 2 100
Mediterranean basin	1.87	25 000

Differences in vegetation structure between regions are in part a consequence of differences in the annual distribution of rainfall. In South Africa the sclerophyllous fynbos community contains an abundance of ericaceous species as an understorey to low broader-leaved shrubs including members of the Proteaceae and Myrtaceae[86].

The Australian heaths are structurally similar, with Epacridaceae replacing Ericaceae. Californian shrublands, known as chaparral, are characterized by Adenostoma (Rosaceae) and a high richness of *Arctostaphylos* species and other members of the Ericaceae. The shrublands, or matorral, of Chile include many

Table 5.10
Estimated plant species richness in the five regions of Mediterranean-type climate

Source: UNEP[78].

Map 5.7
Current forest distribution

Map adapted from the global landcover classification developed by the University of Maryland. The Maryland classification includes 13 classes and was based on AVHRR remote-sensing data with a spatial resolution of 1 km. The map shown was derived by reclassifying the Maryland landcover data to accord with an ecologically based classification of five major forest types. For presentation purposes the data have here been generalized to a 4-km grid.

Source: Data from University of Maryland Global Land Cover Facility. For full description see Hansen[98].

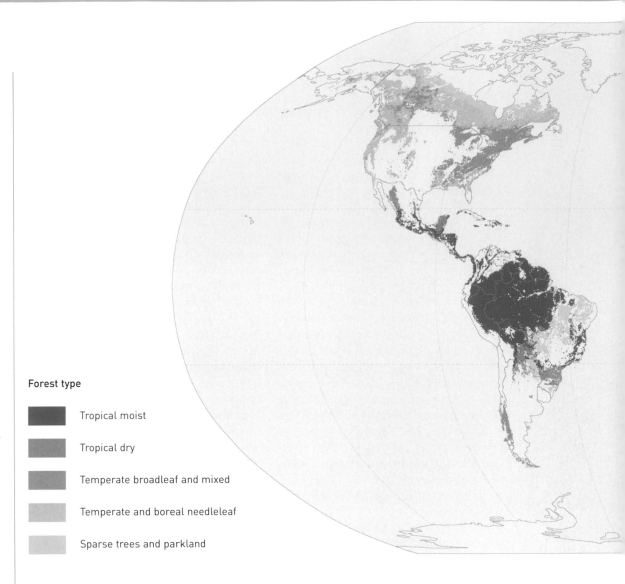

Forest type

Tropical moist

Tropical dry

Temperate broadleaf and mixed

Temperate and boreal needleleaf

Sparse trees and parkland

of the same genera as those of California, while in the Mediterranean basin itself Ericaceae, Cistaceae, Leguminosae and Oleaceae are all important.

Species richness in Mediterranean-type ecosystems, particularly among plants, is generally high – approaching values for moist tropical forest areas – and levels of endemism are also very high. Among the five Mediterranean-type ecosystems, species richness appears highest on the poorer soils of South Africa and southwest Australia (Table 5.10), and lower on the richer soils of California, Chile and the Mediterranean basin[28]. Countries around the Mediterranean Sea hold some 25 000 vascular species (about 10 percent of all vascular plants) of which around 60 percent are endemic to the Mediterranean region.

The remaining four Mediterranean-type ecosystem regions are all considered to hold a disproportionately high floristic diversity in relation to their area[87].

At fine scale, mean plant richness in the fynbos of South Africa is moderate, i.e. around 16 species per m^2, but many species have small ranges, and there is a uniquely high turnover in the species composition of plant communities along ecological and geographical gradients. At landscape scale, richness accordingly rises to very high values, for 2 256 species occur in 471 km^2 on the Cape peninsula and the entire Cape floristic region (including some non-fynbos vegetation) holds some 8 550 species, about 70 percent of which are endemic.

The Mediterranean-type ecosystems in general have a relatively high proportion of

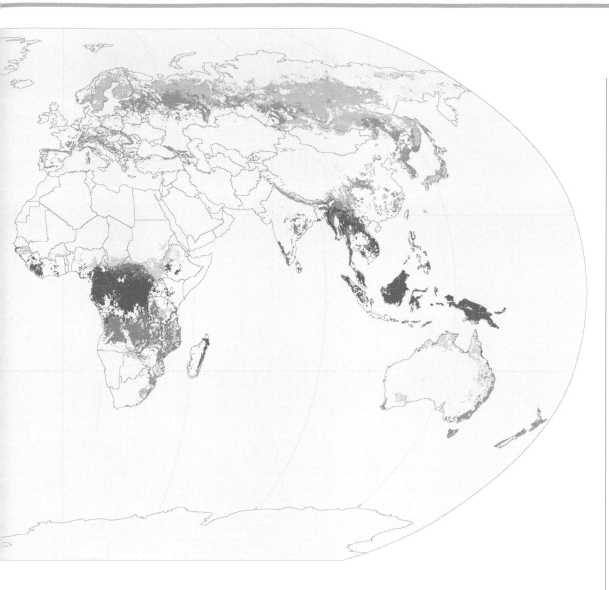

their species categorized as threatened. The Cape flora, largely within a Mediterranean-type ecosystem, occupies only 4 percent of the land area of southern Africa, but accounts for nearly 70 percent of the region's threatened plant species. About one third of the natural vegetation has been transformed by human activity; the remaining natural vegetation is at risk from a number of invasive introduced woody plants, and the effects of an introduced ant (that suppresses native seed-storing ants and thus renders seed liable to destruction by rodents or fire). Around 10 percent of the Californian flora is considered threatened (equivalent to approximately one quarter of all threatened plants in the United States). In Australia, heath habitats, primarily in the southwest Mediterranean-type ecosystem region, rank third after 'woodland' and 'scrub'

in numbers of 'endangered' category plants. Given their much smaller extent, this indicates that a far higher proportion of their flora is threatened than in either woodland or scrub habitats.

Vertebrate diversity tends to be lower than in plants. To take an example, in the Cape Mediterranean-type ecosystem, reptile diversity is only moderate while bird and mammal diversity is relatively low. The absence of large mammals in California and the Mediterranean basin may be linked to overhunting by humans during the late Pleistocene[88].

Several threatened animal species rely on shrubby or scrub habitat. The Iberian lynx *Lynx pardinus*, found in the light woodland and maquis of Spain and Portugal, where habitat loss and hunting have led to decline, is possibly the most threatened cat species.

Map 5.8
Non-forest terrestrial ecosystems

Map adapted from the global landcover classification developed by the University of Maryland. The Maryland classification includes 13 classes and was based on AVHRR remote-sensing data with a spatial resolution of 1 km. The map shown was derived by reclassifying the Maryland landcover data to accord with a highly generalized classification of non-forest terrestrial ecosystem types. For presentation purposes the data have here been generalized to a 4-km grid.

Source: Data from University of Maryland Global Land Cover Facility. For full description see Hansen[98].

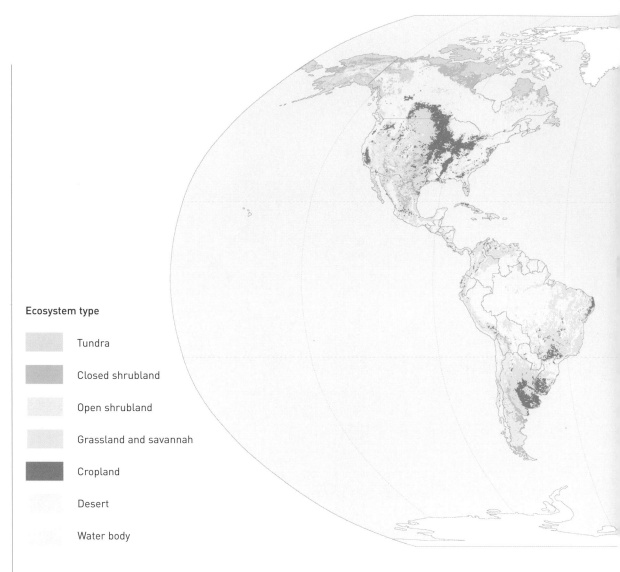

Ecosystem type

Tundra

Closed shrubland

Open shrubland

Grassland and savannah

Cropland

Desert

Water body

The riverine rabbit *Bunolagus monticularis* of South Africa is restricted to a small area of riverine bush in the central Karoo, where it is threatened by further loss of this habitat to agriculture. Among birds, the island cisticola *Cisticola haesitatus* is endemic to the island of Socotra in the western Indian Ocean, where it is threatened by loss of light scrub and grassland habitat, possibly through over-grazing by goats.

Mediterranean-type shrublands are not notably high in either biomass or net primary production. Biomass at mature fynbos sites is typically 15-16 metric tons per hectare, and in chaparral may be twice that[89]. Combined with their relatively low rates of primary production, related to both climate and soil fertility factors, the incidence of fire tends to reduce the accumulation of carbon in these ecosystems

and their soils. Total carbon storage is probably between 100 and 150 metric tons per hectare in Mediterranean-type shrublands[85].

The Mediterranean basin itself has for many centuries been subject to intense human activities, including forest clearance and grazing, such that little genuinely natural vegetation remains. It has been suggested that the plant diversity is locally high because of the number of species that have evolved as components of successional vegetation in response to frequent disturbance. In other Mediterranean-type shrublands, expanding human populations and conversion of land to agricultural or residential use are important pressures. These changes are often accompanied by changing fire and grazing regimes, and both these changes tend to facilitate invasion by non-native plant and animal

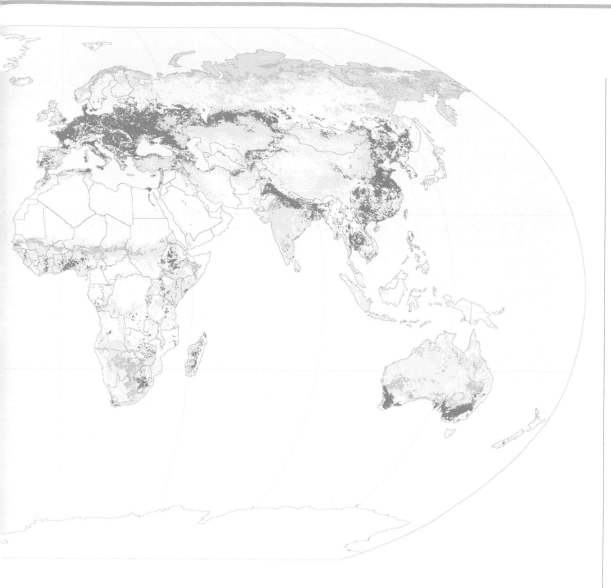

species, which threaten native species populations, especially in California and South Africa[90].

Deserts and semi-deserts
Nearly 10 million km² of the Earth's land area is hyperarid, or true desert, where rainfall is extremely low and unpredictable in space and time, so that in some years none falls at all. These areas have a ratio of rainfall to potential evapotranspiration (P/PET) of less than 0.05[91]. The Sahara desert alone makes up nearly 70 percent of the world hyperarid zone. Other extensive areas are found in the Arabian Peninsula and central Asia, with smaller areas in southwest Africa, the Horn of Africa, western South America and western North America. In semi-deserts, areas with less arid climates, the vegetation is usually more substantial than in deserts, but covers no more than 80 percent of the ground. Temperate deserts and semi-deserts cover nearly 6 million km² in Eurasia and North and South America[92]. Polar regions and some high mountain areas with a permanently cold, dry climate also meet the definition of desert, but have completely different ecological characteristics from true drylands and are not usually considered with them.

Plant cover in desert ecosystems ranges from areas without vegetation to areas with low densities of small shrubs and perennial grasses, often with populations of annuals that vary in density depending on seasonal precipitation. Deserts are often characterized by short periods of relatively high productivity during rainy periods, interspersed with long periods of very low productivity. As

Desert species show a wide range of adaptations to extreme environments.

Dating back some 15 million years, the Namib appears to be Earth's oldest desert.

terrestrial higher taxa are represented, and their species diversity may in some situations be comparable to that of more mesic habitats[93].

True desert species show a wide range of adaptations to the conditions of an extreme environment. Characteristic plants include the Cactaceae in the Americas and the succulent Euphorbiaceae in Africa. Semi-desert species include salt bush *Atriplex*, and creosote bush *Larrea*. Amongst animals, groups that are intrinsically adapted to very low moisture environments include reptiles and many arthropods, although species in a wide range of other groups have also evolved to cope with these conditions. Strategies for survival amongst both plants and animals often include long periods of dormancy (as seeds, in the case of many plants) between rainfall events.

In the particular conditions prevailing in the so-called fog deserts (notably the Atacama desert and the Skeleton Coast desert), different strategies have evolved. Here plants and animals make use of the regular moisture-laden fogs, which roll in from the cold offshore currents, to obtain a low but predictable supply of water.

The often overlooked inland water habitats of deserts may contain a particularly high proportion of locally endemic species; the pools of Cuatro Ciénegas, Mexico, for example, contain numerous mollusk and fish species found nowhere else.

a result, most herbivorous insects develop only one generation per season, leading to a time lag in the interaction between herbivorous insects and their food resources[93]. Consequently, most of the primary productivity in deserts ends up as dry plant material or seeds, which accumulates because microbial decomposition is limited by the low moisture availability. These resources form the basis for populations of detritivores such as termites, darkling beetles and isopods, and seed predators such as ants. These groups can support a rich fauna of predatory arthropods and reptiles as well as omnivorous birds and mammals. It is often assumed that all deserts have low species diversity because of the harsh environmental conditions, but among animals almost all

Many of the larger desert and sub-desert vertebrates are threatened; the openness of these arid areas means that species such as antelopes and other bovids are more conspicuous than forest species and thus more vulnerable to overhunting. Threatened vertebrates include the wild bactrian camel *Camelus bactrianus* with a few remnant populations in the Gobi desert of Mongolia and China, and Przewalski's gazelle *Procapra przewalskii* of China's subdesert steppes, now restricted by overhunting and habitat loss to a few small areas surrounding Lake Quinghai. The Mexican prairie dog *Cynomys mexicanus* is confined to prairies and intermontane basins with herbs and grasses where it is threatened by persecution and continuing

habitat loss. The Addax antelope *Addax nasomaculatus* was originally widespread from the western Sahara to Egypt and Sudan, but as a result of uncontrolled hunting it persists only in small and scattered local populations and is extinct through much of its former range.

Principal threats to desert environments include the activities of domestic livestock such as cattle, which cause soil compaction through trampling, and can damage vegetation and waterholes. Introduced species, such as rabbits in Australia, have been highly damaging to some desert environments. Other human activities that have affected desert habitats include use of off-road vehicles, irrigation and afforestation schemes, and housing projects. In drylands, most adverse impacts that lead to some form of land degradation can be categorized as 'desertification'. Under the UN Convention to Combat Desertification, the latter term is defined explicitly as 'land degradation in arid, semi-arid and dry sub-humid areas resulting from various factors, including climatic variations and human activities'. According to the above definition, hyperarid lands (true deserts) are not susceptible to desertification, because their productivity is already so low that it cannot be seriously decreased by human action. The effects of desertification on arid and semi-arid areas promote poverty among rural people, and by placing greater pressure on natural resources, poverty tends to reinforce any existing trend toward desertification.

REFERENCES

1 Loveland, T.R. *et al.* 1999. An analysis of the IGBP global land-cover characterization process. *Photogrammetric Engineering and Remote Sensing* 65: 1021-1032.

2 ITTO 2001. *Annual review and assessment of the world timber situation 2000.* International Tropical Timber Organization.

3 Salisbury, H.E. 1989. *The great black dragon fire: A Chinese inferno.* Little, Brown and Co., Boston.

4 Farjon, A. and Page, C.N. 1999. *Status survey and conservation action plan: Conifers.* IUCN-the World Conservation Union, Gland.

5 Hoang Ho-dzung 1987. The moss flora of the Baektu mountain area. In: Yang, H. *et al.* (eds). *The temperate forest ecosystem*, pp. 29-31. Institute of Terrestrial Ecology Symposium 20, National Environment Research Council, UK.

6 Väisänen, R., Biström, O. and Heliövaara, K. 1993. Sub-cortical Coleoptera in dead pines and spruces: Is primeval species composition maintained in managed forests? *Biodiversity and Conservation* 2(2): 95-113.

7 Rundel, P.W. and Boonpragob, K. 1995. Dry forest ecosystems of Thailand. In: Bullock, S.H., Mooney, H.A. and Medina, E. (eds). *Seasonally dry tropical forests*, pp. 93-123. Cambridge University Press, Cambridge.

8 Röhrig, E. 1991. Floral composition and its evolutionary development. In: Röhrig, E. and Ulrich, B. (eds). *Ecosystems of the world 7: Temperate deciduous forests*, pp. 17-24. Elsevier, Amsterdam.

9 Ovington, J.D. and Pryor, L.D. 1983. Temperate broad-leaved evergreen forests of Australia. In: Ovington, J.D. (ed.) *Ecosystems of the world 10: Temperate broad-leaved evergreen forests*, pp. 72-102. Elsevier, Amsterdam.

10 Barnes, B.V. 1991. Deciduous forests of North America. In: Röhrig, E. and Ulrich, B. (eds). *Ecosystems of the world 7: Temperate deciduous forests*, pp. 219-344. Elsevier, Amsterdam.

11 Williams, P.H., Gaston, K.J. and Humphries, C.J. 1997. Mapping biodiversity value worldwide: Combining higher-taxon richness for different groups. *Proceedings of the Royal Society, Biological Sciences* 264: 141-148.

12 Thomas, W.W. 1999. Conservation and monographic research on the flora of tropical America. *Biodiversity and Conservation* 8: 1007-1015.

13 Olson, D.M. *et al.* 2001. Terrestrial ecoregions of the world: A new map of life on Earth. *BioScience* 51: 933-938. Also see website http://www.panda.org/resources/programmes/global200/pages/list.htm (accessed March 2002).

14 FAO 2001. *State of the world's forests 2001*. Food and Agriculture Organization of the United Nations, Rome.

15 UNESCO 1973. *International classification and mapping of vegetation*. United Nations Educational, Scientific and Cultural Organization, Paris.

16 Hilton-Taylor, C. (compiler) 2000. *2000 IUCN Red List of threatened species*. IUCN-the World Conservation Union, Gland and Cambridge. Available online at http://www.redlist.org/ (accessed March 2002).

17 Veblen, T.T., Schlegel, F.M. and Oltremari, J.V. 1983. Temperate broad-leaved evergreen forests of South America. In: Ovington, J.D. (ed.) *Ecosystems of the world 10: Temperate broad-leaved evergreen forests*, pp. 5-32. Elsevier, Amsterdam.

18 Ching, K.K. 1991. Temperate deciduous forests in East Asia. In: Röhrig, E. and Ulrich, B. (eds). *Ecosystems of the world 7: Temperate deciduous forests*, pp. 539-556. Elsevier, Amsterdam.

19 Lear, M. and Hunt, D. 1996. Updating the threatened temperate tree list. In: Hunt, D. (ed.) *Temperate trees under threat*, pp. 161-171. International Dendrology Society, UK.

20 Ohba, H. 1996. A brief overview of the woody vegetation of Japan and its conservation status. In: Hunt, D. (ed.) *Temperate trees under threat*, pp. 81-88. International Dendrology Society, UK.

21 Schaefer, M. 1991. The animal community: Diversity and resources. In: Röhrig, E. and Ulrich, B. (eds). *Ecosystems of the world 7: Temperate deciduous forests*, pp. 51-120. Elsevier, Amsterdam.

22 Ovington, J.D. and Pryor, L.D. 1981. Temperate broad-leaved evergreen forests of Australia. In: Ovington, J.D. (ed.) *Ecosystems of the world 10: Temperate broad-leaved evergreen forests*, pp. 73-99. Elsevier, Amsterdam.

23 Röhrig, E. 1991. Biomass and productivity. In: Röhrig, E. and Ulrich, B. (eds). *Ecosystems of the world 7: Temperate deciduous forests*, pp. 165-174. Elsevier, Amsterdam.

24 Zinke, P.J. *et al.* 1984. *Worldwide organic soil carbon and nitrogen data*. Environmental Sciences Division, Publication No. 2212. Oak Ridge National Laboratory, US Department of Energy.

25 Gentry, A.H. 1988. Changes in plant community diversity and floristic composition on environmental and geographical gradients. *Annals of the Missouri Botanical Garden* 75: 1-34.

26 Whitmore, T.C. 1984. *Tropical rain forests of the Far East*. Clarendon Press, Oxford.

27 Grubb, P.J. 1977. Control of forest growth on wet tropical mountains. *Annual Review of Ecology and Systematics* 8: 83-107.

28 Huston, M.A. 1994. *Biological diversity: The coexistence of species on changing landscapes*. Cambridge University Press, Cambridge.

29 Nelson, B. *et al.* 1994. Forest disturbance by large blowdowns in the Brazilian Amazon. *Ecology* 75: 853-858.

30 Tanner, E.V.J., Kapos, V. and Healey, J.R. 1991. Hurricane effects on forest ecosystems in the Caribbean. *Biotropica* 23: 513-521.

31 Pires, J.M. 1957. Noçôes sobre ecologia e fitogeografia da Amazônia. *Norte Agronômico* 3: 37-53.

32 de Oliveira, A.A. and Mori, S.A. 1999. A central Amazonian terra firme forest. I. High tree species richness on poor soils. *Biodiversity and Conservation* 8: 1219-1244.

33 Gentry, A.H. 1988. Tree species richness of upper Amazonian forests. *Proceedings of the National Academy of Sciences* 85: 156-159.

34 Benzing, D.H. 1989. Vascular epiphytism in America. In: Lieth, H. and Werger, M.J.A. (eds). *Ecosystems of the world 14B: Tropical rain forest ecosystems: Biogeographical and ecological studies*, pp. 133-154. Elsevier, Amsterdam.

35 Prance, G.T. 1989. American tropical forests. In: Lieth, H. and Werger, M.J.A. (eds). *Ecosystems of the world 14B: Tropical rain forest ecosystems: Biogeographical and ecological studies*, pp. 99-132. Elsevier, Amsterdam.

36 Jenkins, M. 1992. Biological diversity. In: Sayer, J.A., Harcourt, C.S. and Collins, N.M.H. *The conservation atlas of tropical forests: Africa*, pp. 26-32. Macmillan Publishers, London.

37 Harcourt, C.S. and Sayer, J.A. (eds) 1996. *The conservation atlas of tropical forests: The Americas*. Simon and Schuster, New York.

38 Lynch, J.D. 1979. The amphibians of the lowland tropical forests. In: Duellman, W.E. (ed.) *The South American herpetofauna: Its origin, evolution, and dispersal*, pp. 189-215. Monograph No. 7 of the Museum of Natural History Kansas, Lawrence.

39 Goulding, M., Leal Carvalho, M. and Ferreira, E.G. 1988. *Rio Negro: Rich life in poor waters: Amazonian diversity and foodchain ecology as seen through fish communities.* SPB Academic Publishing, The Hague.

40 World Conservation Monitoring Centre. 1992. Groombridge, B. (ed.) *Global biodiversity: Status of the Earth's living resources*. Chapman and Hall, London.

41 Hölldobler, B. and Wilson, E.O. 1990. *The ants.* Harvard University Press, Cambridge MA.

42 Hamilton, L.S., Juvik, J.O. and Scatena, F.N. 1993. *Tropical montane cloud forests.* East-West Center, Honolulu.

43 Salati, E. and Vose, P.B. 1984. Amazon basin: A system in equilibrium. *Science* 225: 129-138.

44 Mooney, H.A., Bullock, S.H. and Medina, E. 1995. Introduction. In: Bullock, S.H., Mooney, H.A. and Medina, E. (eds). *Seasonally dry tropical forests*, pp. 1-8. Cambridge University Press, Cambridge.

45 Sarmiento, G. 1992. A conceptual model relating environmental factors and vegetation formations in the lowlands of tropical South America. In: Furley, P.A., Proctor, J. and Ratter, J.A. (eds). *Nature and dynamics of forest-savanna boundaries*, pp. 583-601. Chapman and Hall, London.

46 Menaut, J.C., Lepage, M. and Abbadie, L. 1995. Savannas, woodlands and dry forests in Africa. In: Bullock, S.H., Mooney, H.A. and Medina, E. (eds). *Seasonally dry tropical forests*, pp. 64-92. Cambridge University Press, Cambridge.

47 Gentry, A.H. 1995. Diversity and floristic composition of neotropical dry forests. In: Bullock, S.H., Mooney, H.A. and Medina, E. (eds). *Seasonally dry tropical forests*, pp. 146-194. Cambridge University Press, Cambridge.

48 Murphy, P.G. and Lugo, A.E. 1995. Dry forests of Central America and the Caribbean. In: Bullock, S.H., Mooney, H.A. and Medina, E. (eds). *Seasonally dry tropical forests*, pp. 9-28. Cambridge University Press, Cambridge.

49 Ceballos, G. 1995. Vertebrate diversity, ecology and conservation in neotropical dry forests. In: Bullock, S.H., Mooney, H.A. and Medina, E. (eds). *Seasonally dry tropical forests*, pp. 195-220. Cambridge University Press, Cambridge.

50 Dodson, C.H. and Gentry, A.H. 1991. Biological extinction in western Ecuador. *Annals of the Missouri Botanical Garden* 78: 273-295.

51 Janzen, D.H. 1988. Tropical dry forests, the most endangered tropical ecosystem. In: Wilson, E.O. 1988 (ed.) *Biodiversity*, pp. 13-137. National Academy Press, Washington DC.

52 Murphy, P.G. and Lugo, A.E. 1986. Ecology of tropical dry forest. *Annual Review of Ecology and Systematics* 17: 67-88.

53 Bye, R. 1995. Ethnobotany of the Mexican tropical dry forests. In: Bullock, S.H., Mooney, H.A. and Medina, E. (eds). *Seasonally dry tropical forests*, pp. 423-438. Cambridge University Press, Cambridge.

54 Maass, J.M. 1995. Conversion of tropical dry forest to pasture and agriculture. In: Bullock, S.H., Mooney, H.A. and Medina, E. (eds). *Seasonally dry tropical forests*, pp. 399-422. Cambridge University Press, Cambridge.

55 Bullock, S.H. 1995. Plant reproduction in neotropical dry forests. In: Bullock, S.H., Mooney, H.A. and Medina, E. (eds). *Seasonally dry tropical forests*, pp. 277-304. Cambridge University Press, Cambridge.

56 Tukhanen, S. 1999. The northern timberline in relation to climate. In: Kankaapää, Tasanen, S.T. and Sutinen, M.-L. (eds). *Sustainable development in northern timberline forests*, pp. 29-62. Finnish Forest Research Institute Research Papers 734.

57 Solbrig, O.T., Medina, E. and Silva, J.F. 1996. Determinants of tropical savannas. In: Solbrig, O.T., Medina, E. and Silva, J.F. (eds). *Biodiversity and savanna ecosystem processes: A global perspective*, pp. 31-41. Springer, Berlin.

58 Newsome, A.E. 1983. The grazing Australian marsupials. In: Bourlière, F. (ed.) *Ecosystems of the world 13: Tropical savannas*, pp 441-459. Elsevier, Amsterdam.

59 Solbrig, O.T. 1996. The diversity of the savanna ecosystem. In: Solbrig, O.T., Medina, E. and Silva, J.F. (eds). *Biodiversity and savanna ecosystem processes: A global perspective*, pp. 1-27. Springer, Berlin.

60 Lavelle, P. 1983. The soil fauna of tropical savannas. I. The community structure. In: Bourlière, F. (ed.) *Ecosystems of the world 13: Tropical savannas*, pp. 477-484. Elsevier, Amsterdam.

61 FAO 2001. *Global forest resources assessment 2000*. Food and Agriculture Organization of the United Nations, Rome.

62 Williams, M. 1989. Deforestation: Past and present. *Progress in Human Geography* 13: 176-208.

63 Williams, M. 1991. Forests. In: Turner II, B.L. *et al.* (eds). *The Earth as transformed by human action*, pp. 179-201. Cambridge University Press with Clark University, Cambridge.

64 FAO 1999. *State of the world's forests 1999*. Food and Agriculture Organization of the United Nations, Rome.

65 Holdsworth, A.R. and Uhl, C. 1997. Fire in Amazonian selectively logged rain forest and the potential for fire reduction. *Ecological Applications* 7: 713-725.

66 Laurance, W.F. and Bierregaard, R.O. (eds) 1997. *Tropical forest remnants: Ecology, management and conservation of fragmented communities*. University of Chicago Press, Chicago.

67 Smith, T.M., Leemans, R. and Shugart, H.H. 1992. Sensitivity of terrestrial carbon storage to CO_2-induced climate change: Comparison of four scenarios based on general circulation models. *Climatic Change* 21: 367-384.

68 Pounds, J.A., Fogden, M.P.L. and Campbell, J.H. 1999. Biological response to climate change on a tropical mountain. *Nature* 398: 611-615.

69 Bliss, L.C. 1997. Arctic ecosystems of North America. In: Wielgolaski, F.E. (ed.) *Ecosystems of the world 3: Polar and alpine tundra*, pp. 551-683. Elsevier, Amsterdam.

70 Wielgolaski, F.E. 1997. Introduction. In: Wielgolaski, F.E. (ed.) *Ecosystems of the world 3: Polar and alpine tundra*, pp. 1-6. Elsevier, Amsterdam.

71 Chernov, Yu I. and Matveyeva, N.V. 1997. Arctic ecosystems in Russia. In: Wielgolaski, F.E. (ed.) *Ecosystems of the world 3: Polar and alpine tundra*, pp. 361-507. Elsevier, Amsterdam.

72 Diaz, A., Péfaur, J.E. and Durant, P. 1997. Ecology of South American páramos with emphasis on the fauna of the Venezuelan páramos. In: Wielgolaski, F.E. (ed.) *Ecosystems of the world 3: Polar and alpine tundra*, pp. 263-310. Elsevier, Amsterdam.

73 Campbell, J.S. 1997. North American alpine ecosystems. In: Wielgolaski, F.E. (ed.) *Ecosystems of the world 3: Polar and alpine tundra*, pp. 211-261. Elsevier, Amsterdam.

74 Grabherr, G. 1997. The high-mountain ecosystems of the Alps. In: Wielgolaski, F.E. (ed.) *Ecosystems of the world 3: Polar and alpine tundra*, pp. 97-121. Elsevier, Amsterdam.

75 Hedberg, O. 1997. High-mountain areas of tropical Africa. In: Wielgolaski, F.E. (ed.) *Ecosystems of the world 3: Polar and alpine tundra*, pp. 185-197. Elsevier, Amsterdam.

76 Zöckler, C. 1998. Patterns of biodiversity in arctic birds. *WCMC Biodiversity Bulletin* 3: 1-15.

77 Bazilevich, N.I. and Tishkov, A.A. 1997. Live and dead reserves and primary production in polar desert, tundra and forest tundra of the former Soviet Union. In: Wielgolaski, F.E. (ed.) *Ecosystems of the world 3: Polar and alpine tundra*, pp. 509-539. Elsevier, Amsterdam.

78 United Nations Environment Programme 1995. Heywood, V. (ed.) *Global biodiversity assessment*. Cambridge University Press, Cambridge.

79 Risser, P.G. 1988. Diversity in and among grasslands. In: Wilson, E.O. and Peter, F.M. (eds). *Biodiversity*, pp. 176-180. National Academy Press, Washington DC.

80 Coupland, R.T. 1992. Overview of the grasslands of North America. In: Coupland, R.T. (ed.) *Ecosystems of the world 8A: Natural grasslands: Introduction and western hemisphere*, pp. 147-150. Elsevier, Amsterdam.

81 Coupland, R.T. 1993. Review. In: Coupland, R.T. (ed.) *Ecosystems of the world 8B: Natural grasslands: Eastern hemisphere and resumé*, pp. 471-482. Elsevier, Amsterdam.

82 Lavrenko, E.M. and Karamysheva, Z.V. 1993. Steppes of the former Soviet Union and Mongolia. In: Coupland, R.T. (ed.) *Ecosystems of the world 8B: Natural grasslands: Eastern hemisphere and resumé*, pp. 3-60. Elsevier, Amsterdam.

83 Wege, D.C. and Long, A.J. 1995. *Key areas for threatened birds in the neotropics*. Birdlife International, Cambridge.

84 Sauer, J.R. *et al.* 1997. *The North American breeding bird survey results and analysis Version 96.3*. Patuxent Wildlife Research Center, Laurel. Available online at: www.mbrpwrc.usgs.gov/bbs/bbstext.html (accessed April 2002).

85 Adams, J. 1997. Estimates of preanthropogenic carbon storage in global ecosystem types. Available online at http://www.esd.ornl.gov/projects/qen/carbon3.html (accessed April 2002).

86 Cowling, R.M. and Holmes, P.M. 1992. Flora and vegetation. In: Cowling, R.M. (ed.) *The ecology of fynbos: Nutrients, fire and diversity*, pp. 23-61. Oxford University Press, Cape Town.

87 Myers, N. 1990. The biodiversity challenge: Expanded hot-spots analysis. *Environmentalist* 10: 243-256.

88 Davis, G.W., Richardson, D.M., Keeley, J.E. and Hobbs, R.J. 1996. Mediterranean-type ecosystems: The influence of biodiversity on their functioning. In: Mooney, H.A. *et al.* (eds). *Functional roles of biodiversity: A global perspective*, pp. 151-183. John Wiley and Sons, Chichester.

89 Keeley, J.F. 1992. A Californian's view of fynbos. In: Cowling, R.M. (ed.) *The ecology of fynbos: Nutrients, fire and diversity*, pp. 372-388. Oxford University Press, Cape Town.

90 Richardson, D.M., Macdonald, I.A.W., Holmes, P.M. and Cowling, R.M. 1992. Plant and animal invasions. In: Cowling, R.M. (ed.) *The ecology of fynbos: Nutrients, fire and diversity*, pp. 271-308. Oxford University Press, Cape Town.

91 Middleton, N. and Thomas, D. (eds) 1997. *World atlas of desertification*. Second edition. United Nations Environment Programme. Arnold, London.

92 West, N.E. 1983. Approach. In: West, N.E. (ed.) *Ecosystems of the world 5: Temperate deserts and semi-deserts*, pp. 1-2. Elsevier, Amsterdam.

93 Polis, G.A. 1991. Desert communities: An overview of patterns and processes. In: Polis, G.A. (ed.) *The ecology of desert communities*, pp. 1-26. University of Arizona Press, Tucson.

94 Phillips, O.L. *et al.* 1994. Dynamics and species richness of tropical rain forests. *Proceedings of the National Academy of Sciences* 91: 2805-2809.

95 FGDC 1995. FGDC Vegetation Classification Standards. Federal Geographic Data Committee, Reston. Unpublished.

96 UNEP-WCMC 1999. Contribution of protected areas to global forest conservation. In: *International forest conservation: Protected areas and beyond.* A discussion paper for the intergovernmental forum on forests. Commonwealth of Australia, Canberra.

97 Loh, J. (ed.) 2000. *Living planet report 2000.* WWF – World Wide Fund for Nature, Gland.

98 Hansen, M. *et al.* 2000. Global land cover classification at 1 km resolution using a decision tree classifier. *International Journal of Remote Sensing* 21: 1331-1365.

99 Barthlott, W. *et al.* 1999. Terminological and methodological aspects of the mapping and analysis of global biodiversity. *Acta Botanica Fennica* 162: 103-110.

100 Heywood, V.H. (ed.) 1993. *Flowering plants of the world.* B.T. Batsford Ltd, London.

6 Marine biodiversity

MOST OF THE PLANET IS COVERED BY OCEAN waters whose average depth is four times the average elevation of the land, making the open sea by far the largest ecosystem on Earth. Despite this volume, marine net primary production remains similar to or less than that on land because photosynthesis in the sea is carried out by microscopic bacteria and algae restricted to the sunlit surface layers (plants are virtually absent).

The diversity of major lineages (phyla and classes) is much greater in the sea than on land or in freshwaters, and many phyla of invertebrate animals occur only in marine waters. Species diversity appears to be far lower, perhaps because marine waters are physically much less variable in space and time than the terrestrial environment.

Marine fisheries are the largest source of wild protein, derived from fishes, mollusks and crustaceans. The world catch from capture fisheries has grown fivefold over the past five decades, but appears to have declined during the 1990s despite increased fishing effort. More than half of the world's major fishery resources are now in need of remedial management, mainly because of excess exploitation.

THE SEAS

Oceans cover 71 percent of the world's surface. They are on average around 3.8 kilometers (km) deep and have an overall volume of some 1 370 million km³. The whole of the world ocean (all contiguous seas) is theoretically capable of supporting life, so that the marine part of the biosphere is far larger than the terrestrial part. However, as on land, life in the oceans is very unevenly distributed – some parts are astonishingly productive and diverse while others are virtually barren.

With the present configuration of land masses, a major part (37 percent) of the world ocean is within the tropics, and about 75 percent lies between the 45° latitudes. The largest continental shelf areas are in high northern latitudes (Table 6.2), but about 30 percent of the total shelf area is in the tropics. Within the tropics, the shelf is most extensive in the western Pacific (China Seas south to north Australia).

Although knowledge of the functioning of the marine biosphere has increased enormously in the past few decades, overall it remains far less well known and understood than the terrestrial part of the globe. The main reason for this is, quite simply, that

Table 6.1
Area and maximum depth of the world's oceans and seas

Source: Couper[65].

Ocean/Sea	Area (km²)	Depth (meters)
Pacific Ocean	165 384 000	11 524
Atlantic Ocean	82 217 000	9 560
Indian Ocean	73 481 000	9 000
Arctic Ocean	14 056 000	5 450
Mediterranean Sea	2 505 000	4 846
South China Sea	2 318 000	5 514
Bering Sea	2 269 000	5 121
Caribbean Sea	1 943 000	7 100
Gulf of Mexico	1 544 000	4 377
Sea of Okhotsk	1 528 000	3 475
East China Sea	1 248 000	2 999
Yellow Sea	1 243 000	91
Hudson Bay	1 233 000	259
Sea of Japan	1 008 000	3 743
North Sea	575 000	661
Black Sea	461 000	2 245
Red Sea	438 000	2 246
Baltic Sea	422 000	460

	Open ocean	Continental shelf
Total area (million km²)	360.3	26.7
Latitude bands (% of total)		
Polar and boreal (45-90°)	26.6	40.9
Temperate (20-45°)	36.8	28.8
Tropical (0-20°)	36.6	30.3

Table 6.2
Relative areas of continental shelves and open ocean

Source: Adapted from Longhurst and Pauly[1], after Moiseev.

much of it is inaccessible to humans. Study of any part below the top few meters requires specialized equipment and is expensive and time consuming. Knowledge of most of the sea is thus based largely on a range of remote-sensing and sampling techniques and often remains sketchy. As these techniques become more sophisticated, so our understanding of marine ecosystems, particularly those away from the coastal zone, is undergoing constant revision.

Sea water and ocean currents

Sea water is a complex but relatively uniform mixture of chemicals. Most of the 92 naturally occurring elements can be detected dissolved in it, but most only in trace concentrations; the most abundant are sodium (as Na+) and chlorine (as Cl–), which occur at a concentration some ten times higher than the next most abundant element, magnesium. The term used to quantify the total amount of dissolved salts in sea water is salinity, a dimensionless ratio, which generally ranges between 33 and 37 and averages 35. Most of the substances dissolved in sea water are unreactive and remain at relatively stable concentrations; however those that play a part in biological systems can be highly variable in time and space.

The world's sea waters are constantly in motion, at all scales from the molecular to the oceanic. Large-scale ocean circulation plays a vital role in mediating global climate as well as influencing the functioning of marine ecosystems. It is driven by complex interactions between a number of physical variables, notably latitudinal variations in solar radiation (and consequent heating and cooling), precipitation and evaporation, transfer of frictional energy across the ocean

surface from winds, and forces resulting from the rotation of the planet.

Surface currents are largely driven by prevailing winds. The most important features are vast, anticyclonic gyres in the subtropical regions of the world's oceans. These are primarily driven by the westerly trade winds in the Roaring Forties and circulate clockwise in the northern hemisphere and counterclockwise in the southern hemisphere. Also important are the eastward-flowing Antarctic circum-polar current and equatorial current. At smaller scales, eddies and rings are ubiquitous and are analogous to weather systems in the atmosphere; typical oceanic eddies may be 100 km across and persist for a year or more.

The density of sea water increases with increasing salinity and with decreasing temperature until it freezes at around –1.9°C. Sea water of different density does not mix readily, and so the oceans tend to be well stratified vertically, with bodies of less dense sea water (warmer and less saline) sitting on top of cooler, more saline and denser bodies. The stratification is rarely stable over the long term, however, as the influence of climate changes the properties of surface waters and causes various forms of vertical mixing. The single most important factor controlling large-scale deep-water circulation appears to be the generation of cold, high-salinity water near the surface in the Weddell Sea and off Greenland in the North Atlantic. Here, during winter, the sea freezes and the floating ice sheet is virtually free of salt, leaving the underlying water more saline, cold and dense. This sinks to the bottom and moves south along the Atlantic floor, some passing south of the equator and circulating throughout the world ocean, producing what is known as the Great Conveyor – the interlocking system of major circulation currents in the deep sea. Because it originates near the surface it is well oxygenated as well as cold, and is the major reason why aerobic organisms can thrive in the deep sea.

Major zones of upwelling occur along the western boundaries of continents where trade winds blow towards the equator, causing surface waters to be pushed offshore to be replaced by cooler deep waters. There are five major upwelling areas: the Humboldt current region off the Chilean and southern Peruvian coast of South America; the California current region off western North America; the Canary current off the coast of Mauritania in north-west Africa; the Benguela current region off southern Africa; and the northwest Arabian Sea. Marine production is much enhanced in these regions which typically support major pelagic fisheries. Outside these upwelling zones, stratification tends to persist throughout the year in the tropics and subtropics.

MAJOR MARINE ZONES
The continental shelf

Marine waters around major land masses are typically shallow, lying over a continental shelf which may be anything from a few kilometers to several hundred kilometers wide. The most landward part is the littoral or intertidal zone where the bottom is subject to periodic exposure to the air. Water depth here varies from zero to several meters. Seaward of this the shelf slopes gently from shore to depths of one to several hundred meters, forming the sublittoral or shelf zone. Waters below low-tide mark in the continental shelf region are referred to as neritic.

The extent, gradient and superficial geology of continental shelf areas are determined by many factors, including levels of tectonic activity in the Earth's crust. More than 80 percent of the global volume of river-borne sediment is deposited in the tropics (and an estimated 40 percent of it by just two river systems: the Huang He or Yellow River and the Ganges-Brahmaputra)[1], and this is reflected in the extent of shelf areas in parts of the tropics, and in the high turbidity of coastal waters in monsoon regions. Most shelf areas in the tropics are overlain by sands or muds composed of sediment of terrestrial origin (terrigenous deposits).

Shelf regions support the marine communities most familiar to humans, and many of the marine resources of particular value to them. Although mangrove and coral reefs are two of the best known tropical coastal ecosystems, they dominate only a minor part of the world coastline: the former mainly in deltaic or other low-lying coastal plains, and the latter only in shallow waters where

terrestrial sedimentation is very low. Soft-bottom habitats with sparse vegetation are probably the most widespread coastal marine ecosystem type, and virtually the entire seabed away from the coastline is covered in marine sediments (see Box 6.1).

The deep sea

At the outer edge of the shelf there is an abrupt steepening of the sea bottom, forming the continental slope which descends to depths of 3-5 km. The sea bottom along the slope is referred to as the bathyal zone. At the base of the continental slope are huge abyssal plains which form the floor of much of the world ocean. Open waters above these plains make up the oceanic pelagic zone. Given that the oceans cover some 71 percent of the globe, and that the shelf area is relatively narrow, the oceanic pelagic zone is by far the most extensive ecosystem on Earth. The plains are punctuated by numerous submarine ridges and sea mounts which may break the surface to form islands. There are also a number of narrow trenches which have depths of from 7 000 to 11 000 meters (m). These constitute the hadal zone. Ocean

Box 6.1 Life in sediments

Spaces between the particles of marine sediments contain water, air, detritus and organisms, the last divided by size into two main categories. The microfauna includes archeans, bacteria and protoctists. These include a number of primary producers at the base of the food web in shallow waters, and have a role in interstitial sulfur chemistry and oxygenation of sediment. The meiofauna includes sediment-living forms between 0.1 and 1 mm in size, and is a major component of seabed ecosystems, particularly in the deep sea. Four phyla contain only marine meiofauna: the Gastrotricha, Gnathostomulida, Kinorhyncha and Loricifera. Nematodes are typically the most numerous component, with harpacticoid copepods and foraminiferans also important. High concentrations occur around the burrows of deposit-feeding mollusks and polychaetes, with high bacterial numbers and elevated nutrient flux. They play a key role in the flow of nutrients from the microfauna to larger organisms, of which several distinctive species exist in and on the seabed. Sea cucumbers, crinoids, polychaete worms, sea spiders, isopods and amphipods are most abundant. Some taxa are only found here. Pogonophorans occur mostly below 3 000 m and many of the known species are restricted to the sediments of deep ocean trenches (more than 6 000 m). Small-scale variation in food supply, such as the fortuitous appearance of a large animal carcass, may enhance spatial structure and provide opportunity for other species to colonize. A micro-landscape of hills and valleys is created by the burrowing and fecal mounds of echiurid and polychaete worms.

trenches are formed as a consequence of plate tectonic processes where sectors of expanding ocean floor are compressed against an unyielding continental mass or island arc, resulting in the crust buckling downwards (subducting) and being destroyed within the hot interior of the Earth.

CLASSIFYING THE MARINE BIOSPHERE

Much less effort has been devoted to categorizing and mapping units within the marine biosphere than to terrestrial environments. As on land, basic marine habitat types are commonly defined by some combination of structural, climatic and community features (e.g. 'tropical coral reef') but, probably in part reflecting issues of scale and resolution, few global maps of marine ecosystem types are available.

marine ecosystem (LME) units have been defined[4, 5], with a view to improving planning and management. These are large regions of marine space, some 200 000 km^2 or larger, that extend from river basins and estuaries out to the seaward boundary of the continental shelf, with distinct combinations of bathymetry, hydrography and production. With similar aims, the conservation organization WWF has distinguished 43 large marine ecoregions, based broadly on oceanographic and community features, and also intended to be globally representative[6].

THE BASIS OF LIFE IN THE SEAS

In the sea, as on the land, photosynthesis is the driving force behind maintenance of life. Because photosynthesis in nature depends on sunlight, with few exceptions primary pro-

Coral reefs develop within the photic zone because of the symbiotic relationship between some coral species and zooxanthellae.

The principal global scheme intended to classify the entire world ocean within an objective oceanographic system is based on long-term data on sea surface color (obtained by the CZCS radiometer carried by the Nimbus orbiting satellite during 1978-86)[2, 3]. These data reflect chlorophyll concentrations and provide a basis for estimating primary production rates, and changes over time, on a 1-degree grid. These values, together with numerous other data sets, have been used as the basis of a classification of the world ocean into four ecological domains and 56 biogeochemical provinces. Taking a less quantitative approach, some 64 large

ductivity is confined to those parts of the ocean that are sunlit. Water absorbs sunlight strongly so that light intensity decreases rapidly with increasing depth. Red wavelengths are most rapidly absorbed, except where turbidity is high, while blue-green wavelengths penetrate the deepest. Even in the clearest waters the latter are completely absorbed by around 1 km depth, this marking the extreme limit of the so-called photic zone. Photosynthesis is thus limited to the continental shelf area and the first few hundred meters of surface waters (and often much less) of the open ocean which together make up a very small proportion of the total volume of the oceans. The portion of the photic zone where sunlight is strong enough to support appreciable amounts of photosynthesis is the euphotic zone.

Virtually all other marine organisms, including those of the unlit middle depths and the deep sea, are dependent ultimately on growth of primary producers in areas that may be widely distant from them in time and space. The most important exceptions to this are the archeans and bacteria living around hydrothermal vents associated with rift zones in the ocean floor. The water here can be 10°C warmer than in adjacent areas and these microorganisms are able to grow using hydrogen sulfide gas emitted at the vents as an energy source, and they in turn are used by other organisms.

With some exceptions primary production per unit area tends to be lower in marine environments than in terrestrial ones, especially if highly managed terrestrial agricultural systems are considered. This is because over the vast majority of the ocean the euphotic zone is far distant from the lithosphere. The latter provides essential nutrients for life, and these therefore have to be transported to the euphotic zone to allow life processes to continue. There is also usually a steady loss of nutrients and organic compounds from the euphotic zone, owing to the sinking of particles and bodies into the dark regions of the sea where no photosynthesis can take place.

Continued productivity in the open sea is contingent on the replacement of the lost nutrients. The latter may originate either on land, in the form of river outflow, or from marine sediments. Their replacement in the pelagic euphotic zone is dependent on the mixing or vertical movement of the water column. At latitudes higher than 40°, winter mixing allows replenishment of the euphotic zone, particularly in continental shelf areas. However, in permanently stratified subtropical and tropical oceanic waters there is little vertical mixing, and therefore little influx of new nutrients. Productivity in these areas is correspondingly low. In zones of upwelling, surface waters are regularly replaced by nutrient-rich bottom waters and very high levels of productivity can be achieved, at least seasonally.

Until the 1980s it had been believed that photosynthesis in the pelagic ocean was carried out only by single-celled phytoplankton, between 1 and 100 microns in diameter (1 micron = 0.001 mm), and also that vast expanses of open ocean where phytoplankton could not be detected were, in terms of productivity, the marine equivalent of deserts. New observational techniques have since revealed the presence in great abundance of exceptionally small and previously unknown organisms, collectively termed picoplankton. These appear to be predominantly photosynthesizing unicellular cyanobacteria 0.6-1 micron in size, such as Prochlorococcus. Because of their extraordinary abundance (some 100 million cells may be present in 1 liter), and despite their minute size, these organisms play a crucial role in the productivity of open ocean waters[7] and have led to marked upward revisions in estimates of overall marine productivity.

The major role played by microscopic organisms in marine productivity has a number of important implications for marine ecology. Although much remains unknown about the structure and dynamics of pelagic food webs, it seems that a high proportion of marine primary production is used directly by microscopic organisms (both autotrophic and heterotrophic) and cycled back into non-living forms (dissolved carbon dioxide and organic carbon) rather than supporting populations of larger organisms.

Oceanic primary producers divert a high proportion of their energy into reproduction rather than accumulating biomass, in contrast to terrestrial primary producers (plants). As a result of this, average standing biomass per unit area in the oceans has been estimated at around one-thousandth that on land. Population turnover of oceanic primary producers is also several orders of magnitude higher than turnover of major terrestrial primary producers (plants).

The small size and rapid turnover of oceanic primary producers mean that there are no organisms directly analogous to the woody plants that so enrich terrestrial environments by providing structurally complex habitats for other organisms. The nearest

Table 6.3
Marine diversity by
phylum

Notes: Strictly marine groups
shown in bold. Estimates are
from a variety of sources and
in some cases, e.g. Mollusca,
differ in detail from those in
Table 2.1.

equivalents are the large brown algae known as kelp (phylum Phaeophyta), whose structure is less complex and which are much more narrowly distributed. Structurally complex habitats may in contrast be created by animals, particularly corals (phylum Cnidaria) and, to a lesser extent, sponges (phylum Porifera), mollusks and serpulid worms (phylum Annelida).

BIOLOGICAL DIVERSITY IN THE SEAS

It is well known that diversity at higher taxonomic levels (phyla and classes) is much greater in the sea than on land or in freshwater. Of the 82 or so eukaryote phyla currently recognized (see Chapter 2), around 60 have marine representatives compared with around 40 found in freshwater and 40 on land. Amongst animals the preponderance is even higher, with 36 out of 37 phyla having marine representatives (Table 6.3).

Some 23 eukaryote phyla, of which 18 are animal phyla, are confined to marine environments. Most of these are relatively obscure and comprise few species. The major exceptions are the Echinodermata, of which some 7 000 species are known, and the Foraminifera, with around 4 000 known, extant species. A number of other important phyla including the coelenterates (Cnidaria), sponges (Porifera) and brown and red algae (Phaeophyta and Rhodophyta, respectively)

Domain	Kingdom	Phyla		Estimated no. of described marine species
Archaea		2 phyla		?
Bacteria		12 phyla		?
Eukaryota	Protoctista	27 phyla, including:		
		Chlorophyta	Green algae	7 000
		Phaeophyta	Brown algae	1 500
		Rhodophyta	Red algae	4 000
		others incl.		23 000
		Foraminifera		
	Animalia			
		Placozoa		**1**
		Porifera	Sponges	10 000
		Cnidaria	Coelenterates	10 000
		Ctenophora	**Comb jellies**	**90**
		Platyhelminthes	Flatworms	15 000
		Nemertina	Nemertines	750
		Gnathostomulida	**Gnathostomulids**	**80**
		Rhombozoa	**Rhombozoans**	**65**
		Orthonectida	**Orthonectids**	**20**
		Gastrotricha	Gastrotrichs	400
		Rotifera	Rotifers	50
		Kinorhyncha	**Kinorhynchs**	**100**
		Loricifera	**Loriciferans**	**10**
		Acanthocephala	Spiny-headed worms	600
		Entoprocta	Entoprocts	170
		Nematoda	Nematodes	12 000
		Nematomorpha	Horsehair worms	<240

Domain	Kingdom	Phyla		Estimated no. of described marine species
		Ectoprocta	Ectoprocts	5 000
		Phoronida	**Phoronid worms**	**16**
		Brachiopoda	**Lamp shells**	**350**
		Mollusca	Mollusks	?75 000
		Priapulida	**Priapulids**	**8**
		Sipuncula	**Sipunculans**	**150**
		Echiura	**Echiurids**	**140**
		Annelida	Annelid worms	12 000
		Tardigrada	Water bears	few
		Chelicerata	Chelicerates	1 000
		Mandibulata	Mandibulate arthropods	few
		Crustacea	Crustaceans	38 000
		Pogonophora	**Beard worms**	**120**
		Bryozoa	Bryozoans	4 000
		Echinodermata	**Echinoderms**	**7 000**
		Chaetognatha	**Arrow worms**	**70**
		Hemichordata	**Hemichordates**	**100**
		Urochordata	Tunicates and ascidians	2 000
		Cephalochordata	Lancelets	**23**
		Craniata	Craniates	15 000
	Fungi	3 phyla		500
	Plantae	Anthophyta		50
Total				**ca 250 000**

are very largely marine, each with only a small number of non-marine (usually freshwater) species.

The reason for this predominance of marine higher taxa (particularly amongst animals) is believed to be because most of the fundamental patterns of organization and body plan, i.e. the different basic kinds of organism that are distinguished as phyla, originated in the sea and remain there, but only a subset of them has spread to the land and into freshwaters. It is noteworthy that only a third or so of marine phyla are found in the pelagic realm, the remainder being confined to sea bottom (benthic) areas – the habitat where eukaryotic organisms are believed to have evolved.

In contrast, known species diversity in the sea is much lower than on land – some 250 000 species of marine organisms are currently known, compared with more than 1.5 million terrestrial ones. Much of this difference is because of the large number of described terrestrial arthropods, for which there is no marine equivalent. Amongst fishes, almost as many freshwater species as marine are known, despite the fact that freshwater habitats account for only around one ten-thousandth of the volume of marine ones. Similarly, the most diverse known marine habitats (coral reefs) are far less diverse in terms of species number than the moist tropical forests that are often taken as their terrestrial counterparts.

The apparent lower total species diversity of the marine biosphere is likely in part at least to be a result of the physical characteristics of water, particularly its high heat capacity and its ability to mix. Because of these, marine environments (particularly deep-water ones) tend to show much less variation in time and space in their physical characteristics than terrestrial ones. This lack of physical variation seems to result in a similar lack of ecological variation over wide areas.

In contrast to terrestrial faunas, where one phylum – the Mandibulata (insects and relatives) – vastly outnumbers all others in terms of known species, marine species are much more evenly distributed across higher taxa. The largest marine phyla – Mollusca and

Crustacea – each comprise far fewer than 100 000 known marine species, in contrast with the Mandibulata, of which around 1 million terrestrial species have been identified to date. The only major eukaryote phyla (i.e. those with 10 000 or more described species) that are believed to have comparable levels of diversity both on land and in the sea are the Platyhelminthes, Nematoda, Mollusca and Craniata. As on land, vertebrates are by far the best-known group of marine organisms. Of the 50 000 or so described extant species, more than 15 000 may be considered marine (Table 6.4), the overwhelming majority of which are fishes (Table 6.5) and a few tetrapods (Table 6.6).

Marked latitudinal gradients in species richness have been described in a number of groups, for example in benthic isopods and mollusks[8], and it is generally true that coastal waters within the tropics tend, in parallel with terrestrial environments, to be richer in number of species than those at higher latitudes (although many exceptions to this are known, such as penguins, pinnipeds and auks). Although marine biogeography had advanced relatively little since the early 20th century, improved distribution data and more robust methods have recently been developed. These have been applied to definition of coastal biogeographic regions[9], and important areas for marine biodiversity analogous to those delimited on land. For example, 18 centers of endemism within the world tropical reef zone have been identified using data on reef fish, coral, snails and lobsters (Map 6.1); ten centers were regarded as 'hotspots' because of their higher threat score[10].

Class		No. of marine species
Myxini	Hagfishes	43
Cephalaspidomorphi	Lampreys	9
Elasmobranchii	Sharks, skates and rays	900
Holocephali	Chimaeras	31
Actinopterygii	Ray-fin fishes	13 800
Sarcopterygii	Lobe-fin fishes	2
Reptilia	Reptiles	60
Aves	Birds	307
Mammalia	Mammals	110

Table 6.4
Diversity of craniates in the sea by class

Note: Because of changing taxonomy, incomplete information and occupation of multiple habitat types, these estimates are indicative only.

Order	Common fishes	No. of families	No. of genera	No. of species occurring in marine habitats	Order	Common fishes	No. of families	No. of genera	No. of species occurring in marine habitats
Myxiniformes	Hagfishes	1	6	43	Atheriniformes	Silversides	8	47	139
Petromyzontiformes	Lampreys	1	6	9	Beloniformes	Needlefishes, sauries, flyingfishes, halfbeaks	5	38	140*
Chimaeriformes	Chimaeras	3	6	31	Cyprinodontiformes	Rivulines, killifishes, pupfishes, four-eyed fishes, poeciliids, goodeids	8	88	13
Heterodontiformes	Bullhead sharks and horn sharks	1	1	8	Stephanoberyciformes	Gibberfishes, pricklefishes, whalefishes, hairyfish, tapetails	9	28	86
Orectolobiformes	Carpet sharks	7	14	31	Beryciformes	Fangtooths, spinyfins, lanterneyefishes, roughies, pinecone fishes, squirrelfishes	7	28	123
Carcharhiniformes	Ground sharks	7	47	207	Zeiformes	Dories, boarfishes, oreos, parazen	6	20	39
Lamniformes	Mackerel sharks	7	10	16	Gasterosteiformes	Pipefishes, seahorses, sticklebacks, sandeels, seamoths, snipefishes, shrimpfishes, trumpetfishes	11	71	238*
Hexanchiformes	Cow sharks	2	4	5	Synbranchiformes	Swamp-eels	3	12	3
Squaliformes	Dogfishes and sleeper sharks	4	23	74	Scorpaeniformes	Gurnards, scorpionfishes, velvetfishes, flatheads, sablefishes, greenlings, sculpins, oilfishes, poachers, snailfishes, lumpfishes	25	266	1 219*
Squatiniformes	Angel sharks	1	1	12	Perciformes	Perches, basses, sunfishes, whitings, remoras, jacks, dolphinfishes, snappers, grunts, damselfishes, dragonfishes, wrasses, butterflyfishes, etc.	148	1 496	7 371*
Pristiophoriformes	Saw sharks	1	2	5	Pleuronectiformes	Plaice, flounders, soles	11	123	564*
Rajiformes	Rays	12	62	432*	Tetraodontiformes	Triggerfishes, puffers, boxfishes, filefishes, molas	9	100	327*
Coelacanthiformes	Coelacanths	1	1	2	**Total** (conservative working estimate)				**ca 14 000**
Acipenseriformes	Sturgeons	2	6	12					
Salmoniformes	Salmonids	1	11	21					
Stomiiformes	Lightfishes, hatchetfishes, barbeled dragonfishes	4	51	321					
Ateleopodiformes	Jellynose fishes	1	4	12					
Aulopiformes	Greeneyes, pearleyes, waryfishes, lizardfishes, barracudinas, lancetfishes	13	42	219					
Myctophiformes	Lanternfishes	2	35	241					
Lampridiformes	Oarfishes, ribbonfishes, crestfishes, opahs	7	12	19					
Polymixiiformes	Beardfishes	1	1	5					
Ophidiiformes	Pearlfishes, cusk-eels, brotulas	5	92	350*					
Gadiformes	Cods, hakes, rattails	12	85	481					
Batrachoidiformes	Toadfishes	1	19	64*					
Lophiiformes	Anglerfishes, goosefishes, frogfishes, batfishes, seadevils	16	65	297					
Mugiliformes	Mullets	1	17	65*					

Table 6.5
Diversity of fishes in the seas by order

Note: Strictly marine orders in bold; other orders that are mainly marine (more than 50% of species) marked with an asterisk*.

Source: After Nelson[66].

Algae

The macro-algae are superficially plant-like protoctists that lack the vascular tissue used by higher plants to transport water and nutrients. They are almost exclusively aquatic; three of the four principal groups consisting of large-sized species are mainly marine in occurrence. These three, the green, brown and red algae ('seaweeds'), are all cosmopolitan in distribution and occur in a range of environments, although some constituent families have somewhat restricted ranges. There are more marine species of red algae (Rhodophyta) – around 4 000 – than the greens (Chlorophyta, ca 1 000) and browns (Phaeophyta, ca 1 500) combined.

As with pinnipeds and seabirds, the cold and cool temperate regions of the world appear to be surprisingly rich in species. On present incomplete information, the region around Japan (northwest Pacific), the North Atlantic, and the tropical and subtropical western Atlantic hold the most species of marine algae. Southern Australia is not so species rich but appears to have the highest proportion of endemics. There are few species of larger algae in regions of cold-water upwelling; small isolated islands and polar regions also have few species. In contrast, coral reefs support a unique and generally diverse algal flora that includes many crustose coralline algae (more species of which are likely to be discovered). Mangrove areas also support a well-defined algal vegetation. Sandy coastlines hold few species of large algae and often form barriers to seaweed dispersal.

Marine fishes

Fishes are considered a paraphyletic group (see Chapter 2): that is, all living species are thought to share a common ancestor but to share this ancestor with another group (in this case, the tetrapods) not categorized as fishes. Apart from some 50 or so species of generally parasitic lampreys and hagfishes in the superclass Agnatha, fishes are divided into two unequal-sized groups, the cartilaginous fishes (Chondrichthyes) including chimaeras (class Holocephali) and sharks and rays (class Elasmobranchii), and the bony fishes (Osteichthyes), including the 'typical' ray-finned fishes (class Actinopterygii) and the lobe-finned coelacanths and lungfishes (class Sarcopterygii). Some 60 percent of all known living fish species (i.e. around 14 000 species) occur in marine habitats. They range in size from an 8 mm-long goby *Trimmatom nanus* in the Indian Ocean to the 15-m whale shark *Rhincodon typus*, respectively the smallest and largest of all fish species, and occur in virtually all habitats, from shallow inshore waters to the abyssal depths.

The elasmobranchs (sharks, skates and rays) are an overwhelmingly marine group with around 850 living species in ten orders. Although far less diverse than the bony fishes, the cartilaginous fishes include many of the largest fish species, a number of which are top predators in marine ecosystems. There tend to be more shark species at lower latitudes, but

at family level richness tends to be higher on the edge of the tropics (Map 6.2). The bony fishes are a remarkably diverse group, with an enormous range of morphological, physiological and behavioral adaptations. Of the 26 orders of bony fishes with marine representatives (Table 6.5), by far the largest and most diversified is the Perciformes. This is the largest of all vertebrate orders and dominates vertebrate life in the ocean, as well as being the dominant fish group in many tropical and subtropical freshwaters.

As with other groups of organisms, the majority of fishes in the sea are strictly marine, occurring only in salt water. A

Table 6.6
Marine tetrapod diversity

Notes: Birds follow the list of seabirds recognized in Croxall *et al.*[67] with the additional inclusion of four eider ducks and three steamer ducks in the family Anatidae and the red phalarope *Phalaropus fulicaria* (Scolopacidae). Taxonomy follows Sibley and Monroe[68]. Figures in parentheses indicate species that breed largely or entirely inland. Strictly marine families shown in bold.

Class	Order	Family		No. of marine species
Reptilia	Chelonia	**Dermochelyidae**	Leathery turtle	1
		Cheloniidae	Sea turtles	6
	Squamata	Elapidae	Sea snakes and sea kraits	55
		Acrochordidae	File snakes	1
		Iguanidae	Iguanas	1
Aves	Anseriformes	Anatidae		7
	Ciconiiformes	Scolopacidae		1
		Laridae	Gulls, terns, skuas, auks, skimmers	120 (13)
		Phaethontidae	Tropicbirds	3
		Sulidae	Gannets and boobies	9
		Phalacrocoracidae	Cormorants and shags	36 (2)
		Pelecanidae	Pelicans	2
		Fregatidae	Frigatebirds	5
		Spheniscidae	Penguins	17
		Procellariidae	Petrels, albatrosses, shearwaters	115
Mammalia	Cetacea	**Balaenidae**	Right whales	3
		Balaenopteridae	Rorquals	6
		Eschrichtiidae	Gray whale	1
		Neobalaenidae	Pygmy right whale	1
		Delphinidae	Dolphins	32
		Monodontidae	Beluga and narwhal	2
		Phocoenidae	Porpoises	6
		Phystereidae	Sperm whales	2
		Platanistidae	River dolphins	1
		Ziphiidae	Beaked whales	19
	Sirenia	Trichechidae	Manatees	1
		Dugongidae	Dugong	1
	Carnivora	Mustelidae	Otters and weasels	2
		Odobenidae	Walrus	1
		Otariidae	Eared seals	14
		Phocidae	Earless seals	17

Map 6.1
Coral reef hotspots

The location of 18 areas
defined by high endemism
in reef fishes, corals, snails
and lobsters, including the
ten areas identified as
'hotspots' on the basis of
high threat score.

Source: Adapted from Roberts[10].

Endemic-rich areas

Endemic-rich area at higher risk

Endemic-rich area

proportion, however, may also occur in inland waters, often passing a particular part of their life cycle there. Species that spend most of their life in marine waters but ascend rivers to breed, such as many salmonids (family Salmonidae, order Salmoniformes) and sturgeons (family Acipenseridae, order Acipenseriformes), are referred to as anadromous. Those that breed at sea but spend their lives otherwise in freshwater, such as most eels in the family Anguillidae (order Anguiliformes), are referred to as catadromous. Species with a wide salinity tolerance that may occur in marine, brackish and fresh waters (e.g. some sawfishes, family Pristidae, order Rajiformes) are referred to as euryhaline while those with narrow tolerances, be they to marine, brackish or fresh water, are referred to as stenohaline.

Reptiles

Present-day diversity of reptiles in the seas is low. One important reason for this appears to be that modern reptilian kidneys cannot tolerate high salinities and thus the only reptiles that have adapted to marine environments are those which have developed specialized salt-excreting glands. The most thoroughly marine reptiles are undoubtedly the sea snakes in the subfamily Hydrophiinae (family Elapidae). These spend their entire lives in the sea, giving birth to live young there. Although largely air-breathing like other reptiles, they can also absorb some oxygen directly from sea water and are thus able to remain submerged for long periods. Around 50 species are known, widely distributed in tropical parts of the Indo-Pacific region. In addition the little file snake *Acrochordus*

granulatus (family Acrochordidae), from northern Australia and Southeast Asia is also entirely aquatic, but occurs in brackish estuaries as well as sea water.

Five species of sea krait in the subfamily Laticaudinae are also largely marine, feeding mainly on eels. However they return to land to breed, generally on small tropical islands. They, too, are confined to the Indo-Pacific region. One species of lizard, the Galapagos marine iguana *Amblyrhynchus cristatus* (family Iguanidae), feeds underwater on marine algae but spends a considerable proportion of time on land. Several other reptile species regularly enter sea water, most notably a number of homalopsine mangrove snakes (family Colubridae) from the Indo-Pacific and the estuarine crocodile *Crocodylus porosus* (family Crocodylidae) from the same region.

Undoubtedly the most prominent group of marine reptiles is the sea turtles, comprising the leathery turtle *Dermochelys coriacea* in the family Dermochelyidae and six members of the family Cheloniidae. All species are large (ranging from 70-centimeter (cm) adult carapace length in *Lepidochelys kempii* to, exceptionally, 250 cm in *Dermochelys coriacea*) and most are widely distributed in tropical and subtropical waters (Map 6.3). Sea turtles are almost completely marine; only the females emerge to nest on land, mostly within the tropics. One species, the loggerhead *Caretta caretta*, nests largely in temperate areas of the northern hemisphere. Sea turtles typically have a long period to maturity (often up to 25 years in the case of the green turtle *Chelonia mydas*) and a long lifespan. Females often nest only every two or three years. They

Map 6.2
Shark family diversity

Diversity in sharks, based on the distribution of all 30 families plotted as a density surface. Most sharks are coastal in occurrence and for illustration purposes the family density is shown within a band extending out 400 km from the coastline (slightly further than the 200 nautical-mile EEZ limit).

Source: Prepared by UNEP-WCMC; family distributions aggregated from the species range maps published by Compagno[72].

Level of diversity

High

Low

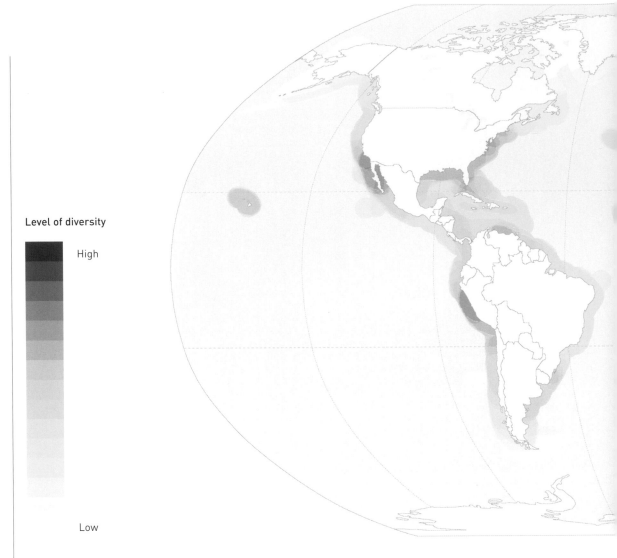

habitually return to the same nesting beaches, sometimes undergoing protracted migrations from feeding grounds. They may lay two or three clutches in a season, sometimes consisting of more than 100 eggs each, depending on the species. Nest, hatchling and juvenile mortality are often high.

Birds
Defining marine birds, or seabirds, is somewhat more problematic than defining marine species in other groups. All birds breed in terrestrial habitats, but a large number (almost all of them non-passerines) obtain all or much of their food from aquatic or littoral habitats. Some of these, including all frigatebirds (Fregatidae), tropicbirds (Phaethontidae), gannets and boobies (Sulidae), penguins (Spheniscidae) and petrels, alba-

trosses and shearwaters (Procellariidae) are indisputably marine, in that they obtain all their food from marine habitats, almost invariably breeding along coastlines and spending most or all of their time when not breeding out at sea. Many others, however, have less clear-cut habits. Some, such as a number of cormorants and shags (Phalacrocoracidae) have both resident inland and marine populations. Others, such as a number of gulls and terns (Laridae) and some ducks and geese (Anatidae), may breed inland but spend the rest of the year living in coastal areas or out at sea. Yet others, such as sandpipers (Scolopacidae) and other waders, typically feed in littoral or intertidal habitats rather than in the sea itself; many of these species also occur inland.

Adopting a somewhat arbitrary division,

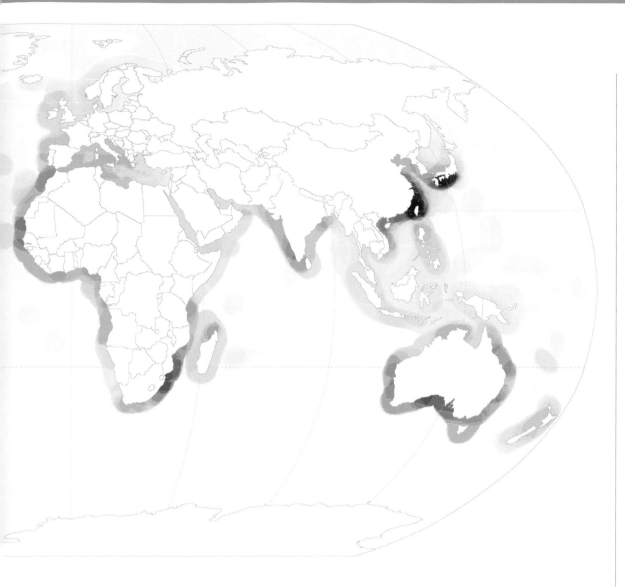

and excluding all wading birds with the exception of the red phalarope *Phalaropus fulicaria* (a truly pelagic species outside the breeding season), over 300 species of birds can be considered wholly or largely marine.

In common with pinnipeds, seabirds show a latitudinal distribution in which diversity is much higher at higher latitudes (temperate and polar regions) than it is in the tropics. Two thirds of all seabirds are confined as breeding species to these latitudes, compared with only 7 percent that are exclusively tropical. This stands in sharp contrast to the pattern found in most major terrestrial groups (see Chapter 5) and many marine groups such as sea turtles, mangroves and reef-building corals (see, respectively, Maps 6.3, 6.4, 6.6) in which species diversity increases dramatically with decreasing latitude. Diversity is also markedly higher in the southern than in the northern hemisphere, with over half of all seabird species breeding in southern temperate and polar latitudes. Dominance of this region is even more marked in the Procellariidae, the family with the greatest number of truly marine species, in which over 60 percent of species breed at these latitudes and half are confined to it (Table 6.7).

Table 6.7
Regional distribution of breeding in seabirds

Notes: Several species breed in more than one latitudinal band so that overall totals exceed actual number of seabirds. Numbers in parentheses indicate approximate number confined to the region.

Breeding distribution	Ocean area (million km²)	Procellariidae	Laridae	Others	Total
Northern temperate and polar	79.5	24 (17)	50 (35)	50 (36)	124 (88)
Northern tropical	75.2	15 (5)	27 (0)	19 (0)	61 (5)
Southern tropical	78.1	25 (8)	30 (4)	26 (5)	81 (17)
Southern temperate and polar	130.1	70 (58)	34 (15)	56 (41)	160 (114)

Map 6.3
Marine turtle diversity

An overview of marine turtle diversity, represented by the number of turtle species nesting in any one area. Each symbol shows the location of a nesting site or area, colored according to the number of species present, up to a maximum of five species in some parts of the tropics.

Source: Prepared using spatial data from a GIS database of marine turtle nesting beaches maintained at UNEP-WCMC.

Number of species nesting in area

5

4

3

2

1

Mammals

Wholly aquatic mammals (those that never normally emerge on to land) are confined to two orders, the Cetacea and the Sirenia. The Cetacea comprises some 78 species, all except five marine, distributed throughout the world's seas. They include the largest living animals – the rorquals in the family Balaenopteridae. All cetaceans are carnivorous; the baleen or whalebone whales (families Balaenidae, Balaenopteridae, Eschrichtiidae and Neobalaenidae) are filter feeders, feeding on organisms several orders of magnitude smaller than they are.

Of four living members of the order Sirenia, only one – the dugong *Dugong dugon* – is exclusively marine, occurring widely in coastal waters of the Indo-Pacific. One other, the Caribbean manatee *Trichechus manatus*, is found in both marine and inland waters while the other two (the Amazonian manatee *Trichechus inunguis* and West African manatee *Trichechus senegalensis*) enter coastal waters marginally if at all. One other species, the very large Steller's sea cow *Hydrodamalis gigas*, survived in waters around Bering and Copper Islands in the North Pacific until the early 18th century. All sirenians are herbivores; marine populations feed mainly on seagrasses.

The remaining marine mammals are all included in the order Carnivora. Two New World otters in the family Mustelidae, the sea otter *Enhydra lutris* from the north temperate Pacific coast and the marine otter *Lutra felina* from the south temperate Pacific coast, feed very largely or exclusively in marine waters; other otter species may frequent coastal areas but are predominantly inland

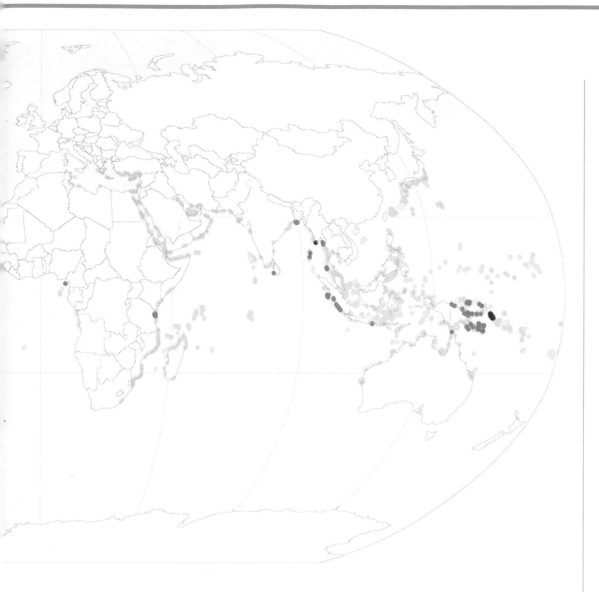

water animals. Members of the three pinni-ped families Odobenidae (the walrus), Otariidae (eared seals) and Phocidae (earless seals) are all largely aquatic, emerging on land to breed and rest, particularly when molting; all are marine with the exception of one or two species of Phocidae (the Baikal seal *Phoca sibirica* and, if the Caspian is regarded as a lake rather than a sea, the Caspian seal *Phoca caspica*). One member of the family Phocidae, the Caribbean monk seal *Monachus tropicalis*, has become extinct this century. All species are carnivorous.

In contrast to most terrestrial mammal families, pinnipeds are considerably more diverse and more abundant at higher rather than lower latitudes. Of the 32 extant or recently extant species, only five occur within the tropics (two marginally). Part of the explanation for this undoubtedly lies in the greater availability of suitable habitat at higher latitudes: as noted above, 70 percent of continental shelf waters and just over 60 percent of the world's marine area are found outside the tropics. However, this in itself is unlikely to account for the entire difference. It is probable that the greater productivity of shelf waters at high latitudes, discussed above, and of upwelling areas at mid-latitudes (e.g. the Benguela current off the western coast of South Africa and the Humboldt current off Chile and Peru) plays a major part.

The isolated character of many island breeding sites, such as the Galapagos group, in temperate and sub-polar parts of the southern hemisphere may also have en-couraged speciation of pinnipeds here.

COASTAL AND SHALLOW WATER COMMUNITIES
Mangroves

Mangrove woodland is indeed a truly hybrid terrestrial/marine ecosystem, unique in that terrestrial organisms can occur in the canopy and marine species at the base[11]. Mangroves, or mangals, are a diverse collection of shrubs and trees (including ferns and palms) which live in or adjacent to the intertidal zone and are thus unusual amongst vascular plants in that they are adapted to having their roots at least periodically submerged in sea water. A wide variety of organisms is associated with mangroves including a number of epiphytic, parasitic and climbing plants, and large numbers of crustaceans, mollusks, fishes and birds[12].

Mangrove species are generally divided into those found only in mangrove habitats and those that may also be found elsewhere but which are nevertheless an important component of mangrove habitats. Both groups come from a wide range of families. The former includes around 62 species and seven hybrids in some 22 genera (Table 6.8). The appearance of mangroves is far from uniform: they vary from closed forests 40-50 m high to widely separated clumps of stunted shrubs less than 1 m high[13]. They are only able to grow on shores that are sheltered from wave action, and are particularly well developed in estuarine and deltaic areas; they may also extend some distance upstream along the banks of rivers, e.g. some 300 km up the Fly River in Papua New Guinea.

Mangrove communities are largely restricted to the tropics between 30°N and 30°S, with extensions beyond this to the north in Bermuda and Japan, and to the south in Australia and New Zealand[14] (Map 6.4). They occur over a larger geographical area than coral reefs (see below) and, unlike reefs, are well developed along the western coasts of the Americas and Africa. They have a more restricted distribution than coral reefs in the South Pacific.

There are two main centers of diversity. The eastern group occurs in the Indo-Pacific (the Indian Ocean and western part of the Pacific Ocean) and is the most species rich[15, 16]. The western group is centered around the Caribbean and includes mangrove communities along the west coast of the Americas and Africa.

Global mangrove area is believed to slightly exceed 180 000 km[2], divided regionally as shown in Table 6.9. Mangroves occur in over 100 countries (including dependent territories) but exist in very small areas in many of these. Four countries (Indonesia, Brazil, Australia and Nigeria) between them account for over 40 percent of the world's mangrove area, and Indonesia alone possesses nearly one quarter of the global mangrove area[16].

Although it is known that mangrove ecosystems in most parts of the world have been extensively degraded and cleared, it is difficult to obtain reliable data on the global extent of mangrove loss over time. Mangroves by their very nature occupy highly dynamic and unstable environments so that even without human action the location and extent of mangrove cover would be constantly changing. One assessment[17] suggested that more than 50 percent of the world's mangrove forest cover had been destroyed.

Mangroves stabilize shorelines and de-

Table 6.8
Diversity of mangroves

Note: Where two figures are given, second figure indicates number of hybrids. Families composed solely of mangrove species are shown in bold.

Source: Adapted from Duke[69] and Spalding et al.[16].

Order	Family	No. of species
Filicopsida	Adiantaceae	3
Plumbaginales	Plumbaginaceae	2
Theales	**Pelliceraceae**	1
Malvales	Bombacaceae	2
	Sterculiaceae	3
Ebenales	Ebenaceae	1
Primulales	Myrsinaceae	2
Fabales	Leguminosae	2
Myrtales	Combretaceae	4+1
	Lythraceae	1
	Myrtaceae	1
	Sonneratiaceae	6+3
Rhizophorales	Rhizophoraceae	17+2
Euphorbiales	Euphorbiaceae	2
Sapindales	Meliaceae	2+1
Lamiales	**Avicenniaceae**	8
Scrophuliariales	Acanthaceae	2
	Bignoniaceae	1
Rubiales	Rubiaceae	1
Arecales	Palmae	1

crease coastal erosion by reducing the energy of waves and currents and by holding the bottom sediment in place with their roots. They also act as windbreaks and provide protection from coastal storms. They are generally highly productive ecosystems and are important habitats for crustaceans, shellfish and finfishes. Most of the larger commercial penaeid shrimps are mangrove dependent; these and other species are harvested both on a subsistence basis and commercially, and may provide a major source of income in some countries.

As well as providing habitat for adults of many species of finfish and invertebrates, mangroves serve as spawning and nursery areas for many others, often of major economic importance. The wood provides building material, used locally in houses, as fence poles and to build fish traps, and is also harvested on a large scale for production of pulp and particle board. In many areas mangroves are also an important source of fuel, as firewood and charcoal. Mangrove foliage may provide an important source of fodder for domestic livestock in some countries, particularly during dry seasons when other sources of greenery are in short supply.

Salt marshes

Salt marshes are coastal communities of rooted salt-tolerant (halophytic) plants of terrestrial origin, dominated by grasses, herbs and dwarf shrubs. They share many characteristics with mangroves but replace them geographically in higher latitudes, except for some overlap at the extremes of mangrove distribution (in the Gulf of Mexico, Japan, southern Australia and northern New Zealand). Globally, salt marshes are estimated to cover around 350 000 km² in total[18]. They tend to develop in sheltered areas of mud and silty sand that are flat and slow draining; as plant growth and sedimentation elevates the marshland, so the period of tidal submergence decreases and a network of creeks develops. The two genera which are most prominent as pioneer saltmarsh plants are *Salicornia*, or samphire, and *Spartina*, or cord grass. Species of *Puccinella*, *Scirpus* and *Juncus* are also common.

Salt marshes are highly productive, around 2 500 g C per m² per year. Occurring in highly seasonal latitudes, salt marshes rapidly take up and accumulate nutrients during the growing season. In the autumn, when plants die or become dormant, the uptake of nutrients is greatly diminished and dead organic matter may be transported out of the marsh. The physical features of salt marshes mean that pollutants are not easily or rapidly flushed out. Other threats include infilling, especially around the North Sea[19], and increased erosion through channel dredging.

Rocky shores

Rocky shorelines are generally exposed to oceanic swells and extreme wave action (except for more sheltered fjordlands) but are topographically variable, occurring as wide platforms, steep cliffs or other formations, depending on local geology[20]. They provide a unique habitat for plant growth, with a stable substrate for attachment, and shallow well-lit water that tends to be turbulent and rich in nutrients. Consequently macro-algae communities, dominated by kelps (*Laminaria, Ecklonia, Macrocystis*) and fuccoids (*Fucus, Ascophyllum*) flourish, mainly in temperate regions but also in areas of the tropics where seasonal upwellings of cold water occur, such as Chile and the southern Arabian coast[21]. The net primary productivity of kelp forests is comparable to tropical rainforests, and *Macrocystis pyrifera*, the giant kelp, can attain growth of up to 45 cm a day.

Two physical factors – water movement and desiccation – have influenced the diversity of intertidal rocky shore species. Strong wave action has favored use of crevices, dense aggregations of individuals, and the evolution of strong attachment devices (algal holdfasts,

Table 6.9
Current mangrove cover

Source: Spalding *et al.*[16].

Region	Mangrove area (to nearest thousand km²)	% of total
South and Southeast Asia	75 000	42
Americas	49 000	27
West Africa	28 000	16
Australasia	19 000	10
East Africa and Middle East	10 000	6
Total	**181 000**	

Map 6.4
Mangrove diversity

The location of current mangrove forest, together with contours representing gradients of mangrove species richness. Note that graphic presentation at this scale enormously exaggerates actual forest area.

Source: Reproduced with modification, from Spalding[73].

Mangrove locality and species richness

●   Mangrove forest

High diversity

Low diversity

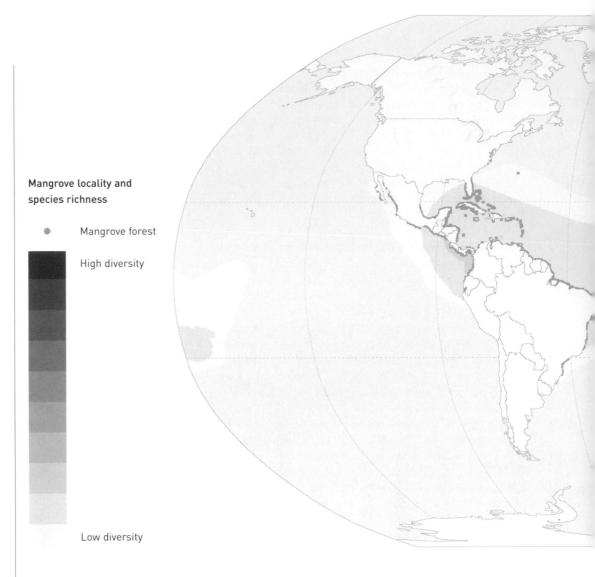

cementation of barnacles, the byssus threads of mussels and the adhesive feet of gastropods and echinoderms). The need to avoid being washed off the rocks means that most organisms on rocky shores are sessile or have limited motile ability. Competition for space is therefore intense and organisms inhabiting rocky shores occupy well-defined zones. It also means that they can be particularly vulnerable to disturbances such as oil spills, especially if these occur during calm periods which extend the residence time of pollutants. However, because they are generally exposed and high-energy environments, rocky shores tend to be less vulnerable than most to pollution. These features also limit some destructive human activities, such as construction, responsible for degrading other marine habitats.

Seagrasses

Seagrasses are a mixed group of flowering plants (not true grasses) that are adapted to live submerged in shallow marine and estuarine environments at a wide range of latitudes. About 58 species are recognized by many authorities, in four families, all within the monocotyledons. They occur from the littoral region to depths of 50 or 60 m but are most abundant in the immediate sublittoral area. There are more species in the tropics than in the temperate zones, and of the 12 seagrass genera seven are confined to tropical seas and five to temperate seas[22]. Most seagrass species are similar in external morphology, with long thin leaves and an extensive rhizome root system which enables them to fasten to the substrate. A variety of substrates are occupied from sand and mud to granite

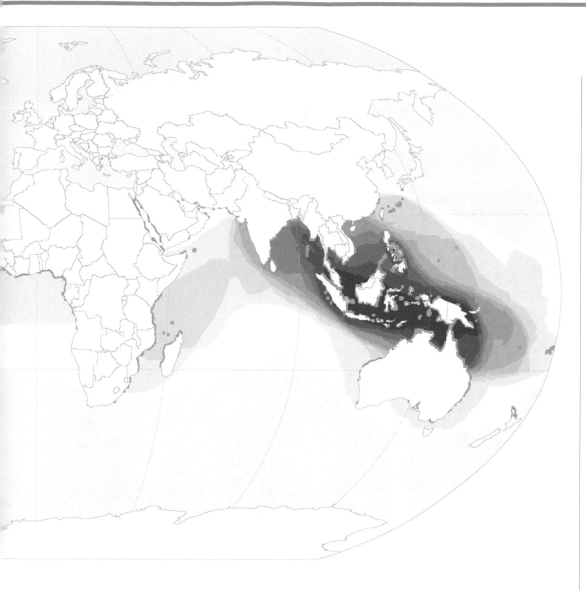

rock, but the most extensive beds occur on soft substrates[11].

While species diversity (see Map 6.5) is highest in Southeast Asia (as with mangroves and corals) there are two important centers in temperate regions: in mainland Japan and in southwest Australia. Secondary centers of diversity include East Africa, the Red Sea and the Mediterranean. Although seagrasses themselves are not a diverse group, they support considerable diversity in some assoc- iated species, including an estimated 450 species of epiphytic algae[22].

In many areas seagrasses form extensive but simple communities, referred to as sea- grass beds or seagrass meadows. Seagrass beds have high productivity and contribute significantly to the total primary production of inshore waters. Seagrasses are particularly important in nutrient-poor (oligotrophic) waters because they can extract some nutri- ents from sediments, unlike other marine primary producers. Many commercially impor- tant species are dependent on seagrass beds, often as nursery habitat, providing shelter from predators and adverse sea conditions. These include mollusks (such as the queen conch *Strombus gigas*), shrimp, lobster, holo- thurians and many finfish (such as grunts, Haemulidae; rabbitfish, Siganidae; emperors, Lethrinidae; and snappers, Lutjanidae). A small but important number of threatened species depend on seagrasses, including sirenians, the green turtle *Chelonia mydas* and many species of seahorse (Syngnathidae). In addition to such direct values, seagrass beds also play an important role in binding sediments, providing some protection from

Map 6.5
Seagrass species diversity

This first global map of seagrass diversity indicates the extent of seagrass habitat (inventory sites) and shows contours of species richness. Diversity contours not shown for parts of West Africa because species inventory incomplete.

Source: Preliminary plot, compiled using multiple sources, from a seagrass atlas in preparation at UNEP-WCMC.

Seagrass locality and species richness

● Seagrass

High diversity

Low diversity

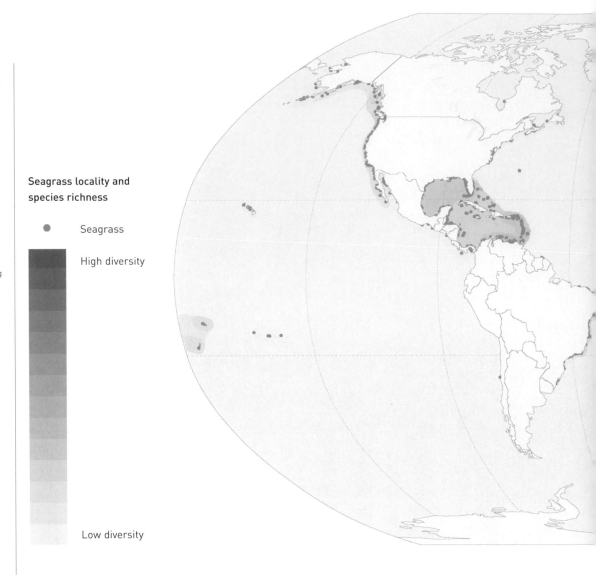

coastal erosion, and may help remove excess nutrients and toxins from coastal waters.

A preliminary estimate[23] suggested that seagrass beds extend over some 600 000 km² globally, but a precise figure is not yet available. Although seagrasses can be highly dynamic ecosystems, with individual seagrass beds undergoing significant shifts in distribution over relatively short periods, there is considerable evidence that there have been major net losses over the past century. Extensive losses along the Atlantic coasts of Europe and North America in the 1930s followed disease caused by a marine slime mold; some evidence suggests that other environmental impacts (possibly increased turbidity in coastal waters) may have increased susceptibility to disease. More recently, nutrient loading has continued to increase in coastal waters worldwide, often leading to enhanced growth of phytoplankton, epiphytes and macroalgae, all of which can outcompete seagrasses for available sunlight. Sedimentation also has a major impact, reducing the passage of light through the water column and physically smothering seagrass plants. To a lesser degree, toxic pollutants and physical disturbance from activities such as trawling and dredging have also played a role in seagrass losses[24].

Shallow tropical coral reefs

The term 'coral reef' applies to a variety of calcium carbonate structures developed by stony corals. They are tropical shallow-water ecosystems, typically with high biodiversity and largely restricted to coastal seas between the latitudes of 30°N and 30°S[25]. They are

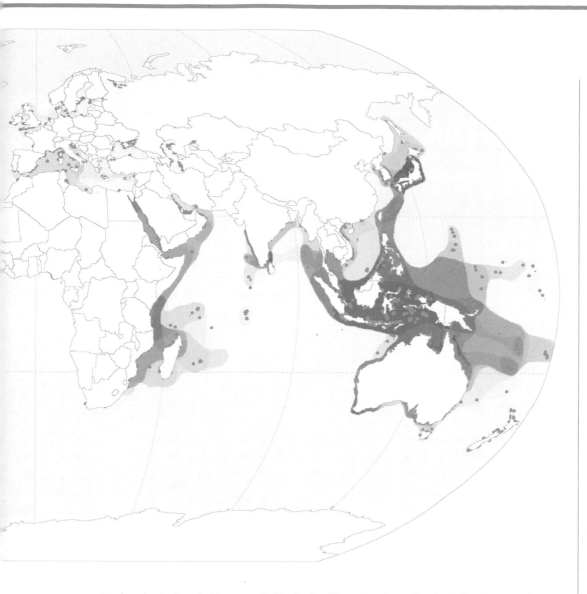

most abundant in shallow, well-flushed marine environments characterized by clear, warm, low-nutrient waters that are of oceanic salinity[25]. There are two basic categories: shelf reefs, which form on the continental shelves of large land masses, and oceanic reefs, which are surrounded by deeper waters and are often associated with oceanic islands. Within these two categories there are a number of reef types: fringing reefs, which grow close to shore; barrier reefs, which develop along the edge of a continental shelf or through land subsidence in deeper waters, and are separated from the mainland or island by a relatively deep, wide lagoon; and atolls, which are roughly circular reefs around a central lagoon and are typically found in oceanic waters, probably originating from the fringing reefs of long-submerged islands.

Two other less clearly defined categories are patch reefs, which form on irregularities on shallow parts of the seabed, and bank reefs, which occur in deeper waters, both on continental shelf and in oceanic waters[12].

The global extent of coral reefs is not known with certainty. A recent estimate, derived by measuring the total reef extent in a comprehensive set of national maps, suggests a world total of 284 300 km². This is equivalent to a little more than 1 percent of the total world ocean shelf area. New information and mapping techniques may increase the accuracy of such estimates, but the total is unlikely to exceed 300 000 km². A regional breakdown is provided in Table 6.11. Although coral reefs occur in around 110 countries and territories, just five countries – Indonesia, Australia, Philippines,

France (overseas departments and territories in the Indian Ocean and Pacific) and Papua New Guinea – account for over half of the global total.

Five different orders within the phylum Cnidaria (coelenterates) include species with calcified skeletons, or 'stony corals'. Many of these are reef building (hermatypic) and some are solitary (ahermatypic). The great majority of hermatypic corals belong to the order Scleractinia, the true stony corals, although not all scleractinians are reef-builders. Together with sea anemones and sea fans, the Scleractinia make up the class Anthozoa. Most scleractinian coral polyps have symbiotic algae (zooxanthellae) within their tissues; these use the nitrates, phosphates and carbon dioxide produced by the coral, and through photosynthesis generate oxygen and organic compounds that provide much of the polyps' nutrition. The zooxanthellae give corals their color and, because they photosynthesize, restrict the corals that contain them to the photic zone[26]. Corals without zooxanthellae typically do not form reefs and can exist in deeper colder waters (see deep-water reefs, below). Because of difficulties with synonymy and in defining species, it is difficult to estimate precisely how many extant species of reef-building coral there are. A recent synthesis[27] deals with around 800 zooxanthellate scleractinians (see Table 6.10); another[28] lists more than 1 300 scleractinians in all, and 260 calcified hydrozoans.

Not all reefs are constructed primarily by corals. Within the red algae (phylum Rhodophyta) and the green algae (phylum Chlorophyta) in particular several genera grow heavily calcified encrustations which bind the reef framework and in places are the main contributors to shallow reef growth.

Coral reefs are among the most productive and diverse of all natural ecosystems. The main center of diversity for reef-building corals is Southeast Asia, with an estimated minimum of 450 species found associated with reefs around the Philippines, Borneo, Sulawesi and associated islands. This area is part of a single, vast, Indo-West Pacific biogeographic province that extends from the Red Sea in the west to the Pitcairn Islands in the east. Many coral genera and a significant number of species are found throughout the region, although overall diversity in the province decreases on leaving this center. In the east of this region, the central and eastern Pacific forms a series of somewhat distinct subregions, characterized by a number of genera and species (particularly in Hawaii) not found further west. This area also shares many species with the Indo-West Pacific province but overall has much lower diversity than most of the latter. The Atlantic, including the Caribbean and the Gulf of Mexico, forms a distinct province with few species in common with the Indo-West Pacific. It is also very depauperate compared with most of the latter (Map 6.6).

Deep-water reefs

Beyond the coral reefs of shallow tropical waters, a few species of coral that lack zooxanthellae form deep-water reefs on hard substrates in high-current areas associated with topographic rises, such as ridges and pinnacles. The best known is *Lophelia pertusa*, a colonial coral that forms structures ranging from patches a few meters in width to reefs many hundreds of meters in size, at depths of 100-3 000 m and temperatures of

Table 6.10
Diversity of stony corals in the order Scleractinia

Note: Includes only the reef-forming scleractinians with zooxanthellae.

Source: Veron[27].

Family	Genera	Species
Acroporidae	4	262
Agaricidae	6	43
Astrocoenidae	4	13
Caryophylliidae	1	1
Dendrophyllidae	4	14
Euphyllidae	5	14
Faviidae	24	125
Fungiidae	13	56
Meandriniidae	7	8
Merulinidae	5	12
Mussidae	13	50
Oculinidae	4	15
Pectiniidae	5	28
Pocilloporidae	3	30
Poritidae	5	92
Rhizangiidae	1	1
Siderastreidae	6	28
Trachyphyllidae	1	1

4-8°C. Although best known through trawling and oil exploration activities in the northeast Atlantic, these deep-water reefs are global in occurrence.

The complex matrix of living and dead branches of *Lophelia* increases spatial heterogeneity above that of the surrounding seabed and provides a habitat for many species. Boring sponges, anemones, bryozoans, gorgonians, polychaetes, barnacles and bivalves occur in large numbers[29] and their diversity is comparable to some shallow-water tropical systems[30]. Although large aggregations of fish are associated with *Lophelia* reefs, and reef areas support higher catches than adjacent seabed[31], the species diversity of fish and coral is much lower than in tropical coral reefs (some 23 species of fish have been recorded on *Lophelia* reefs in the northeast Atlantic).

The reefs are delicate structures easily destroyed by demersal fishing gear which routinely operate to depths of 2 000 m. The total destruction of some Norwegian reefs has already been documented and an estimated 30-50 percent of others have been damaged. Slow growth, in the region of 4-25 mm per year, severely limits the ability of *Lophelia* to recover. The trawl fishery for orange roughy *Hoplostethus atlanticus* and oreos *Allocyttus niger* on sea mounts south of Tasmania has been responsible for substantial destruction of *Solenosmilia variabilis* reefs, with some reduced to more than 90 percent bare rocks[32]. If recovery ever occurs it will take hundreds of years.

OCEANIC PELAGIC COMMUNITIES

There is a fundamental distinction between the processes and patterns observed in open oceans, dominated by global winds and large-scale vertical and horizontal movement of water masses, and those observed nearer to coasts, where shelf bathymetry, coastal winds and local input of nutrients, pollutants and sediments generate a diversity of smaller-scale phenomena.

The oceanic pelagic zone is dominated by the activity of plankton in the euphotic surface waters. Plankton are by definition drifting or weakly swimming organisms, comprising a

wide range of small to microscopic animals, protoctists and bacteria. Free-swimming pelagic organisms, predominantly fishes but also cetaceans and cephalopod mollusks (squid), are collectively termed nekton. These organisms, when adult, are predators of plankton or smaller nekton. They in turn – as vertically migrating fishes or larvae, and as dead organic material – provide food for deep-sea and benthic organisms. With few exceptions, the only other food source for creatures in the aphotic zone is the 'rain' of organic matter, such as feces, molted crustacean exoskeletons, and a variety of other organic material derived from plankton in the surface waters of the ocean. Plankton and larger free-swimming organisms tend strongly to concentrate along major circulation currents (gyres), contact zones and upwellings, and this can give rise to significant local variation in diversity.

The marked vertical gradients within the pelagic zone – of light, temperature, pressure, nutrient availability and salinity – lead to vertical structuring of pelagic species assemblages. Several zones based on changes in species composition with depth have been recognized, including epipelagic (usually taken as from the surface to a depth of 200-250 m and including the euphotic zone); mesopelagic, which underlies the epipelagic zone to a depth of 1 000 m or so; and below this the bathypelagic which changes in a somewhat less well-defined fashion to abyssopelagic at around 2 500-2 700 m depth.

Region	Shallow reef area (to nearest thousand km²)	% of world total
Indo-Pacific	261 200	91.9
Red Sea, Gulf of Aden	17 400	6.1
Arabian Sea, Persian Gulf	4 200	1.5
Indian Ocean	32 000	11.3
Southeast Asia	91 700	32.3
Pacific	115 900	40.8
Wider Atlantic	21 600	7.6
Caribbean	20 000	7.0
Atlantic (excl. Caribbean)	1 600	0.6
Eastern Pacific	1 600	0.6
Total	**284 300**	

Table 6.11
Coral reef area

Source: Spalding *et al.*[70].

Map 6.6
Coral diversity

The location of coral reefs, together with contours representing gradients of species richness among reef-building scleractinian coral species. Note that graphic presentation at this scale enormously exaggerates actual reef area.

Source: Reproduced by permission, with modification, from Veron[74], revised by Veron, pers. comm. November 2001.

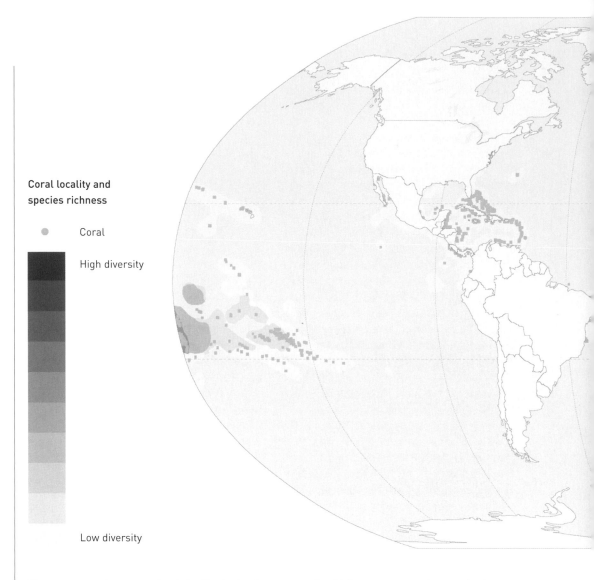

Coral locality and species richness

●   Coral

High diversity

Low diversity

These zones, however, tend to fluctuate in time and space. As well as seasonal changes in water characteristics, many components of the epipelagic and mesopelagic nekton undergo marked diel migrations (i.e. on a 24-hour cycle), ascending to surface waters at night to feed and descending, sometimes over 1 km, during the day. Many species of nekton, particularly cetaceans and larger fishes, are also highly migratory, ranging over enormous expanses of ocean in more or less regular and predictable patterns.

It has generally been assumed that biomass in the pelagic zone everywhere below the euphotic zone is low. However, recent studies have indicated that biomass of tropical mesopelagic animals may be surprisingly high. Study of the mesopelagic fauna has been limited to date, as it requires the use of

expensive high-seas research vessels; knowledge of taxonomy, distribution and biology of most of the species concerned remains very incomplete[1]. One study[33] recognized around 160 fish genera in 30 families as important components of the fauna. Most species are small (less than 10 cm in length) and often bizarrely shaped. Estimates based on a variety of surveys carried out indicate that global biomass of this stock may be large: a figure of 650 million tons (some six to seven times total current marine fisheries landings) has been suggested, although this should be regarded with extreme circumspection[1].

From available data, it would appear that the mesopelagic biomass is greatest in the northern Indian Ocean, and particularly in the northern Arabian Sea, one of the five major upwelling zones. Surveys here indicated

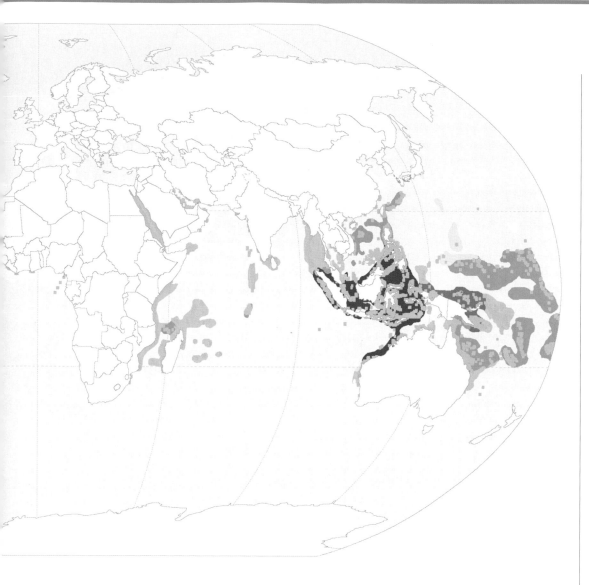

extremely high biomass (25-250 g per m²) in the Gulf of Aden and Gulf of Oman as well as off the western coastline of Pakistan. These figures are around an order of magnitude higher than those recorded elsewhere in the tropics, indicating either a great overestimate for the northern Indian Ocean or an under-estimate elsewhere. Alternatively this region genuinely is ten times as productive as the rest of the tropical ocean system. Although this appears as yet unresolved, it is nevertheless apparent that there is substantial global mesopelagic fish biomass and that the Arabian Sea is particularly rich in these species.

DEEP-SEA COMMUNITIES

Approximately 51 percent of the Earth's surface is covered by ocean over 3 000 m in depth, so deep-sea communities are preva-lent over a significant proportion of the planet. All deep-sea habitat is in the aphotic zone, well below the distance sunlight can pene-trate. As deeper and deeper levels are reached biomass falls exponentially[34].

Despite their enormous volume, the deep oceans appeared to be relatively simple ecosystems and to make little contribution to global species diversity, but discoveries during the past decade or so have shown that in some regions species diversity in the benthic community increases with increasing depth. This was revealed by novel sampling techniques, principally the epibenthic sled[35]. The rate of discovery of new species and the proportion of species currently known from only one sample both suggest that a great number remain to be discovered[36, 37]. As with arthropods in tropical moist forests, estimates for the number of unknown species vary widely, with some suggestions that there may be as many as 10 million undescribed species in the deep sea[38]. Others consider that the true figure is more likely to be around 500 000[39].

Ocean trenches

Ocean trenches are typically close to land masses and tend to have high rates of sedimentation, a significant amount of which is of organic origin and an important available food source for trench communities. Several trenches also underlie highly productive cold-water upwelling zones, the organic fallout from which contributes greatly to productivity there. The water within trenches generally originates from the surrounding bottom water, which is derived from cold surface water at high polar latitudes and is relatively well oxygenated[40].

Trenches tend to be isolated linear systems that because of their seismic activity form a habitat that is unstable and unpredictable compared with the relative environmental stability of the adjacent abyssal plains[40]. Trench faunas are not rich in species but are often high in numbers of endemic species. There are some 25 genera restricted to the ultra-abyssal (hadal) zone, representing some 10-25 percent of the total number of genera present, and two known endemic hadal families: the Galatheanthemidae (Cnidaria) and Gigantapseudidae (Crustacea). The latter family contains a single species: *Giganta-pseudes adactylus*. The greatest number of endemic species known from a single trench is a sample of 200 from the Kurile-Kamchatka trench; this may be compared with 10 endemic species known from the Ryukyu and Marianas trenches.

Hydrothermal vents

Hydrothermal vent communities were first discovered in 1977, at a depth of 2 500 m on the Galapagos Rift, but contribute to what might be one of Earth's most archaic eco-systems. They are now known to be assoc-iated with almost all known areas of tectonic activity at various depths. These tectonic regions include ocean-floor spreading centers,

subduction and fracture zones and back-arc basins[41]. Cold bottom water permeates through fissures in the ocean floor close to ocean floor spreading centers, becomes heated at great depths in the Earth's crust and finds its way back to the surface through hydrothermal vents. The temperature of vent water varies greatly, from around 23°C in the Galapagos vents to around 350°C in the vents of the East Pacific Rise, and they may be rich in metalliferous brines and sulfide ions[40]. Most species live out of the main flow at temperatures of around 2°C, the ambient temperature of deep-sea water. The biomass of vent communities is usually high compared with other areas of similar depth, and dense colonies of tube worms, clams, mussels and limpets typically constitute the major components.

Hydrothermal vent communities are of particular interest because they flourish in the dark at high pressures and low temperatures[42], and unique because they are supported by chemolithoautotrophic archeans and bacteria, notably *Thiomicrospira* species (phylum Proteobacteria), which form dense microbial carpets in the rich hydrothermal fluid and derive their energy chiefly from oxidizing hydrogen sulfide[41,42]. Many of the eukaryote vent species filter-feed on these microorganisms, whilst others rely on sym-biotic sulfur bacteria for energy[43].

The overall species diversity at vents is low compared with other deep-sea soft sediment areas[42], but endemism is high. More than 20 new families or subfamilies, 50 new genera and nearly 160 new species have been recorded from vent environments, including brine and cold seep communities[41, 44]. Vent communities are separated by gaps of between 1 and 100 km, and although they may persist only for several years or decades, sites of vent activity move relatively slowly allowing dispersal of vent organisms[41].

Two further seep patterns are known. Cold sulfide and methane-enriched groundwater seeps occur near the base of the porous limestone of the Florida escarpment, as well as in the Gulf of Mexico and elsewhere. These support a community similar in taxonomic composition to the hydrothermal vents of the east Pacific. Tectonic subduction zone seeps are more diffuse and lower in temperature than hydrothermal vent seeps and are rich in dissolved methane. They are known to occur off Oregon and in the Guaymas basin in the Gulf of California, and cold seeps occur at a depth of 1 000 m in Sagami Bay near Tokyo and in the subduction zones of the trenches off the east coast of Japan.

HUMAN USE OF AND IMPACT ON THE OCEANS

The seas provide many biological resources used by humans. A wide range of animal species, notably fishes, mollusks and crustaceans, contribute to marine fisheries and these provide by far the most important source of wild protein, of particular importance to many subsistence communities around the world. Marine algae are also an increasingly important foodstuff, notably in the Far East, with current annual world production of around 2 million metric tons. Marine organisms are also proving extremely fruitful sources of pharmaceuticals and other materials used in medicines. Relatively minor although locally important uses include exploitation of coastal resources for building materials (e.g. coral limestone, mangrove poles) and other industrial products (e.g. tannins from mangroves).

Access to marine resources is not equitably distributed amongst the world's nations. Most obviously, some 39 states are landlocked, i.e. have no seaboard (five of these border the inland Caspian Sea). Those that do have seaboards show great variation in length of coastline, and area of territorial waters and exclusive economic zones (EEZs), both absolutely and relative to their land areas. They also show great variation in their capacities to exploit marine resources, both on the high seas and within their territorial waters and EEZs.

Human activities, directly and indirectly, are now the primary cause of changes to marine biodiversity. Approximately one third of the world's human population lives in the coastal zone (within 60 km of the sea) and indications are that this proportion will rise during the 21st century[45]. Pressures exerted by the human population on the marine biosphere are substantial and increasing.

Many species of fishes, mollusks and crustaceans provide humans with their largest source of wild protein.

Most identified threats relate to coastal and inshore (continental shelf) areas. However, threats to the oceanic realm are undoubtedly increasing: fisheries and their attendant physical effects, such as habitat alteration owing to dredging and trawling, have entered deeper continental slope waters having previously been largely confined to the epipelagic zone, and deep-water oil and gas mining is planned. Even abyssal and hadal areas are susceptible to human impact. A small, steady increase in abyssal temperature of 0.32°C in 35 years has been attributed to global climate change brought about by human activities. Ocean waste dumping and the potential for deep-water mining and mineral extraction are also causes for concern, as are the changes in biomass and species composition in the waters above these regions[46].

The following five activities have been identified as the most important agents of present and potential change to marine biodiversity at genetic, species and ecosystem levels[46]:

- fisheries operations;
- chemical pollution and eutrophication;
- alteration of physical habitat;
- invasions of exotic species;
- global climate change.

WORLD MARINE CAPTURE FISHERIES
World marine capture fisheries have grown fivefold in the past half-century, with annual landings increasing from nearly 18 million metric tons between 1948 and 1952 to around 87 million metric tons during the period 1994-97, with a fall caused mainly by El Niño in 1998 and a subsequent recovery in 1999. Marine capture fisheries made up just over 70 percent of current recorded global production of aquatic resources in the late 1990s, the

Just five species of finfish, among them the Atlantic herring, account for a quarter of the global catch.

remainder being accounted for by inland capture fisheries (see Chapter 7) and aquaculture[47, 48]. With capture fisheries apparently remaining more or less stable, the increase in total marine production during the 1990s was due to a continuing increase in aquaculture production.

Composition of marine fisheries
Marine fisheries encompass a wide range of organisms, including algae, invertebrate animals in various phyla and vertebrates including fishes (often termed finfishes in fisheries analysis), reptiles, mammals and birds (although by convention the last of these groups is not normally considered in fisheries analysis).

The Food and Agriculture Organization of the United Nations (FAO) recognizes in total just under 1 000 'species items' (species, genera or families) that feature at least periodically in national catch statistics. However, globally important marine fisheries are confined to relatively few groups, with over 80 percent of landings by weight being finfishes and virtually all the remainder mollusks and crustaceans.

In terms of major species groups, by far the most important are the herrings and anchovies in the order Clupeiformes, which in 1998 accounted for over 22 million metric tons, or around 25 percent of marine landings. These are followed by cod, hake and haddock (Gadiformes), and jacks and mullets (some Perciformes and Mugiliformes), with production of more than 9 million and nearly 8 million metric tons respectively. The most important invertebrate group overall is cephalopod mollusks (squid, cuttlefish and octopus) of which some 3.4 million metric tons were reported landed.

In terms of individual species, for several years during the 1990s just five species of finfish, anchoveta *Engraulis ringens*, Alaska pollock *Theragra chalcogramma*, Chilean jack mackerel *Trachurus murphyi*, Atlantic herring *Clupea harengus* and chub mackerel *Scomber japonicus*, together made up around one quarter of marine landings. Each accounted for over 2 million metric tons annually, and among several others around this level, Japanese anchovy *Engraulis japonicus* and Skipjack tuna *Katsuwonus pelamis* were increasingly important. In most years by far the most important single species is the anchoveta *Engraulis ringens*, whose fishery (off the west coast of South America) was nearly 13 million metric tons in 1994, constituting by far the largest single-species fishery the world has ever seen; but just under 8 million in 1997 and less than 2 million in 1998.

Distribution of marine fisheries

The geographical distribution of marine fisheries is determined both by the distribution of harvestable fish stocks and by a range of complex socioeconomic factors. The former is largely determined by variations in productivity, which, as noted above, are themselves largely determined by nutrient availability, so that overall the most productive fisheries areas are on continental shelves at higher latitudes and in upwelling zones at lower latitudes. As a generalization, the latter are associated with pelagic fish stocks and the former more with demersal or semi-demersal (deep-water or bottom-dwelling) stocks, although pelagic stocks play an increasingly important role even here.

As might be expected purely on the basis of its size, the Pacific Ocean is by far the most important major fisheries area, accounting for over 60 percent of marine landings. The northwest Pacific alone – an area with extensive continental shelf development – accounts for nearly half this total.

The various upwelling zones are not all of equal importance in fisheries. That associated with the Humboldt current off Peru and Chile is the single most productive, while those associated with the California, Benguela and Canary currents are of somewhat lesser importance, although each is still a major fisheries area. The Arabian Sea upwelling appears to be anomalous, in that it evidently supports major populations of mesopelagic (i.e. middle-depth) rather than epipelagic species. Not only are the former generally considered of low value, with an identified market only as animal feed, but capture and processing requires expensive, advanced technology. They thus remain virtually unexploited at present and are considered along with the Antarctic krill stocks to be the major unexploited fisheries resource left.

Trends in marine fisheries

National fishery statistics are collated by the FAO. These data are the principal source of information on global fishery trends, although it is widely acknowledged that they are variable in quality. During the 1950s and 1960s, total landings increased steadily as

1996   1998

Million metric tons

Alaska pollock *(Theragra chalcogramma)*
4.5
4.0

Atlantic herring *(Clupea harengus)*
2.3
2.4

Japanese anchovy *(Engraulis japonicus)*
1.3
2.1

Chilean jack mackerel *(Trachurus murphyi)*
4.4
2.0

Skipjack tuna *(Katsuwonus pelamis)*
1.6
1.9

Chub mackerel *(Scomber japonicus)*
2.2
1.9

Anchoveta *(Engraulis ringens)*
8.9
1.7

Largehead hairtail *(Trichiurus lepturus)*
1.3
1.4

Atlantic cod *(Gadus morhua)*
1.3
1.2

Yellowfin tuna *(Thunnus albacares)*
1.1
1.2

Blue whiting *(Micromesistius poutassou)*
0.6
1.2

Figure 6.1
Species contributing most to global marine fisheries

Source: FAO[75].

new stocks were discovered, while improved fishing technology and an expansion of fishing effort enabled fuller exploitation of existing stocks of both pelagic and demersal species. Long-range fleets increased in size during this period and, as traditional fishing grounds in the North Atlantic and North Pacific became fully exploited, moved into new fishing grounds closer to the tropics and in the southern hemisphere. By concentrating their efforts in the richest ocean areas, these fleets were largely responsible for the rapid increase in world catches.

At the beginning of the 1970s, the Peruvian anchoveta fishery alone contributed some 20 percent of marine fisheries production. These stocks collapsed around 1972, at the same time as the important South African pilchard fishery in the Atlantic, seemingly in association with an ENSO (El Niño Southern Oscillation) event. There was a sharp drop in overall marine fisheries production, after which the global catch increased more slowly than before, reaching the early 1970s level by the end of the decade. Landings of most demersal fish stocks remained relatively constant, however, implying that they were close to full exploitation. Long-range fleets continued to expand in importance.

The 1980s once again saw a period of continuous growth (averaging 3.8 percent a year) in reported world landings. As in the 1970s landings of demersal stocks were generally static or declining so that shoaling pelagic species provided most of the increase in fish production. In fact, just three pelagic species (Peruvian anchoveta, South American sardine *Sardinops sagax*, and Japanese

sardine *Sardinops melanostictus*) and one semi-demersal species (Alaska pollock) accounted for 50 percent of the increase in world landings during the 1980s[47]. Most of this increase appears to have been because of favorable climatic effects on stock sizes rather than new fishery developments or improved management practices[47].

Following a sharp decline at the end of the 1980s, FAO data indicate slow net growth in marine capture fisheries through the 1990s. Four of the five most important fishes in fisheries in the late 1990s are pelagic, the exception being the Alaska pollock. This dominance of pelagic over demersal species is reflected in overall fisheries figures, with pelagic landings well over twice demersal landings globally. This contrasts sharply with the situation in the early 1950s when pelagic landings were only some 30 percent greater in volume than demersal landings (Figure 6.2).

This increasing dependency on pelagic fish stocks is symptomatic of a major crisis in global marine fisheries. In general demersal fishes are more valuable per unit weight than pelagic species so that all else being equal the former are preferentially harvested. The increased importance of the latter in the past 40 years is indicative of the growing overexploitation of fisheries stocks worldwide – as valuable demersal stocks have been depleted so attention has turned to the intrinsically less valuable pelagic stocks.

Of some 441 fishery stocks for which status data were available in 1999, FAO considered that only 4 percent were underexploited, and with a further 21 percent assessed as moderately exploited, around 25 percent of stocks analyzed were above the level of abundance thought to correspond to maximum sustainable yield (MSY) level (or have a fishing capacity below this level). The remaining 75 percent of stocks were considered to require strict control of fishing capacity and fishing effort in order for them to recover to MSY biomass[49]. The proportion of stocks in this condition has increased between 1974, when first systematically reviewed, and 1999 (Figure 6.3). In terms of ocean regions, the situation worsened steadily in the North Atlantic and North Pacific until the early

Figure 6.2
Marine fisheries landings by major group

Note: 1984-99 data are for capture fisheries only; pre-1984 data include aquaculture.

Source: FAO[40, 75].

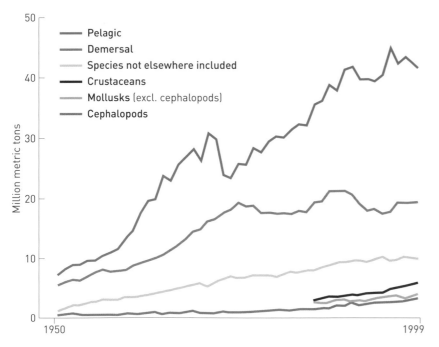

1990s, when there were signs of possible stabilization, mainly in the former. Stocks appear still to be in decline in the tropical and southern parts of these oceans, with some possible stabilization in the tropical Atlantic.

Recent analysis of fishery data suggests that the widespread overexploitation of marine fisheries is probably more serious than the global catch statistics indicate because mis-reporting by countries with particularly large fisheries, coupled with wide fluctuation in Peruvian anchoveta stocks (linked with El Niño events), can produce spurious trends at global level. When more realistic estimates of the catch in China, for example, are substituted for reported catch figures believed to be incorrect, the global catch appears to have declined by 0.36 million metric tons annually since 1988, rather than increased by 0.33 million metric tons (Figure 6.4). The declining trend is much steeper if the pelagic Peruvian anchoveta are excluded[50].

There are three major reasons for the declining state of many marine fisheries. First, and most fundamental, most fisheries have traditionally been regarded as an 'open access' resource, so that, in effect, it pays any one fisher to harvest as much as possible at any given time because, if they do not, somebody else will. Secondly, technological innovations have made fishing much more efficient. Thirdly, there has been high investment in the world's commercial fishing fleet (partly a consequence of the nature of fisheries as an open access resource but also for complex socioeconomic and political reasons).

Bycatches and discards
The effects of overfishing are compounded by the wastefulness of many marine capture fisheries. FAO estimated in 1994 that global marine fisheries bycatch and discards amounted to 18-40 million metric tons (mean 27 million) (Map 6.7). This represented just over 25 percent of the annual estimated total catch (i.e. landings represent around 75 percent of actual catch). Although figures are not available, it is generally assumed that the great majority of discards die. Further losses are caused by the mortality of animals which escape from fishing gear during fishery

operations, but it is impossible at present to estimate the importance of this. Shrimp fishing produces the largest volume of discards (around 9 million metric tons annually).

Bycatches include non-target, often low-value or 'trash' species, as well as undersized fish of target species. Non-target species may include marine mammals, reptiles (sea turtles) and seabirds, as well as finfishes and invertebrates. Of particular concern in recent years has been mortality of marine mammals, especially dolphins, in pelagic drift nets, of sea turtles in shrimp trawls and more recently of diving seabirds, especially albatrosses, in long-line fisheries. Discarding

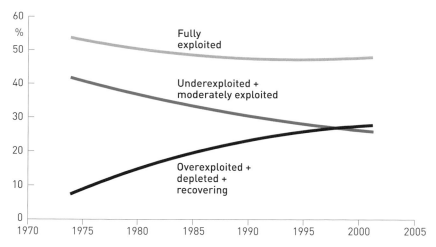

Figure 6.3
Global trends in the state of world stocks since 1974

Source: Adapted from Figure 40 in FAO[49].

Figure 6.4
Trends in global fisheries catch since 1970

Source: Adapted from Watson and Pauly[50].

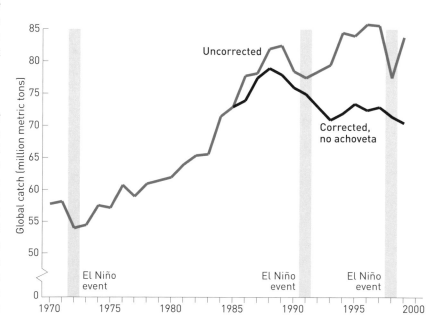

Map 6.7
Marine fisheries catch and discards

The location of the fishery areas recognized by FAO for statistical purposes is shown on this map, with symbols representing the approximate late 1990s yield from capture fisheries and the volume of discarded catch, most of which is presumed not to survive. Each symbol represents approximately 1 million metric tons.

Source: Data from FAO[48].

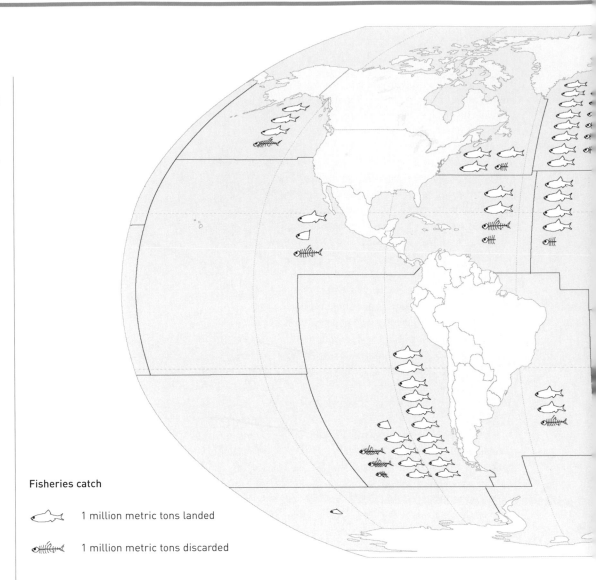

Fisheries catch

1 million metric tons landed

1 million metric tons discarded

may be a side-effect of management systems intended to regulate fisheries (e.g. non-transferable quotas may cause discarding of over-quota catch; species-specific licensing may cause discard of non-licensed but still commercially valuable species).

Solutions to bycatches and discards will be found essentially through improvement in the selectivity of fishing gear and fishing methods. Much of the research in this has been carried out in higher latitudes and is not readily transferable to multispecies tropical fisheries, where the tropical shrimp trawls still produce high rates of bycatch. Improved use of bycatch either as fishmeal or human food is also a possibility; however, this does not address the problem of mortality of potentially threatened species (sea turtles, seabirds, cetaceans), nor the

wasteful capture of immature specimens of harvestable species.

A further problem in the efficient use of marine resources is post-harvest loss. It is almost impossible to estimate this accurately, but FAO believes it to exceed 5 million metric tons per year (i.e. around 5 percent of harvest). Most significant are physical losses of dried fish to insect infestations and loss of fresh fish through spoilage. These problems are particularly significant in developing countries.

AQUACULTURE
One major response to the growing crisis in marine capture fisheries has been the rapid rise in various forms of aquaculture (Figure 6.5). The latter may be defined as the rearing in water of organisms (animals, plants and algae) in a process in which at least one

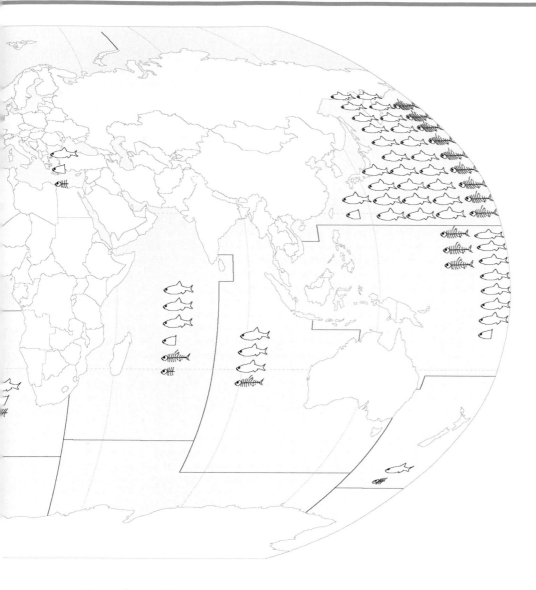

phase of growth is controlled or enhanced by human action. The animals used are generally finfishes, mollusks and crustaceans, although a number of other groups such as sea squirts (Tunicata), sponges (Porifera) and sea turtles are cultured in small quantities. Seaweeds of various kinds are also cultured, some in large amounts. Most of the species grown in any quantity are low in the food chain, being either primary producers, filter-feeders or finfishes that in their adult stages are either herbivores or omnivores.

FAO notes that aquaculture is the world's fastest growing food production sector, annual output having increased at an average rate of some 10 percent in the period 1984-98 (compared with less than 2 percent for capture fisheries) (see Figure 6.5). In 1999 aquaculture provided around one quarter of recorded global fisheries production. Of the total 32.9 million metric tons recorded in 1999, almost 20 million originated inland, and nearly 13 million were produced in marine and brackish environments[49]. In 1996, some 7.7 million metric tons of algae and plants were produced, almost all of this seaweed, chiefly Japanese kelp *Laminaria japonica*, nori *Porphyra tenera* and wakame *Undaria pinnatifida*. The first of these was, in terms of volume, the most important of all aquaculture species, with around 4.4 million metric tons produced.

In marine and brackish (usually estuarine) environments, by far the most important animal group in terms of volume is the mollusks, whose 1997 recorded production of some 8.6 million metric tons made up more than 75 percent of all animal production in

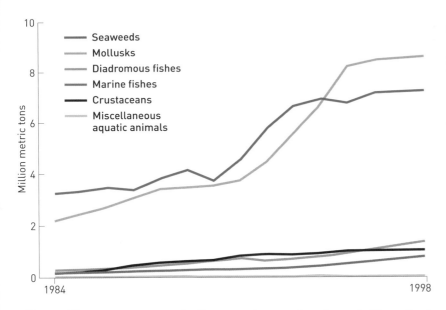

**Figure 6.5
Marine aquaculture
production**

Source: FAO[40, 75].

these environments. Around 50 mollusk species are produced in significant quantity, almost all bivalves. As with most culture systems, production is heavily skewed to a small number of species, with 65 percent of production composed of just three: the Pacific cupped oyster *Crassostrea gigas*, Japanese carpet shell *Ruditapes philippinarus* and Yesso scallop *Pecten yessoensis*. The Far East dominates production, with around 75 percent of that recorded taking place in China and most of the remainder in Japan.

Although production of marine crustaceans accounts for only 10 percent or so by volume of marine and brackish water animal aquaculture, it has disproportionately high economic importance, and is also the sector that has given rise to most environmental concerns. Between 1984 and 1998 annual production grew nearly sixfold, from less than 200 000 to over 1 million metric tons. The great majority of production takes place in tropical and subtropical Asia and is dominated by *Penaeus* species; globally this genus produces over 90 percent of aquaculture crustacean supply by weight. Three species of *Penaeus* account for around three quarters of crustacean production. The giant tiger prawn *P. monodon* is the most widely cultivated and accounts for nearly half; the whiteleg shrimp *P. vannamei* is cultured in the Americas and accounts for around 15 percent of estimated global supply (around 70 percent of this

originating in Ecuador); and the fleshy prawn *P. chinensis* is cultured in China and currently accounts for around 10 percent of production, having declined considerably since the early 1990s when around 200 000 metric tons were produced annually. Other marine crustaceans cultivated include other *Penaeus* species, some *Metapenaeus*, and spiny lobsters *Panulirus*. These groups, however, make an insignificant contribution to global supply.

Growth in crustacean aquaculture has been fuelled by the high value of the product: the market in 1996 was estimated to be worth nearly US$7.5 billion, or around one quarter of the total value of marine and brackish water aquaculture[48]. The great majority of production takes place in low-income countries – the five countries producing over 100 000 metric tons annually being China, Thailand, Indonesia, Ecuador and Bangladesh – and is aimed mainly at the export market (primarily to Europe, the United States and Japan) and to a lesser extent at the domestic luxury market. Pressure is high to produce maximum returns on investment so that increasingly intensive farming methods are used. These are widely acknowledged to be having adverse social and environmental impacts in the countries of production, as well as leading to increasing difficulties in maintaining supply, owing to the spread of major diseases. The last of these accounts for the major decline in the Chinese fleshy prawn industry during the 1990s. Impacts include:

- loss of mangrove habitat;
- abstraction of freshwater;
- introduction of pathogens and other damaging non-native species;
- escape of cultured non-native species;
- pollution;
- diversion of low-quality or cheap fish food resources (may lead to more efficient use of bycatches and trash fish but also to more indiscriminate catch fisheries);
- diversion of effort from other forms of aquaculture (notably milkfish *Chanos chanos*).

The aquarium trade
Up to 2 million people worldwide (about half in the United States and a quarter in Europe) are

thought to keep marine aquariums, most of which are stocked with wild-caught species. In 1997 a total of 1 200 metric tons of coral was traded internationally, with 56 percent imported by the United States and 15 percent by the European Union. Approximately half of this was live coral for aquariums, a tenfold increase on the amount of live coral traded in the late 1980s[51]. Qualitative estimates of trade suggest that 14-30 million fish may be traded per year, representing some 1 200 species, about two

thirds of which are from coral reefs. Aquarium species are typically gathered by local fishers using live capture techniques or chemical stupefactants (such as sodium cyanide) which are non-selective and adversely affect the health of specimens as well as killing non-target organisms. Inappropriate shipping methods and poor husbandry along the supply chain often cause high mortality among the fish and invertebrates collected.

While the current impacts of the aquarium trade remain poorly known, the industry has considerable potential to contribute to sustainable development. It is relatively low in volume but very high in value – a kilo of aquarium fish from one island country was valued at almost US$500 in 2000, whereas reef fish harvested for food were worth only US$6[52]. Aquarium species are a high-value source of income in many coastal communities with limited resources, with the actual value to the fishers determined largely by market access. In Fiji many collectors pay

an access fee to the villages to collect on their reefs, but by selling directly to exporters they can have incomes many times the national average. By contrast, in the Philippines there are many middlemen, and collectors themselves typically earn only around US$50 per month. Targeting mostly non-food species, aquarium fisheries could in principle provide an alternative economic activity for low-income coastal populations and an important source of foreign exchange for national economies, as well as an economic incentive for the sustainable management of reefs. The application of international certification schemes may provide an important tool for achieving this.

OTHER MAJOR IMPACTS ON THE MARINE BIOSPHERE
Alteration of physical habitat
Physical alteration of habitats through human action chiefly affects coastal and inshore areas. Impacts here can be severe, although few attempts have been made to quantify them on a global basis. Major causes include coastal development, particularly landfilling and construction of groynes and jetties, aquaculture, dredging of channels for navigational purposes, extraction of materials such as sand and coral, stabilization of shorelines, and destructive fishing methods such as beam-trawling, use of explosives and *muro-ami* (using rocks on ropes to drive fishes into nets). Upstream activities, such as deforestation and dam construction can greatly alter sediment loads in rivers, affecting patterns of sediment deposition in estuarine areas.

Chemical pollution and eutrophication
Human activities have increased inputs of a huge range of organic and inorganic chemicals into marine ecosystems. Such inputs may enter by direct discharge (e.g. in sewage outflow pipes), via river and stream outflow, as land runoff, through the atmosphere or from seagoing vessels. Because virtually all such input originates on land, as with most other human impacts on marine ecosystems, areas most affected are coastal and inshore regions, particularly enclosed or semi-enclosed water bodies. In oceanic regions,

The aquarium trade targets mostly non-food species and many invertebrates such as crabs, anemones and shrimp.

mixing of the enormous volume of sea water generally ensures that inputs become rapidly diluted.

Major categories of input include nutrients of various kinds (e.g. nitrates and nitrites, phosphates, dissolved organic matter), persistent organic pollutants (POPs), including a range of chlorinated hydrocarbons, and heavy metals such as cadmium (Cd), copper (Cu), mercury (Hg), lead (Pb), nickel (Ni) and zinc (Zn). Quantifying these inputs and assessing their impact is problematic, particularly because many occur naturally in sea water.

Many POPs and heavy metals can act as toxins above certain concentrations, inducing mortality or morbidity or impairing reproductive success, particularly in cases where they become increasingly concentrated towards the top of food chains. Their overall impact on marine ecosystems remains uncertain. More easily observable is the impact

Worldwide, human activities have increased inputs of nitrogen and phosphorus in rivers and coastal waters fourfold.

of eutrophication resulting from the increased input of organic and inorganic nutrients (particularly nitrogen and phosphorus) into coastal waters, mainly through fertilizer runoff and sewage disposal. It is believed that human intervention has increased river inputs of nitrogen and phosphorus worldwide into coastal areas by more than fourfold over background levels. These inputs lead to increases in productivity in coastal waters, often in the form of algal blooms. These blooms may themselves be noxious; they also typically cause the euphotic zone to reduce in vertical extent and are implicated in the

development of hypoxic (low dissolved oxygen concentration) and anoxic (zero dissolved oxygen) zones. A shallowing of the euphotic zone may cause die-off of photosynthesizing benthic algae in shallow-water areas. This has occurred, for example, in the Black Sea where the euphotic zone had decreased from 50-60 m vertical extent in the early 1960s to around 35 m by 1990, leading to a decrease of up to 95 percent in living biomass of benthic macrophytic algae such as Phyllophora, formerly an important harvested resource.

Hypoxia and anoxia result from the activities of oxygen-respiring bacteria below the euphotic zone feeding on accumulated dead algae and other organisms and waste matter raining down from above. Hypoxia results in the emigration of mobile aerobic species and mortality of sedentary ones. This may have catastrophic impact on local fisheries. Most hypoxic zones vary in extent through the year and from year to year and some are only seasonal, disappearing when winter mixing causes re-oxygenation of bottom waters. They may be very extensive – the hypoxic zone to the west of the Mississippi delta covered some 16 000 km^2 in 1997, having covered some 9 000 km^2 in 1989. Over 50 such zones have been identified worldwide to date; some appear to be at least in part induced by natural phenomena while others are believed entirely anthropogenic.

Invasions of exotic species

As on land, the breakdown of biogeographic barriers in the sea appears to be having a major, and increasing, impact on marine ecosystems. The chief source is the deliberate or accidental translocation of organisms, but the construction of marine corridors between previously isolated areas has had a major impact on geographically restricted areas. These factors have resulted in a rapid spread of alien species in all the world's oceans.

Translocation can result from deliberate introduction of harvestable species, accidental escapes from aquaculture and aquarium operations, transport in ballast water of ships and release of fouling organisms that adhere to the hulls of ships and boats. The extent of introductions of this kind has yet to be fully assessed but is certainly large. In many cases (see Box 6.2) the introduced species appear to be having a major impact on native biota, although in general it is difficult to separate the effects of a particular species from general ecosystem deterioration.

Deliberate introductions include the planting of mangroves and *Nipa* palms along coastlines, and the cultivation of fish, crustaceans, mollusks and algae in many coastal regions. Atlantic salmon that have escaped from aquaculture are reportedly affecting wild stocks in the northeast Pacific, and their pathogens have themselves moved into the wild populations of closely related species. Ballast water is commonly pumped into the hold of ships as a means of controlling balance and position in the water, and is liable to be flushed out far distant from where it was taken on. It has been estimated that on any one day the ballast water of the world's ocean fleets contains around 10 000 different species[53].

The principal artificial corridors are the Suez Canal (opened in 1869) and the Panama Canal (opened in 1914). The 165-km long Suez Canal is a continuous seawater channel with a water level at the Red Sea end some 1.2 m higher than at the Mediterranean end, leading to a constant northward flow of water. It is estimated that to date some 400-500 marine species have migrated through the canal (so-called Lessepsian migrants[54], after Ferdinand de Lesseps who planned the canal) and established themselves in the Mediterranean, while a far smaller number have moved in the other direction. New species are believed to arrive in the Mediterranean at the rate of four to five annually. Because the Panama Canal has a separate freshwater section, some 25 m above sea level, migration through it has been limited to date.

Global climate change

Human-induced climate change is liable to impact directly on marine and coastal areas by warming (particularly of the surface layers), by sea-level rise (associated both with thermal expansion and the melting of terrestrial ice caps and glaciers), and through change in the gases dissolved in surface waters. These impacts are well understood,

and measurable changes are already apparent. A more complex array of secondary changes may also occur, including changes to ocean stratification and surface mixing, changes to patterns of surface current, and perhaps to global systems such as the El Niño Southern Oscillation.

Tropical coral reefs appear particularly sensitive to temperature change. Reef-building corals are adapted to stable thermal conditions and in most areas appear to be growing close to their upper temperature limit. Temperatures little more than 1°C above the normal maximum for a period of a few weeks are sufficient to drive a stress response known as coral bleaching. During a particularly strong El Niño event in 1998, warmer waters around the Seychelles and the Maldives induced a bleaching event in which 60-90 percent of all corals in the area died, equivalent to 5 percent of the world coral reef area. Although this event was linked to an extreme climatic perturbation, it occurred at a time of rising global temperatures and provides an indication of the impact of potential future climate change[58]. More subtle changes associated with the gradually changing background conditions, particularly temperatures, have been recorded in the

Box 6.2 Marine introductions

Leidy's comb jelly *Menopsis leidyi* was introduced from the American Atlantic into the waters of the Black Sea in 1982, presumably in ballast water. Unchallenged by natural predators, this species proliferated to a 1988 peak estimated at around 1 000 million metric tons wet weight (about 95 percent of the entire wet weight biomass in the Black Sea). The species depleted the natural zooplankton stocks, with subsequent algal blooms and decline of the fishing industry in the Black Sea[55].

The Asian clam *Potamocorbula amurensis* spread through the northern San Francisco estuary (United States) following its introduction, possibly in ballast water, in 1986. The species reaches high density, up to 2 000 individuals per m^2, and has caused sharp declines in the abundance and extent of several plankton species; its impact on fisheries is not yet clear. With more than 200 introduced species, this bay may be the most invaded aquatic habitat in North America.

The green alga *Caulerpa taxifolia* is thought to have escaped from an aquarium in the western Mediterranean and is spreading rapidly in the coastal waters of Spain, France and Italy, with severe impact on the native seagrass beds and on coastal fisheries[56]. It has recently been reported in the coastal waters of California[57].

distribution of pelagic seabirds along the Californian coast[59]; the faunal composition of intertidal communities[60]; and penguin distribution in the Antarctic[61].

The mean sea level has risen 18 cm during the past 100 years and further increases could have a severe impact on coastal communities. Rising sea levels will lead to the inundation of some coastal lands, whilst in many other areas they will alter patterns of coastal erosion, and they may increase groundwater salinization. While many intertidal habitats are highly adaptable, the growing human presence in most coastal areas will prevent the natural migration of these habitats, leading to overall losses of saltmarsh, mangrove, or even beach and rocky shore habitats.

The biomass of the world's oceans is very low compared with terrestrial environments, but because of the rapid turnover in oceanic carbon cycles, marine phytoplankton (cyanobacteria and algae) play an important role in removing dissolved carbon dioxide from solution, and are intermediates in the transport of organic carbon to the deep ocean. Once in the deep ocean, this carbon is effectively removed from exchange with the atmosphere for millennia. In this way, marine photosynthesizers help to buffer the rising concentration of carbon dioxide in the atmosphere, but this service, and marine primary productivity, may be affected if ocean temperatures rise (warmer waters hold less carbon dioxide in solution) and if the broad patterns of ocean circulation change significantly[62]. It is thought that increasing atmospheric temperatures may affect the generation of cold, oxygen-rich bottom waters beneath Arctic and Antarctic ice sheets, with major implications for deep-sea biota and for global patterns of seawater circulation, in particular the Great Conveyor, driven by bottom water generated in the North Atlantic.

THE CURRENT STATUS OF MARINE BIODIVERSITY

Because they are usually much less readily observed, marine species are in general much more difficult to monitor and assess than terrestrial ones. Assessment is based on sampling and, in the case of harvested species, often on the basis of catch rates, although the latter may vary in response to a wide range of factors in addition to changes in the population of the species concerned. An exception lies with those groups such as pinnipeds, sea turtles and seabirds that nest or breed on land. Because many of these tend to be colonial species and because they tend to breed in open habitats (beaches, cliff tops, ice sheets), they may be easier to monitor than many other species, either terrestrial or aquatic. In the case of large, commercially valuable fish stocks, monitoring at large scale has in some cases been carried out for many years, so that estimates of the stock level are obtainable.

Threatened and extinct species

The only major marine species groups (classes or above) that have been comprehensively assessed in terms of threatened species status to date are mammals and birds. In addition, sea turtles and a number of fish families and genera (e.g. the sturgeons in the order Acipenseriformes and the seahorses Hippocampus in the order Syngnathiformes) have also been assessed. Other threatened marine species have been identified on more of an ad hoc basis. Data are summarized in Table 6.12.

Relatively speaking, far fewer marine species are known to have become extinct since 1600 than either terrestrial or fresh-

Much less readily observed, marine species are in general more difficult to monitor and assess than terrestrial ones.

water ones. Cataloged extinctions comprise two marine mammals (the Caribbean monk seal *Monachus tropicalis* and Steller's sea cow *Hydrodamalis gigas*) and five seabirds (three island petrels, Pallas's cormorant *Phalacrocorax perspicillatus* and the great auk *Alca impennis*). In addition five coastal or island duck species have disappeared at various times from the late 17th century onwards; however there is in most cases insufficient information to determine whether these species were predominantly marine or terrestrial.

As a gross generalization, marine species appear to be somewhat less extinction prone as a result of mankind's activities than freshwater or terrestrial ones. There are arguably two main reasons for this. First, because of the size of the world ocean and the fact that people do not actually live in it, the marine biosphere remains as a whole considerably more buffered from human intervention than terrestrial and inland water areas. Second, marine species on the whole appear to be more widespread than terrestrial or inland water ones. In the open ocean, there are vast areas with apparently similar habitat conditions and there are few barriers to dispersal so that many species have circumglobal distributions. In addition, many forms that as adults are sessile (e.g. sponges and corals) or sedentary (many mollusks and crustaceans) have planktonic larvae that are often widely dispersed in water currents. For this reason, many coral reef species, for example, are found in suitable habitat throughout the Indo-Pacific region. In addition, many of the most heavily exploited fish species have high fecundity (in the case of some tunas amounting to several million eggs in a single spawning), so that they have at least potentially high population growth rates unparalleled in terrestrial vertebrates.

There are of course significant exceptions to all these. Coastal regions in many parts of the world, and enclosed or semi-enclosed marine areas such as the Baltic, Black and Yellow Seas, are often under intense pressure from a range of human activities. A number of marine species do appear to have restricted ranges (e.g. the Hawaiian coral reefs are relatively rich

in species found nowhere else while many southern hemisphere seabirds are apparently confined to a small number of breeding sites) and significant numbers have low or very low reproductive rates (many chondrichthyine fishes, marine mammals and seabirds).

Marine species which are easily exploitable or have high economic value may suffer catastrophic declines if exploitation is not strictly controlled.

Until recently by far the most important human activity affecting marine species was uncontrolled exploitation. Where species are either easily exploitable or are highly sought-after (i.e. have high unit value), or both, they may suffer catastrophic declines. This is the case with sea turtles and a number of marine mammals and birds that are or have been harvested principally at their terrestrial breeding sites (which are often colonial), as well as with the great whales and the dugong, which although strictly marine are air-breathing and therefore spend some time at the sea surface (when they may be spotted). Most of these species have relatively low reproductive rates, so that even if they are ultimately afforded protection population recovery rates may be very slow.

A recent synthesis of a wide range of information, including paleoecological and archeological data relating to early human communities, suggests that overfishing of larger vertebrates and mollusks is characteristic of indigenous and colonial human use of coastal ecosystems, and the first of what is typically a series of impacts[63]. Massive losses in biomass and abundance appear to have occurred, on a scale largely unsuspected, and seemingly amounting to the loss of entire trophic levels of consumer organisms, with radical consequences for ecosystem status. Overfishing is likely to be followed by impacts of pollution, mechanical habitat loss, introduced species and climate change. Loss of filter-feeding organisms that maintain water quality is liable to be followed by eutrophication, hypoxia and disease, as exemplified by conditions in Chesapeake Bay following the collapse of the oyster fishery in the early 20th century. Synergistic interactions of this kind are making the effective, long-term management of marine resources one of the major – and most intractable – problems currently facing humankind.

Land-breeding species may also be susceptible to other threats, such as predation,

Table 6.12
Taxonomic distribution and status of threatened marine animals

Note: Only the birds and mammals have been comprehensively assessed for species at risk; numbers refer to species (units such as subspecies and geographic populations appear in the Red List database but these are not tabulated here).

Source: Status categories from 2000 Red List database, www.redlist.org[71] (accessed February 2002).

Phylum and class or order	Common name	'Critically endangered'	'Endangered'	'Vulnerable'
Cnidaria				
Anthozoa	Stony corals			2
Mollusca				
Bivalvia	Bivalves			4
Gastropoda	Gastropods	1	2	6
Craniata – fishes				
Carcharhiniformes	Ground sharks	1	3	5
Hexanchiformes	Cow sharks			1
Lamniformes	Mackerel sharks			3
Orectolobiformes	Carpet sharks			2
Pristiformes	Sawfishes	2	5	
Rajiformes	Rays		2	1
Rhinobatiformes	Guitar fishes	1		1
Squaliformes	Dogfishes and sleeper sharks			1
Squatiniformes	Angel sharks		1	2
Coelacanthiformes	Coelacanths	1		
Acipenseriformes	Sturgeons	2	10	6
Batrachoidiformes	Toadfishes			5
Clupeiformes	Herrings and anchovies		1	1
Gadiformes	Cods, hakes, rattails	1		2
Gasterosteiformes	Sticklebacks			1
Lophiiformes	Anglerfishes, etc.	1		
Ophidiiformes	Pearlfishes, cusk-eels, brotulas			1
Perciformes	Perches, etc.	7	5	33
Pleuronectiformes	Plaice, flounders, soles		1	1
Salmoniformes	Salmonids	1	4	6
Scorpaeniformes	Gurnards, scorpionfishes, etc.	1	2	1
Siluriformes	Catfishes		1	
Syngnathiformes	Pipefishes, seahorses, etc.	1	1	35
Tetraodontiformes	Triggerfishes, etc.			3

coastal development and pollution. It is noteworthy in this context that the family Procellariidae contains nearly three times as many threatened species (36 out of 115 species, or 28 percent) as the average bird family, in which 11 percent of species are threatened, and nearly six times as many critically endangered species as would be expected at random. It is almost certainly the tendency of these birds to nest on islands, whose biotas have in general suffered enormously more from mankind's influence in the past few centuries (see Chapter 4), rather than their seagoing habits that has led to this.

For truly marine species (chiefly finfishes and invertebrate animals) the situation appears somewhat different. Even when these have been exploited to the point of stock collapse, as has occurred for example with the cod *Gadus morhua* stocks off Newfoundland in the North Atlantic, the species concerned do not appear to have become imminently threatened with biological extinction. This is in part because once stocks are reduced below a certain level it is often no longer economically

Phylum and class or order	Family	Common name	'Critically endangered'	'Endangered'	'Vulnerable'
Craniata – Reptilia					
Squamata	Iguanidae	Iguanas			1
Chelonia	Dermochelyidae	Leathery turtle	1		
	Cheloniidae	Sea turtles	2	3	1
Craniata – Aves					
Anseriformes	Anatidae	Ducks	1	1	1
Charadriiformes	Alcidae	Auks, puffins			4
	Charadriidae	Plovers	1	2	1
	Laridae	Gulls, terns, skuas, auks, skimmers	1	1	5
Ciconiiformes	Ardeidae	Egrets, herons			1
Pelecaniformes	Fregatidae	Frigatebirds	1		1
	Pelecanidae	Pelicans			1
	Phalacrocoracidae	Cormorants and shags			8
	Sulidae	Gannets and boobies	1		1
Procellariiformes	Diomedeidae	Albatross	2	2	12
	Hydrobatidae	Strom petrels	1		1
	Pelecanoididae	Diving petrel		1	
	Procellariidae	Petrels, shearwaters	10	6	20
Sphenisciformes	Spheniscidae	Penguins		3	7
Craniata – Mammalia					
Carnivora	Mustelidae	Otters, etc.		2	
	Otariidae	Eared seals		1	5
Cetacea	Phocidae	Earless seals	1	1	1
	Balaenidae	Right whales		1	
	Balaenopteridae	Rorquals		3	1
	Delphinidae	Dolphins		1	1
	Monodontidae	Beluga			1
	Phocoenidae	Porpoises	1		1
	Physeteridae	Sperm whales			1
Sirenia	Dugongidae	Dugong			1
	Trichechidae	Manatees			3

viable to continue harvesting them. Generally, the residual population at this stage is still large enough to allow recovery if harvesting ceases, particularly in the case of species with high fecundity and therefore high potential intrinsic rates of increase. Exceptions to this are species that have low fecundity, particularly if they also have a long period to maturity, with limited ranges, and which may either have high unit value or be caught as bycatches.

In the case of bycatches, because the fishery is not directed at the species concerned, its intensity will not decrease as population levels decrease so that it may theoretically be possible at least locally to extirpate species, particularly if they are habitually caught before they reach maturity. Examples include several sawfish species

whale populations that have not apparently recovered as rapidly as projected following the cessation of their harvest.

The marine living planet index

An impression of the overall trend in a large sample of species for which population indicators are available can be derived from the WWF living planet index[64]. This approach is designed to represent the change in the 'average species' in the sample from one five-year interval to the next, starting in 1970. The marine sample represents 217 aquatic and coastal species of mammals, birds, reptiles and fishes, and the overall trend is for a significant decline in population levels over the last three decades of the 20th century (Figure 6.6). The sample is dominated by the stocks and species that humans have an interest in monitoring, most of the fishes among these being of commercial importance as a fisheries resource. These should also be stocks that humans have an interest in managing as well as possible. That the index has declined in every five-year interval since 1970 is evidence that such management is failing, as confirmed by the picture painted above of global marine capture fisheries.

Assessing the status of marine and coastal ecosystems

Threatened species inventories and the marine living planet index can give a very general overall impression of the status of marine biodiversity. Assessing marine ecosystem 'health' is much more problematic. However, snapshots can be obtained from examining particular ecosystems, such as mangroves and coral reefs[26]. In the former, an overall assessment can be made on the basis of the area destroyed or severely degraded. In the latter areal measures are more problematic, in part because reef extent is much more difficult to measure than mangrove extent and, of greater importance, because the vast majority of a reef is composed of non-living calcareous deposits. Measures of the change in extent of these give little insight into the state of the living component of the reef. For this reason other measures, such as estimates of incidence of coral disease, may be feasible.

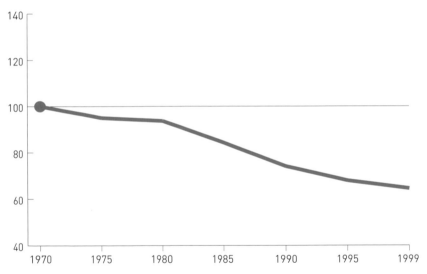

Figure 6.6
Marine population trends

Note: A simplified representation of the average population change in a sample of 217 marine species, see text.

Source: Loh[64].

(family Pristidae). These are large, slow-growing, predominantly inshore species that give birth to relatively small numbers of live young. Population densities appear to be naturally low and animals are widely caught as bycatch in inshore fisheries before they are large enough to reproduce. As a result five species are classified as 'endangered' and two as 'critically endangered'. In addition, it is possible that trophic shifts may occur when populations of some species are severely reduced, inhibiting recovery of these populations when exploitation ceases. This has been suggested in the case of some great

REFERENCES

1   Longhurst, A.R. and Pauly, D. 1987. *Ecology of tropical oceans.* ICLARM (International Center for Living Aquatic Resources Management) Contribution No. 389. Academic Press Inc., California.

2   Longhurst, A.R. 1995. Seasonal cycles of pelagic production and consumption. *Progress in Oceanography* 36: 77-167.

3   Longhurst, A.R. 1998. *Ecological geography of the sea.* Academic Press, San Diego.

4   Sherman, K. and Busch, D.A. 1995. Assessment and monitoring of large marine ecosystems. In: Rapport, D.J., Guadet, C.L. and Calow, P. (eds). *Evaluating and monitoring the health of large-scale ecosystems.* Springer-Verlag, Berlin, published in cooperation with NATO Scientific Affairs Division. NATO Advanced Science Institutes Series I: *Global Environmental Change*, 28: 385-430.

5   Available online at http://www.edc.uri.edu/lme/default.htm (accessed February 2002).

6   Olson, D.M. and Dinerstein, E. 1998. The Global 200: A representation approach to conserving the earth's distinctive ecoregions. *Conservation Biology* 12: 502-515.

7   Vaulot, D. 1995. Marine biodiversity at the micron scale. *International Marine Science Newsletter* 75/76. United Nations Educational, Scientific and Cultural Organization, Paris.

8   Rex, M.A., Stuart, C.T. and Coyne, G. 2000. Latitudinal gradients of species richness in the deep-sea benthos of the North Atlantic. *Proceedings of the National Academy of Sciences* 97: 4082-4085.

9   Adey, W.H. and Steneck, R.S. 2001. Thermogeography over time creates biogeographic regions: A temperature/space/time-integrated model and an abundance-weighted test for benthic marine algae. *Journal of Phycology* 37: 677-698.

10 Roberts, C.M. *et al.* 2002. Marine biodiversity hotspots and conservation priorities for tropical reefs. *Science* 295: 1280-1284.

11 Nybakken, J. 1993. *Marine biology: An ecological approach.* 3rd edition. HarperCollins College Publishers, New York.

12 World Conservation Monitoring Centre 1992. Groombridge, B. (ed.) *Global biodiversity: Status of the Earth's living resources.* Chapman and Hall, London.

13 Finlayson, M. and Moser, M. (eds) 1991. *Wetlands.* Facts on File Ltd, Oxford.

14 Woodroffe, C.D. and Grindrod, J. 1991. Mangrove biogeography: The role of Quarternary environmental sea-level change. *Journal of Biogeography* 18: 479-492.

15 Tomlinson, P.B. 1986. *The botany of mangroves.* Cambridge University Press, Cambridge.

16 Spalding, M.D., Blasco, F. and Field, C.D. (eds) 1997. *World mangrove atlas.* International Society for Mangrove Ecosystems, Okinawa.

17 Lasserre, P. 1995. Coastal and marine biodiversity. *International Marine Science Newsletter* 75/76: 13-14.

18 Risser, P.G. 1986. Spatial and temporal variability of biospheric and geospheric processes: Research needed to determine interactions with global environmental change. Report of a workshop. International Council of Scientific Unions Press, Paris.

19 Ducrotoy, J.P., Elliott, M. and de Jonge, V.N. 2000. The North Sea. In: Sheppard, C. (ed.) *Seas at the millennium: An environmental evaluation*, Vol. I, pp. 43-64. Elsevier Science, Amsterdam.

20 Lewis, J.R. 1977. Rocky foreshores. In: Barnes, R.S.K. (ed.) *The coastline*, pp. 147-158. Wiley, London.

21 Wilson, S.C. 2000. Northwest Arabian Sea and Gulf of Oman. In: Sheppard, C. (ed.) *Seas at the millennium: An environmental evaluation*, Vol. II, pp. 17-34. Elsevier Science, Amsterdam.

22 Phillips, R.C. and Meñez, E.G. 1988. Seagrasses. *Smithsonian Contributions to the Marine Sciences* 34: 104.

23 Hemminga, M.A. and Duarte, C.M. 2000. *Seagrass ecology.* Cambridge University Press, Cambridge.

24 Short, F.T. and Wyllie-Echeverria, S. 2000. Global seagrass declines and effects of climate change. In: Sheppard, C. (ed.) *Seas at the millennium: An environmental evaluation, Vol. III: Global issues and processes.* Elsevier Science, Amsterdam.

25 Wilkinson, C.R. and Buddemeier, R.W. 1994. Global climate change and coral reefs: Implications for people and reefs. Report of the UNEP-IOC-ASPEI-IUCN Global Task Team on the Implications of Climate Change on Coral Reefs.

26 Hulm, P. and Pernetta, J.C. 1993. *Reefs at risk: Coral reefs, human use and global climate change. A programme of action.* IUCN-the World Conservation Union, Gland.

27 Veron, J. 2000. *Corals of the world.* 3 vols. Australian Institute of Marine Science, Queensland, Australia.

28 Cairns, S.D., Hoeksema, S.B. and van der Land, J. 1998. List of extant stony corals. Available online at http://www.nodc.noaa.gov/col/projects/coral/hardcoral/Hardcoralmain.html (accessed February 2002).

29 Mortensen, P.B. *et al.* 1995. Deep water bioherms of the scleractinian coral *Lophelia pertusa* at 64°N on the Norwegian shelf: Structure and associated megafauna. *Sarsia* 80: 145-158.

30 Rogers, A.D. 1999. The biology of *Lophelia pertusa* and other deep-water reef-forming corals and impacts from human activities. *International Review of Hydrobiology* 84(4): 315-406.

31 Fossa, J.H. *et al.* 1999. Effects of bottom trawling on *Lophelia* deep water corals in Norway. Poster presented to the ICES (International Council for Exploration of the Seas) workshop on the ecosystem effects of fishing, February 1999.

32 Koslow, J.A. and Gowlett-Holmes, K. 1998. The seamount fauna of southern Australia: Benthic communities, their conservation and impacts of trawling. Report to Environment Australia and the Fisheries Research Development Corporation, FRDC Project 95/058.

33 Gjøsaeter, J. and Kawaguchi, K. 1980. *A review of the world resources of mesopelagic fish.* FAO Fisheries Technical Paper 193: 1-151.

34 Rowe, G.T. 1983. Biomass and production of the deep-sea macrobenthos. In: Rowe, G.T. (ed.) *Deep-sea biology, Vol. 8 The sea*, pp. 453-472. John Wiley and Sons Inc., New York.

35 Hessler, R.R. and Sanders, H.L. 1967. Faunal diversity in the deep-sea. *Deep-Sea Research* 14: 65-78.

36 Angel, M.V. 1991. Biodiversity in the deep ocean. A working document for the UK Overseas Development Administration. Unpublished ms.

37 Grassle, J.F. 1991. Deep-sea benthic biodiversity. *Bioscience* 41(7).

38 Grassle, J.F. and Maciolek, N.J. 1992. Deep-sea species richness: Regional and local estimates from quantitative bottom samples. *American Naturalist* 139: 313-341.

39 May, R.M. 1992. Bottoms up for the oceans. *Nature* 357: 278-279.

40 Angel, M.V. 1982. *Ocean trench conservation.* Commission on Ecology Papers No. 1. IUCN-the World Conservation Union, Gland.

41 Gage, J.G. and Tyler, P.A. 1991. *Deep-sea biology: A natural history of organisms at the deep-sea floor.* Cambridge University Press, Cambridge.

42 Grassle, J.F. 1986. The ecology of deep-sea hydrothermal vent communities. *Advances in Marine Biology* Vol. 23. Academic Press.

43 Hecker, B. 1985. Fauna from a cold sulphur-seep in the Gulf of Mexico: Comparison with hydrothermal vent communities and evolutionary implications. *Bulletin of the Biological Society of Washington* 6: 465-473.

44 Grassle, J.F. 1989. Species diversity in deep-sea communities. *TREE (Trends in Ecology and Evolution)* 4(1).

45 UNCED 1992. *The global partnership for environment and development. A guide to Agenda* 21. United Nations Conference on Environment and Development, Geneva.

46 Committee on Biological Diversity in Marine Systems 1995. *Understanding marine biodiversity: A research agenda for the nation.* National Academy Press, Washington DC.

47 FAO 1990. *Review of the state of world fishery resources.* Marine Resources Service, Food and Agriculture Organization of the United Nations, Rome.

48 FAO 1999. *The state of world fisheries and aquaculture 1998.* Food and Agriculture Organization of the United Nations, Rome.

49 FAO 2000. *The state of world fisheries and aquaculture 2000.* Fisheries Department, Food and Agriculture Organization of the United Nations, Rome. Available online at http://www.fao.org/DOCREP/003/X8002E/X8002E00.htm (accessed April 2002).

50 Watson, R. and Pauly, D. 2001. Systematic distortions in world fisheries catch trends. *Nature* 414: 534-536.

51 Green, E. and Shirley, F. 1999. *The global trade in coral.* World Conservation Press, Cambridge.

52 Edwards, A.J. and Shepherd, A.D. 1992. Environmental implications of aquarium fish collection in the Maldives, with proposals for regulation. *Environmental Conservation* 19: 61-72.

53 Carlton, J.T. 1999. Invasions in the sea: Six centuries of re-organising the Earth's marine life. In: Sandlund, O.T., Schei, P.J. and Viken, A. (eds). *Invasive species and biodiversity management,* pp. 195-212. Kluwer Academic Publishers, Netherlands.

54 Goldschmid, A. 1999. Essay about the phenomenon of Lessepsian migration. Colloquial Meeting of Marine Biology I, Salzburg 1999. Available online at http://www.sbg.ac.at/ipk/avstudio/pierofun/lm/Lesseps.htm (accessed April 2002).

55 GESAMP 1997. Opportunistic settlers and the problem of the ctenophore *Mnemiopsis leidyi* invasion in the Black Sea. IMO/FAO/UNESCO-IOC/WMO/WHO/IAEA/UN/UNEP Joint Group of Experts on the Scientific Aspects of Marine Environmental Protection. *Rep. Stud. GESAMP,* 58: 84.

56 Anon 1997. *Medwaves* 34: 11-13. News Bulletin of the Mediterranean Action Plan, Athens.

57 Dalton, R. 2000. Researchers criticize response to killer algae. *Nature* 406: 447.

58 Spalding, M., Teleki, K. and Spencer, T. 2001. Climate change and coral bleaching. In: Green, E. *et al.* (eds). *Impacts of climate change on wildlife,* pp. 40-41.

59 Veit, R.R. *et al.* 1997. Apex marine predator declines 90% in association with changing oceanic climate. *Global Change Biology* 3: 23-28.

60 Sagarin, R.D. *et al.* 1999. Climate-related changes in an intertidal community over short and long time scales. *Ecological Monographs* 69: 465-490.

61 Ainley, D., Wilson, P. and Fraser, W.R. 2001. Effects of climate change on Antarctic sea ice and penguins. In: Green, E. *et al.* (eds). *Impacts of climate change on wildlife,* pp. 26-27.

62 Baliño, B.M., Fasham, M.J.R. and Bowles, M.C. 2001. *Ocean biogeochemistry and global change: JGOFS research highlights 1988-2000.* International Geosphere-Biosphere Programme Science No. 2. International Council of Scientific Unions, Stockholm.

63 Jackson, J.B. *et al.* 2001. Historical overfishing and the recent collapse of coastal ecosystems. *Science* 293: 629-638.

64 Loh, J. (ed.) 2000. *Living planet report 2000.* WWF – World Wide Fund for Nature, Gland.

65 Couper, A.D. 1993. *Times atlas of the oceans.* Times Books.

66 Nelson, J.S. 1994. *Fishes of the world.* John Wiley and Sons Inc., New York.

67 Croxall, J.P., Evans, P.G.H. and Schreiber, R.W. 1994. *Status and conservation of the world's seabirds.* ICBP Technical Publication No. 2. ICBP (now BirdLife International), Cambridge.

68 Sibley, C.G. and Monroe, Jr, B.L. 1990. *Distribution and taxonomy of birds of the world.* Yale University Press, New Haven and London.

69 Duke, N.C. 1992. Mangrove floristics and biogeography. In: Robertson, A.I. and Alongi, D.M. (eds). *Tropical mangrove ecosystems. Coastal and estuarine Studies* 41, p. 329. American Geophysical Union, Washington DC.

70 Spalding, M.D., Ravilious, C. and Green, E. 2001. *World atlas of coral reefs*. University of California Press, Berkeley.

71 IUCN 2000. 2000 IUCN Red List database available online at www.redlist.org (accessed February 2002).

72 Compagno, L.J.V. 1984. *Sharks of the world*. An annotated and illustrated catalogue of shark species known to date. FAO Fisheries Synopsis No. 125 Vol. 4 (Parts 1-2).

73 Spalding, M.D. 1998. Biodiversity patterns in coral reefs and mangrove forests: Global and local scales. Unpublished PhD thesis, University of Cambridge.

74 Veron, J.E.N. 1995. *Corals in space and time: The biogeography and evolution of the Scleractinia*. University of New South Wales Press.

75 FAO 2002. http://www.fao.org/DOCREP/003/X8002E/x8002e04.htm#P1_6 (accessed April 2002).

7 Inland water biodiversity

INLAND WATERS MAKE UP A MINUTE PROPORTION (much less than a hundredth of 1 percent) of the world's water resource. Despite this, they encompass a wide range of habitat types and contain a disproportionately high fraction of the world's biodiversity. Freshwater is also a vital resource for human survival. In consequence, inland water ecosystems are placed under many, often conflicting, pressures, with increasingly adverse consequences for their biodiversity. There are indications that, overall, a higher proportion of inland water species are in decline than marine or terrestrial forms.

INLAND WATERS

The hydrosphere is estimated to contain about 1 386 million cubic kilometers (km³) of water, almost all of which (97.5 percent) is saline water making up the world ocean, leaving some 35 million km³ (2.5 percent) of freshwater. A major proportion (about 69 percent) of this freshwater is locked up in the form of ice and permanent snow, where it is unavailable to living organisms. The Earth's liquid freshwater is mostly in subterranean groundwaters, with a small proportion in soils and wetlands, and the smallest proportion of all – about 0.01 million km³ or 0.3 percent of all freshwater – makes up the world's lakes and rivers on which inland water biodiversity depends (Table 7.1).

There are very large regional differences in the concentrations of water in all its forms (e.g. about twice as much atmospheric water in equatorial as in temperate latitudes), and in the occurrence of different types of inland waters (e.g. South America has an enormous concentration of river water but few large lakes, while the converse is true for Africa). Over the oceans, evaporation exceeds input from rivers and rainfall, and over land precipitation exceeds evaporation. This excess on land amounts to about 43 000 km³ annually and this represents the global runoff that replenishes the world's rivers, lakes and marine waters, and which humans draw on, together with groundwater, to meet their domestic, agricultural and industrial needs.

Chapter 5, on terrestrial ecosystems, outlines some of the principal differences between terrestrial and aquatic environments as they determine the living conditions of organisms in them. In general, freshwater systems and the organisms within them are far more strongly affected by daily and seasonal changes as a result of weather patterns and climate conditions than are marine aquatic environments, in parallel with change in the broader terrestrial environment. For example, upland streams may receive an influx of cold meltwater in spring, and be subject to strong insolation during midsummer; the volume, speed and transparency of river water is liable to change radically following rainfall in

**Table 7.1
Components of the hydrosphere**

Source: Anon[54]; Shiklomanov[55].

	Area (million km²)	% total area	Volume (million km³)	% total water	% freshwater
Earth's surface	510				
Land	149	29			
World ocean	361	71	1 351	97.5	–
Freshwater	–	–	34.65	2.5	100
Ice and permanent snow	16	–	23.8	1.76	68.9
Groundwater	–	–	10.4	0.77	29.9
Wetlands, soil water, permafrost	2.6	–	0.3	0.02	0.9
Lakes and rivers	1.5	–	0.01	0.0007	0.3

Some coastal or high altitude lakes and lagoons combine high concentrations of dissolved chemicals with high water temperatures, sometimes as high as 70°C. Few kinds of species thrive in these harsh alkaline or saline conditions. Communities typically include cyanobacteria, diatoms, a few small invertebrate animals such as brine shrimps (*Artemia*), with flamingos often being the only large organisms present. Flamingos, of which five living species are recognized, are filter-feeders, with the bill and mouthparts specialized to extract small particles from water. All share the same basic morphology, but tend to extract prey of different sizes: e.g. in sub-Saharan Africa the lesser flamingo feeds almost exclusively on cyanobacteria and other microorganisms, while the greater flamingo (often found in the same lakes), feeds mainly on small macroscopic invertebrates, such as brine shrimps and other crustaceans. Fishes are often absent from lakes where flamingos are abundant, and flamingos are rarely abundant where fishes are common. This may be because the two compete with each other for their food supply but, in the inhospitable lakes where they abound, flamingos have no real competitors. Their early adaptation to an extreme environment that no other large animal was capable of exploiting may explain why they have survived relatively unchanged in morphology for so long (fossils are known from the Eocene, about 50 million years before the present).

the catchment basin. The soil and surface geology of watershed areas have a strong influence on the chemical composition of both river and lake waters and, for example, buffer or reinforce the effect of acid rain.

Inland water habitats

Although the terms 'inland water' and 'freshwater' are often used more or less interchangeably they are not equivalent. A considerable number of inland waters are saline, some much more so than sea water. Conversely, waters of the deltaic regions of some major river systems (most notably the Amazon) may be fresh a considerable distance out to sea.

Despite their vastly smaller extent, inland aquatic habitats show far more variety in their physical and chemical characteristics than marine habitats. They encompass systems as varied as the world's great lakes and rivers, small streams and ponds, temporary puddles, thermal springs and even the minute pools of water that collect in the leaf axils of certain plants, such as bromeliads. Chemically they range from almost pure water to highly concentrated solutions of mineral salts, toxic to all but a few specialized organisms (see Box 7.1).

Inland water habitats can be divided into running or lotic and standing or lentic systems. They may also be divided into permanent water bodies, periodically (usually seasonally) inundated, and ephemeral or transient. Each of these has its own distinct set of ecological characteristics.

There is not necessarily a rigid dividing line between an inland aquatic habitat on the one hand and a terrestrial or marine habitat on the other. Any temporarily inundated area, such as a river floodplain, is effectively a hybrid or transitional system, being at some times essentially aquatic, at other times terrestrial. Similarly there are many areas that consist of shifting mosaics of land and shallow water, or areas of saturated vegetation, such as sphagnum moss bogs that are strictly neither land nor water. These transitional areas are often collectively termed 'wetlands'. Similarly, estuarine areas are transitional areas between inland and marine systems.

Lotic systems: rivers and catchment basins

A river system is a complex but essentially linear body of water draining under the influence of gravity from elevated areas of land toward sea level. The typical drainage system consists of a large number of smaller channels (streams, rills, etc.) at higher elevation merging as altitude falls into progressively fewer but larger channels, which in simplest form discharge by a single large watercourse. Most such systems discharge into the coastal marine environment. Some discharge into lakes within enclosed inland basins; a few watercourses in arid regions enter inland basins where no permanent lake exists.

The source area of all the water passing through any given point in the drainage system is the catchment area for that part of the system. In parallel with the hierarchical aggregation of tributaries of the major river system, sub-catchments aggregate into a single major catchment basin; this is the entire area from which all water at the final discharge point of the system – i.e. usually the sea – is derived. Strictly, the watershed is the line of higher elevation dividing one catchment basin from another, but this term is increasingly used as a synonym of catchment.

The speed and internal motion of river water depends largely on water volume and the shape of its channel. These factors typically differ greatly through the river system, from narrow, steep and fast upland feeder streams to broad, level and slow downstream reaches. In virtually all river systems water volume also varies seasonally. Some rivers in arid or semi-arid catchments flow for only part of the year, or in extreme cases only once every several years.

Large rivers may span many degrees of latitude and pass through a wide range of climatic conditions within their catchments. Variations in water flow and underlying geology also create a wide range of habitats within any river and often within a short distance. Different organisms are typically adapted to different parts of any given river system.

River systems can change course radically as a result of deposition and erosion of their channel, and the uplift and erosion of watershed uplands. Despite the dynamic physical state of these systems, large rivers rarely disappear, and although direct evidence is scarce, indications are that some have been in continuous existence for tens of millions of years. This is consistent with the fact that running waters include representatives of almost all taxonomic groups found in freshwaters, and that several invertebrate taxa occur only in running waters or attain greatest diversity there. By far the largest river catchment in the world is the Amazon (with the Ucayali) in South America which covers just under 6 million km^2 and is nearly 60 percent larger than the next largest, the Congo in central Africa. Unsurprisingly, the former is the major repository of the world's freshwater biodiversity. Between them the 20 largest river catchments cover around 45 million km^2, or about one third of the world's ice-free land surface[1].

Lentic systems
Lakes and ponds
The great majority of existing lakes, of which around 10 000 exceed 1 km^2 in extent, were formed as a result of glacial activity, with most of the rest a result of tectonic activity[2]. Tectonic lakes are formed either as a result of faults caused by deep crustal movements or by volcanism. In the case of the former, a lake may form in a depression caused by a single fault, or in a depressed area between two or more faults – these being graben lakes – or in a rift valley. Most volcanic lakes form in craters or calderas of volcanoes while a few (usually short-lived) may form behind dams caused by lava flows. Glacial lakes occupy basins caused by the scouring action of ice masses.

Most of the world's existing lakes are glacial and geologically very young, dating from the retreat of continental ice sheets at the start of the Holocene, around 11 500 years before present. All such lakes are expected to

The majority of lakes were formed as a result of glacial activity.

Table 7.2
Physical and biodiversity features of major long-lived lakes

Notes: Lakes ordered by volume. A few other lakes have notable endemism among fishes, mollusks, crustaceans or other groups. Among these are lakes Inle (Myanmar), Lanao (Philippines), Malili (Indonesia) and the Cuatro Cienegas basin (Mexico) – but their ages are not yet firmly established. Qualitative remarks (e.g. 'very high', 'low') in the Biodiversity column are related to long-lived lakes, not to lake systems in general. All biodiversity data are approximate and subject to change with new survey data or different taxonomic opinion.

i Evidence indicates that the lake dried out completely, or nearly so, around the late Pleistocene, 10-12 000 years ago[61].

Source: Collated from data in Martens *et al.*[56]. Fish estimates for East African Lakes from Snoeks[62].

Lake	Country	Age (million years)	Max. depth (m)	Vol. (km³)	Biodiversity
Baikal Largest, deepest, oldest extant freshwater lake (20% of all liquid fresh surface water on Earth)	Russia	25-30	1 637	23 000	Very high sp. richness; exceptional endemism in fishes and several invertebrate groups Total animal spp.: 1 825, endemic: 982 Fishes: 56 spp., 27 endemic
Tanganyika	Burundi Tanzania Zambia DR Congo (former Zaire)	20	1 470	18 880	Very high sp. richness; high endemism, especially high among cichlid fishes Total animal spp.: 1 470, endemic: 632 Fishes: 325 spp., including 250 cichlids of which 98% endemic
Malawi	Malawi Mozambique Tanzania	>2	780	8 400	Very high sp. richness; high endemism, especially high among cichlid fishes Fishes: 845 spp., including 800 cichlids of which 99% endemic
Victoria World's second largest freshwater lake (area)	Kenya Tanzania Uganda	>4?[i]	70	2 760	High sp. richness, especially of fishes; exceptional endemism among cichlid fishes – many fish endemics depleted or extirpated following introduction of Nile perch Fishes: 545 spp., including 500 cichlids of which 99% endemic
Titicaca One of world's highest altitude lakes	Bolivia Peru	3	280	890	Moderate sp. richness and endemism (highest among fishes) Total animal spp.: 533, endemic: 61 Fishes: 29 spp., 23 endemic
Biwa	Japan	4	104	674	Moderate sp. richness and endemism (highest in gastropod mollusks and fishes) Total animal spp.: 595, endemic: 54 Fishes: 57 spp., 4 endemic
Ohrid Fed mainly by subterranean karst waters	Albania Macedonia (FYR)	3	295	50	Moderate sp. richness; exceptional endemism in several groups (planarians, oligochaetes, gastropod mollusks, ostracod crustaceans) Fishes: 17 spp., up to 10 endemic

fill slowly with sediment and plant biomass, and to disappear within perhaps the next 100 000 years, along with any isolated biota. Lakes may also be caused by the dissolution of soluble rocks, most notably limestone in karst regions which is gradually dissolved by dilute acids in water running through it, and by changes in the course of rivers in floodplain regions, which result in ox-bow and scroll lakes.

Only about ten existing lakes are known with certainty to have origins much before the Holocene (Table 7.2)[2], and most of these occupy basins formed by large-scale subsidence of the Earth's crust, dating back to at most 20 million (Lake Tanganyika) or 30 million (Lake Baikal) years before the present.

There is good evidence that some extinct lake systems in the geological past were very large and very long-lived under different climatic and tectonic conditions. In general, the long-lived lakes are of particular interest in terms of biodiversity because they tend to be rich in species of several major groups of animals and many of these species are restricted to a single lake basin.

Wetlands

As indicated above, the distinction between a wetland, an aquatic system and a terrestrial system may be essentially arbitrary. However, a number of mixed shallow-water and terrestrial habitat types share several characteristics and are habitually grouped as wetlands. Wetlands in this sense are typically heterogeneous habitats of permanent or seasonal shallow water dominated by large aquatic plants and broken into diverse microhabitats[3]. The four major broad habitat types are:

Bogs

Bogs are peat-producing wetlands in moist climates where organic matter has accumulated over long periods. Water and nutrient input is entirely through precipitation. Bogs are typically acid and deficient in nutrients and are often dominated by sphagnum moss.

Fens

Fens are peat-producing wetlands that are influenced by soil nutrients flowing through the system and that are typically supplied by mineral-rich groundwater. Grasses and sedges, with mosses, are the dominant vegetation. Fens are typically more productive and less acidic than bogs.

Marshes

Marshes are inundated areas with herbaceous emergent vegetation, commonly dominated by grasses, sedges or reeds. They may be either permanent or seasonal and are fed by ground or river water, or both.

Swamps

Swamps are forested freshwater wetlands on waterlogged or inundated soils where little or no peat accumulation occurs. As with marshes, they may be either permanent or seasonal.

Biogeography and important areas

Freshwater lineages that originated in continental water systems may show general patterns of distribution similar to terrestrial groups, corresponding more or less to broad biogeographic realms. Lineages of marine origin may remain restricted to peripheral systems corresponding to the area where the ancestral forms moved into freshwater.

Unlike many terrestrial species, which can disperse widely in suitable habitat, the spatial extent of the range of strictly freshwater species tends to correspond to present or formerly continuous river basins or lakes. These species include fishes and most mollusks and crustaceans. Watersheds between river basins are the principal barriers to their dispersal between systems, and their ranges are extended mainly by physical changes to the drainage pattern (e.g. river capture following erosion or uplift can allow species formerly restricted to one system to move into another), or by accidental transport of eggs by waterbirds, or by flooding.

In many instances, the range within a system will also be restricted by particular habitat requirements (variations in water turbulence or speed, shelter, substrate, etc.). These frequently differ at different stages in the life cycle (for example in fishes the conditions and sites required for egg deposition and development, for early growth of fry,

Continent	Area name	Taxonomic group		
Africa	L. Malawi	Fishes	Mollusks	
Africa	L. Tanganyika	Fishes	Mollusks	Crabs
Africa	L. Victoria	Fishes	Mollusks	
Africa	Madagascar	Fishes	Mollusks	Crabs
Africa	Niger-Gabon	Fishes		Crabs
Africa	Upper Guinea	Fishes	Mollusks	Crabs
Africa	Lower Congo	Fishes	Mollusks	Crabs
Australia	SE Australia & Tasmania	Fishes	Mollusks	Crayfish
Australia	SW Australia	Fishes		Fairy shrimp
Eurasia	SE Asia and lower Mekong River	Fishes	Mollusks	Crabs
Eurasia	Balkans (southwest)	Fishes	Mollusks	
Eurasia	L. Baikal	Fishes	Mollusks	
Eurasia	L. Biwa	Fishes	Mollusks	
Eurasia	L. Inle	Fishes	Mollusks	
Eurasia	L. Poso	Fishes	Mollusks	
Eurasia	Malili Lakes	Fishes	Mollusks	
Eurasia	Sri Lanka	Fishes		Crabs
Eurasia	Western Ghats	Fishes	Mollusks	Crabs
North America	East Mississippi drainage (Ohio, Cumberland, Tennessee rivers)	Fishes	Mollusks	Crayfish
North America	Mobile Bay drainage	Fishes	Mollusks	Crayfish
North America	Western USA	Fishes	Mollusks	Fairy shrimp
South America	L. Titicaca	Fishes	Mollusks	
South America	La Plata drainage	Fishes	Mollusks	
South America	Amazon basin	Fishes	?	Crabs

Table 7.3
Partial list of global hotspots of freshwater biodiversity

Notes: This table lists areas of special importance for diversity in fishes and either mollusks or crustaceans or both. See text and Appendix 6. Six of the seven long-lived lakes in Table 7.2 also appear here.

Source: See sources cited in Appendix 6.

and for feeding and breeding of adults are often different).

Many cave or subterranean freshwater aquatic species (e.g. of fishes, amphibians and crustaceans) have restricted ranges, perhaps consisting of a single cave or aquifer, and limited opportunities for dispersal, depending on the surrounding geology and the consequent morphology of the water system occupied.

Analysis of data from some 151 river basins indicates that there is a strong correlation between the spatial extent of a river catchment and the number of fish species therein. The 'size' of a river can be represented by the area of the basin or by the volume of water flowing through the river system in any given period; the latter is a better predictor of fish species richness than is basin area. When area is taken into account,

there is also a strong relationship between species richness and the latitude of the basin. Recent analysis suggests that latitude may be a surrogate measure for energy availability and productivity within the basin[4, 5], factors known to be well correlated with variation in terrestrial diversity (Chapter 5). No taxonomic class restricted to inland waters has yet been mapped globally at species level but, at a higher taxonomic level, a density surface of freshwater fish families has been developed (see Map 7.1) with a view to providing an indication of global variation in inland water diversity analogous to those available for terrestrial groups (Chapter 5).

A recent analysis of areas important for the maintenance of global freshwater biodiversity[1] was based on the expert view of a number of regional and taxonomic specialists. The analysis was designed to make effective use of readily available information and, although preliminary, yielded the first global overview of freshwater biodiversity hotspots[1]. Maps 7.2, 7.3 and 7.4 show, respectively, important areas for freshwater fishes, mollusks and selected crustacean groups. Further details of all these areas can be found in Appendix 6. Table 7.3 lists the sites and areas that have been identified as of special importance for more than one of the above groups. It is not a comprehensive global listing because it omits several large but imprecisely defined areas of known high diversity, and it omits diverse taxa not covered in the assessment (e.g. amphipods, copepods); nor does it mention sites of key importance mainly for one group of animals. Although the Amazon basin is a vast region rather than an identifiable site, it has such an exceptional diversity of fishes that it could not reasonably be excluded from a list of globally important areas.

More detailed continental reviews are now also available for Asia, including discussion of taxonomy, hotspots and policy[6], for Latin America[7] and North America[8]. In order to help prioritize investment, the conservation organization WWF has selected 53 freshwater ecoregions, based on a combination of biogeographic region, water body type, biodiversity and representativeness[9, 10]. Perhaps more so than in other biomes, these freshwater ecoregions have a firm objective basis because they correspond broadly with catchment basins.

BIOLOGICAL DIVERSITY IN INLAND WATERS

At high taxonomic levels the diversity of freshwater organisms is considerably lower than on land or in the sea. Only one extant eukaryote phylum (Gamophyta – green conjugating algae) is apparently confined to freshwater habitats. The number of species overall is low in absolute terms in comparison with marine and terrestrial groups, but species richness in relation to habitat extent is relatively high. For example, about 10 000 (40 percent) of the 25 000 known fish species are freshwater forms[1]. Given the distribution of water on the Earth's surface this is equivalent to one fish species for every 15 km^3 in freshwaters compared with one for every 100 000 km^3 of sea water. This high diversity of freshwater fishes relative to habitat extent is undoubtedly promoted by the extent of isolation between freshwater systems. Many lineages of fishes and invertebrates have evolved high diversity in certain water systems, and in some cases, species richness and endemism tend to be positively correlated between different taxonomic groups[11].

As is the case with terrestrial habitats, species richness increases strongly toward the equator, so that in most groups of organisms, there are many more species in the tropics than in temperate regions, although in a few specific cases (e.g. freshwater crayfish) this appears to be reversed.

Protoctists

The larger algae comprise some 5 000 species in three major groups (the green, brown and red algae), the great majority of which are marine or brackish water forms ('seaweeds'). The green algae Chlorophyta includes one order of around 80 species (Ulotrichales) that is mainly freshwater. However, one major group usually associated with the green algae – the stoneworts (charophytes) – is almost entirely freshwater. The stoneworts include some 440 species, most of which are endemic at continent level or below; they tend to be very sensitive to nutrient enrichment and have declined in many areas[12].

Fungi

There are more than 600 species of freshwater fungi known, currently more from temperate regions than from the tropics, although probably only a small fraction of existing species have been described, and the tropics have been little sampled[13]. Virtually all described freshwater fungi are ascomycotes with few basidiomycotes and zygomycotes having been identified. They occur wherever vascular plant material is available as a substrate. They appear to be important as parasites, endotrophs and saprotrophs of emergent aquatic macrophytes, as decomposers of submerged allochthonous woody debris, and as a food resource for invertebrates[13-15]. Most are very small, with sporomes (fruiting bodies) less than 0.5 mm in diameter.

Plants

Wetland or aquatic species occur with some frequency in the non-vascular plant phyla, which generally prefer moist habitats, and among the ferns and allies. Mosses in the order Sphagnales (a single family Sphagnaceae and genus *Sphagnum*) often grow submerged, and are key components of peat bogs. Many groups of damp-loving (hygrophilous) terrestrial mosses (e.g. *Thamnium*, *Bryum*, *Mnium*) have aquatic forms. Several genera of Bryales are aquatic or have aquatic species. A number of

Species richness in inland waters, in relation to habitat extent, is relatively high.

Map 7.1
Freshwater fish family diversity

Family richness of typical bony fishes (Actinopterygii) in inland waters, plotted as a world density surface. It is based on generalized range maps of 157 families. Color depth represents the number of families, up to a maximum of 44, potentially present at any point. Two families of cartilaginous fishes (Elasmobranchii) that together have a very few inland water species are omitted. About ten or so families of bony fish that occur in coastal and estuarine waters, but do not extend significantly into inland freshwaters, are omitted. Several families range more or less widely in inland waters and also occur in coastal and estuarine waters around the continents, but this peripheral part of the range is in most cases not represented.

Source: Produced by UNEP-WCMC using range maps prepared from information in Berra[60].

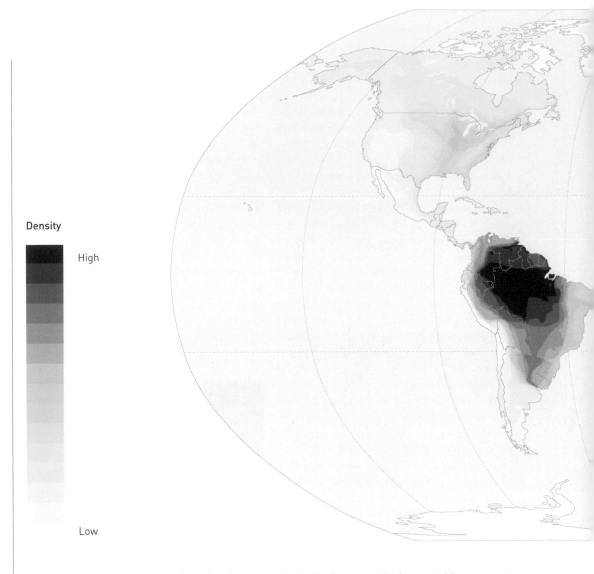

Density

High

Low

liverwort species growing otherwise in wet terrestrial situations may also live submerged, sometimes at considerable depths. Truly aquatic liverworts include several species of *Riccia* and *Ricciocarpus natans* (Ricciaceae; Marchantiales) that live free-floating on the surface of eutrophic lakes. At least 16 species of *Riella* in the Riellaceae (Jungermanniales) are aquatic forms characteristic of temporary waters in semiarid regions, reaching highest diversity in northern and southern Africa. Among the lycophytes, most of the 60 or so species of *Isoetes* (family Isoeteacae) are aquatic, some of great limnological importance, and *Stylites* is an endemic member of the littoral community of Andean lakes. Sphenophytes (horsetails) often occur in moist situations, including around water margins. *Equisetum*

fluviatile, for example, is a notable emergent littoral form of north temperate lakes.

The Filicinophyta include several aquatic forms. The genus *Ceratopteris* (four species, family Pteridaceae) has the only truly aquatic (floating) homosporous ferns; some are cultivated ornamentals, others are edible. A few other species, e.g. *Microsorum pteropus* (Polypodiaceae) and *Microlepia speluncae* (Dennstaedtiaceae) can grow in water. Among heterosporous ferns, the family Marseliaceae comprises three genera and 55-75 species which are either amphibious or fully aquatic. All members of the families Salviniaceae (one genus and about ten species) and Azollaceae (one genus and around six species) are floating aquatic ferns. The latter supports the symbiotic nitrogen-fixing *Anabaena azollae* (phylum Cyanobacteria).

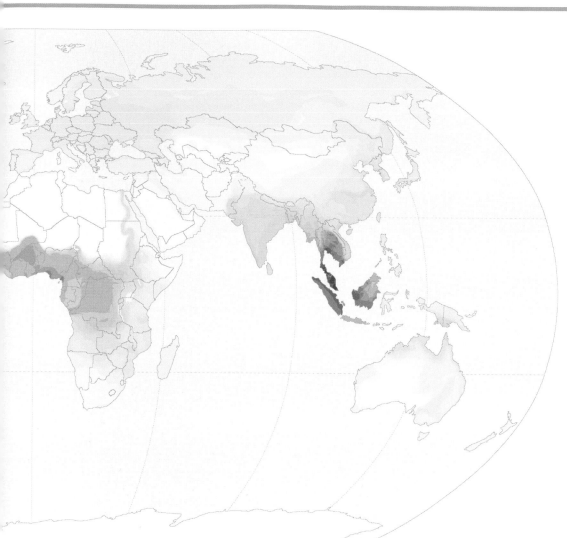

Vascular plants are essentially terrestrial forms, and existing aquatic species are derived from terrestrial ancestors; several different lineages include aquatic species and this transition has therefore occurred several times. It has been estimated that at most 1 percent of angiosperms, i.e. up to 2 700 species, are aquatic[16]. Around 14 angiosperm families consist largely or exclusively of inland water forms (Ceratophyllaceae, Hippuridaceae, Hydrostachydaceae, Nymphaeaceae, Podostemaceae, Trapaceae, Butomaceae, Hydrocharitaceae, Lemnaceae, Limnocharitaceae, Najadaceae, Pontederiaceae, Potamogetonaceae, Zannichelliaceae).

Most inland water plant species are relatively widespread, ranging over more than one continental land mass; many are cosmopolitan, occurring around the world and on remote islands. Of the widespread forms, some are essentially northern temperate species extending to a great or lesser extent into the tropics; some are mainly tropical[16]. The Podostemaceae is particularly noteworthy for its many monotypic genera, and a large number of narrowly endemic species, in at least one instance with several forms restricted to different stretches of a single river. Tropical South America, Madagascar, Sri Lanka, India, Myanmar and Indonesia hold such localized species[16].

Inland water animals

Animal species are considerably more diverse and numerous in inland waters than plants. Most of the major groups include terrestrial or marine species as well as freshwater forms. Apart from fishes, important groups with inland water species include crustaceans (crabs, crayfishes, shrimps, as well as planktonic forms such as filter-feeding Cladocera and filter-feeding or predatory Copepoda), mollusks (including mussels Bivalvia, and snails Gastropoda), insects (including stoneflies Plecoptera, caddisflies Trichoptera, mayflies Ephemoptera), sponges, flatworms, polychaete worms, oligochaete worms, numerous parasitic species in various groups, and numerous microscopic forms. Information is incomplete for many groups, but crustaceans and mollusks have speciated profusely in certain freshwater systems, with a tendency to form local endemic species. Because of the feeding mode – attached bottom-living filter-feeders – bivalves can help maintain water quality but tend to be susceptible to pollution (their larvae are parasitic on fishes). The diversity and ecological role of microorganisms and micro-invertebrates in freshwater sediments have been reviewed[17].

Insects with an aquatic larval phase but a winged adult phase are often restricted to particular river basins (even if adults disperse widely, they may not find suitable habitat), but in general are much less restricted in this way than entirely aquatic species. A relatively large number of species, particularly of crustaceans, occupy temporary pools and have a stage that is desiccation-resistant and can undergo long-range passive dispersal between drainage basins; some such species are thus widely distributed.

Insects

As on land, insects (phylum Mandibulata) are as far as is known by far the most diverse group of organisms in inland waters. The true number of aquatic insects remains unknown; data for three relatively well known areas (Europe, Australia and North America) and extrapolations for possible global totals are included in Table 7.4. In contrast to terrestrial faunas, where beetles (order Coleoptera) are the most diverse, flies and their relatives (order Diptera) appear to be by far the most abundant group in inland water habitats, although also one of the less fully known.

In terms of life histories, there are two main groups of aquatic insects: those in which the adult stage and the active immature stages are passed in water (in some cases with a terrestrial pupal stage); and those in which, after a nymphal or larval stage in water, the adult stage is spent on land or in the air. The great majority of Diptera, and therefore most aquatic insects, form part of the latter group. Included amongst their number are several of enormous economic significance to humans, of which the most significant are almost certainly mosquitoes of the genus *Anopheles*, intermediary hosts of the malaria parasite.

Most aquatic insects are benthic, living in or on the bottom; a small number are planktonic and live suspended in the water column. Around half of the aquatic Hemiptera and a few other insects and non-insect invertebrates live on the water surface (epipleuston).

Fishes

Around 40 percent of known fish species occur in freshwater: almost exactly 10 000 species are confined to freshwater, and a further 1 100 or so occur in freshwater but are not confined to it (Table 7.5). These last include species, such as many salmonids, that grow in the seas but ascend rivers to spawn (anadromous), and others, such as eels, that grow in inland waters but spawn at

Table 7.4
Insects of inland waters

Notes: Data refer to number of species; all estimates approximate.

i Partially aquatic as adult and sometimes as nymph.
ii Number in parentheses refers to fully aquatic Nepomorpha.
iii All these species are parasitoids as larvae.

Source: Collated from data in Hutchinson[57].

Order	Australia	North America	Europe	World
Ephemeroptera	84	614	224	2 250
Odonata	302	415	127	4 875
Plecoptera	196	578	387	2 140
Orthoptera[i]	–	ca 20	–	ca 20
Blattodea[i]	–	0	–	ca 10
Hemiptera	236	404	129 (81)[ii]	3 200
Megaloptera	26	43	6	300
Neuroptera	58	6	9	ca 100
Coleoptera	730	1 655	1 072	5 000
Hymenoptera[iii]	–	55	74	ca 100
Diptera	1 300	5 547	4 050	>20 000
Trichoptera	478	1 340	895	7 000
Lepidoptera	–	–	5	ca 100

sea (catadromous). Freshwater fishes are taxonomically diverse, although not as diverse as marine ones. Thirty-four of the 57 or so extant orders of fishes have at least one strictly freshwater species, while a further two, the sawfishes (Pristiophoriformes) and

tarpons (Elopiformes) have species that occur in freshwater but are not confined to it[18]. This compares with 38 orders that have at least one marine species.

Of the orders of fishes with freshwater species, ten are entirely freshwater and

Table 7.5
Fish diversity in inland waters, by order

Source: Nelson[18] (differs in detail from the later taxonomy of Eschmeyer[58]).

Order (Strictly freshwater orders in bold)	Common fishes	No. of families	No. of genera	No. of species	No. of strictly freshwater species	No. of species using freshwater	Approximate % strictly freshwater	Approximate % using freshwater
Petromyzontiformes	Lampreys	1	6	41	32	41	78	100
Carcharhiniformes	Ground sharks	7	47	208	1	8	0	4
Pristiophoriformes	Sawfishes	1	2	5	0	1	0	20
Rajiformes	Rays	12	62	456	24	28	5	6
Ceratodontiformes	Australian lungfish	1	1	1	1	1	100	100
Lepidosireniformes	Lungfishes	2	2	5	5	5	100	100
Acipensiformes	Sturgeons	2	6	26	14	26	54	100
Amiiformes	Bowfin	1	1	1	1	1	100	100
Anguilliformes	Eels	15	141	738	6	26	1	4
Atheriniformes	Silversides	8	47	285	146	171	51	60
Batrachoidiformes	Toadfishes	1	19	69	5	6	7	9
Beloniformes	Needlefishes, sauries, flyingfishes, halfbeaks	5	38	191	51	56	27	29
Characiformes	Characins	10	237	1 343	1 343	1 343	100	100
Clupeiformes	Herrings and anchovies	5	83	357	72	80	20	22
Cypriniformes	Carp, minnows, loaches	5	279	2 662	2 662	2 662	100	100
Cyprinodontiformes	Rivulines, killifishes, pupfishes, poeciliids, goodeids	8	88	807	794	805	98	100
Elopiformes	Ladyfishes and tarpons	2	2	8	0	7	0	88
Esociformes	Pikes and mudminnows	2	4	10	10	10	100	100
Gadiformes	Cods, hakes, rattails	12	85	482	1	2	0	0
Gasterosteiformes	Pipefishes, sticklebacks, sandeels, etc.	11	71	257	19	41	7	16
Gonorhynchiformes	Milkfish and beaked sandfishes	4	7	35	28	29	80	83
Gymnotiformes	Knifefishes	6	23	62	62	62	100	100
Mugiliformes	Mullets	1	17	66	1	7	2	11
Ophidiiformes	Pearlfishes, cusk-eels, brotulas	5	92	355	5	6	1	2
Osmeriformes	Smelts	13	74	236	42	71	18	30
Osteoglossiformes	Bonytongues	6	29	217	217	217	100	100
Perciformes	Perches, basses, sunfishes, whitings, etc.	148	1 496	9 293	1 922	2 815	21	30
Percopsiformes	Trout-perches, pirate perch, cavefishes	3	6	9	9	9	100	100
Pleuronectiformes	Plaice, flounders, soles	11	123	570	6	20	1	4
Polypteriformes	Bichirs	1	2	10	10	10	100	100
Salmoniformes	Salmonids	1	11	66	45	66	68	100
Scorpaeniformes	Gurnards, scorpionfishes, velvetfishes, etc.	25	266	1 271	52	62	4	5
Semionotiformes	Gars	1	2	7	6	7	86	100
Siluriformes	Catfishes	34	412	2 405	2 280	2 287	95	95
Synbranchiformes	Swamp-eels	3	12	87	84	87	97	100
Tetraodontiformes	Triggerfishes, puffers, boxfishes, filefishes, molas	9	100	339	12	20	4	6

another five are very largely so (with more than 80 percent of their known species in freshwaters). A further 13 are very largely marine with a small proportion of freshwater species (<10 percent) while the remainder have significant numbers of both marine and freshwater species. Over 80 percent of freshwater species are confined to just four orders: the carps and their relatives (Cypriniformes); the characins (Characiniformes); the catfishes (Siluriformes); and the perches and their relatives (Perciformes). The first three of these are wholly or almost entirely freshwater, while the last, the largest order of fishes with nearly 40 percent of known species, is unusual in having significant numbers of both marine and freshwater species.

Amphibians

The great majority of the 5 000 or so living amphibian species have aquatic larval stages and, as none is known to occur in sea water, all these are dependent on inland waters of various kinds for continued survival of populations. In some cases such water bodies may be temporary pools or puddles, or water in the leaf axils of plants. Relatively few species are fully aquatic (Table 7.6). Although the number of fully aquatic species in each of the three extant orders is roughly similar (ca 20-30 in each), these represent very different proportions of each order, being less than 1 percent of anurans, around 5 percent of caudate amphibians and more than 10 percent of caecilians.

Aquatic caudate amphibians are neotenic, that is retain features of the larval stage, most notably external gills. In addition to the fully aquatic amphibians (several of which can survive for short periods in damp conditions out of water), many other species may lead largely aquatic lives or may, as in the case of the Mexican axolotl *Ambyostoma mexicanum*, have completely aquatic neotenic populations.

Reptiles

Very few completely aquatic inland water reptiles are known. The three file-snakes in the family Acrochordidae are live-bearing and may pass their entire lives in water, often in coastal and estuarine areas as well as

freshwaters. Virtually all other reptiles of inland waters are egg-laying and return to land at least to nest; most also spend a proportion of their time on land, often basking on banks or logs. The two most aquatic orders are the Crocodilia and the Chelonia. All the 22 or so extant species of the former are

predominantly aquatic and occur in freshwaters, although one or two may also be found in marine areas. Of the latter, some two thirds of the 250 or so extant species are largely or predominantly aquatic, and a further 30 or so species may be considered amphibious.

Aside from the file-snakes and the homalopsine snakes, a number of other snake species are at least semi-aquatic. These include several genera of natricine snakes, including the North American *Nerodia* and the Asiatic *Sinonatrix*, as well as the anacondas *Eunectes* (family Boidae) and the water cobra *Boulengeria annulata* (family Elapidae). Amongst lizards, no wholly aquatic species is known. However, many can swim proficiently, often using water as a means of escape from predators, and a number are semi-aquatic. These last include several Australian and Old World monitors *Varanus* spp. (family Varanidae), water dragons *Hydrosaurus* and *Physignathus* (Agamidae), the crocodile lizard *Shinisaurus* from south China (Xenosauridae), the Bornean earless monitor *Lanthanotus* (Lanthanotidae), and a number of New World teiids (Teiidae).

Virtually all reptiles of inland waters return to land, at least to nest.

Class and order	Family		No. of freshwater species
AMPHIBIA			
Caudata	**Amphiumidae**	Congo eels	3
	Cryptobranchidae	Giant salamanders and hellbenders	3
	Plethodontidae	Lungless salamanders	4[i]
	Proteidae	Mudpuppies and olm	6
	Sirenidae	Sirens	3
Anura	**Pseudidae**	Paradox frogs	3
	Pipidae	Clawed frogs and pipid toads	28
Gymnophiona	**Typhlonectidae**	Typhlonectid caecilians	19[ii]
REPTILIA			
Chelonia	**Carettochelidae**	Pig-nosed soft-shelled turtle	1
	Trionychidae	Soft-shelled turtles	23
	Platysternidae	Big-headed turtles	1
	Chelydridae	Snapping turtles	2
	Dermatemydidae	Central American river turtle	1
	Chelidae	Austro-american side-necked turtles	37
	Kinosternidae	Mud and musk turtles	12
	Pelomedusidae	Side-necked turtles	22
	Emydidae	Pond and river turtles	64
Crocodilia	**Alligatoridae**	Caimans and alligators	8
	Crocodylidae	Crocodiles	12
	Gavialidae	Gharial and false gharial	2
Squamata	Acrochordidae	File-snakes	3
	Colubridae	Colubrid snakes	40[iii]
AVES			
Anseriformes	Anseranatidae	Magpie goose	1
	Dendrocygnidae	Whistling-ducks	9
	Anatidae	Ducks, geese and swans	141
Gruiformes	Heliornithidae	Limpkin and sungrebes	3
	Rallidae	Rails, gallinules and coots	11[iv]
Ciconiiformes	Jacanidae	Jacanas	8
	Laridae	Gulls, terns, skuas, auks, skimmers	22 (13)[v]
	Podicipedidae	Grebes	21
	Anhingidae	Anhingas	4
	Phalacrocoracidae	Cormorants and shags	5 (2)[v]
	Phoenicopteridae	Flamingos	5
	Pelecanidae	Pelicans and shoebill	6
	Gaviidae	Divers	5
Passeriformes	Cinclidae	Dippers	5
MAMMALIA			
Monotremata	**Ornithorhynchidae**	Platypus	1
Didelphiomorpha	Didelphidae	Opossums	1
Insectivora	Tenrecidae	Tenrecs and otter shrews	4
	Soricidae	Shrews	3
	Talpidae	Moles and desmans	2
Rodentia	**Castoridae**	Beavers	2
	Muridae	Voles and mice	32
	Hydrochaeridae	Capybara	1
Cetacea	**Platanistidae**	River dolphins	5
Sirenia	**Trichechidae**	Manatees	3
Carnivora	Mustelidae	Mustelids, otters	13
	Viverridae	Viverrids	3
	Phocidae	Earless seals	2
Artiodactyla	Hippopotamidae	Hippopotamus	1

**Table 7.6
Tetrapod diversity in inland waters**

Notes: Entirely freshwater families in bold. Taxonomy based on the same vertebrate sources cited in Table 2.1. Bird groups here differ to some extent from Table 7.9 which uses a more traditional arrangement of bird higher taxa.

i Genera *Leurognathus*, *Haideotriton*, *Typhlomolge* only.
ii Sometimes included in the Caecilidae.
iii Subfamily Homalopsinae.
iv Genus *Fulicula* (coots) only.
v Figures in parentheses indicate those species of the total that breed largely or entirely inland also included as seabirds in Table 6.6.

Map 7.2
Major areas of diversity of inland water fish

This map represents an informal synthesis of documented expert opinion on globally important areas for freshwater fish diversity, taking into account species richness and endemism. Two categories are shown: discrete areas and systems known to be of high diversity, and areas where diversity is globally important but less concentrated.

Note: For numbered locations see Appendix 6.

Source: Compiled with the help of members of IUCN/SSC specialist groups and other ichthyologists; first published in WCMC[1].

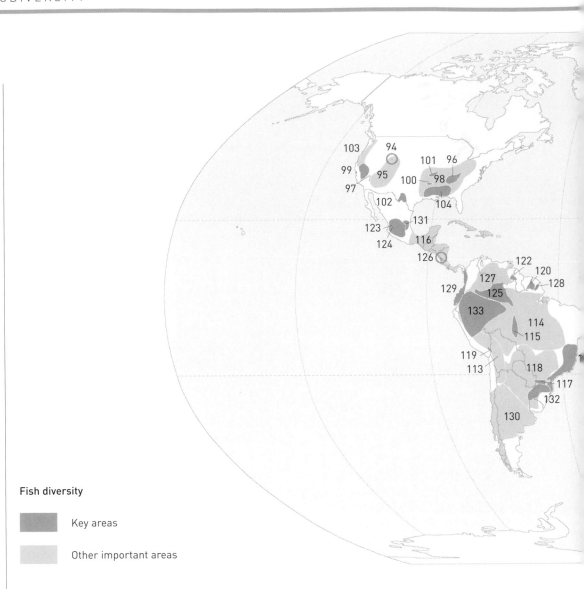

Fish diversity

Key areas

Other important areas

Birds
Unlike mammals, there are no wholly aquatic birds, because all species lay eggs that cannot survive prolonged immersion in water; however, a much higher proportion of bird than mammal species is associated with inland water ecosystems. As with other tetrapod groups, it is impossible to separate rigidly inland water species from primarily terrestrial forms or from seabirds. Table 7.6 includes bird species that are highly adapted to aquatic ecosystems and largely or exclusively inhabit inland waters, rather than marine or coastal areas. Nearly 60 percent of the roughly 250 species belong to one family – the Anatidae – all of whose members are more or less associated with aquatic habitats, although some (such as most goose species) feed very largely on land. It is noteworthy

that, among the Passeriformes or perching-birds – by far the most species-rich order (accounting for over half of all bird species) – only one small family, the dippers (Cinclidae), can be considered truly aquatic in habits.

In addition to these, there are a large number of wading bird species associated with inland or coastal wetland and littoral habitats. These include all or most members of the following families: Anhimidae (screamers – three species); Eurypygidae (sunbittern – one species); Gruidae (cranes – 15 species); Rallidae (rails, gallinules and coots – 142 species); Scolopacidae (sandpipers and their relatives – 88 species); Rostratulidae (painted-snipe – two species); Charadriidae (plovers and their relatives – 89 species); Ardeidae (herons, bitterns and egrets – 65 species); Scopidae (hammerkop – one species);

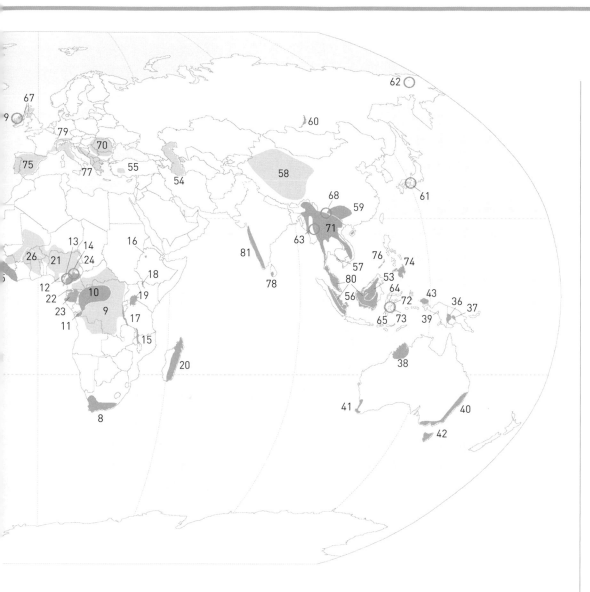

Threskiornithidae (ibises and spoonbills – 34 species); and Ciconiidae (storks – 26 species). There are also a considerably smaller number of non-wading birds that feed largely on fishes and other aquatic animals and are adapted to diving or surface-snatching. Among these are many kingfishers (families Alcedinidae, Dacelonidae and Cerylidae), the fish-owls (*Ketupa* and *Scotopelia* spp., family Strigidae), fish-eagles (*Haliaeetus* and *Ichthyophagus* spp., family Accipitridae) and a few other raptors.

Mammals
Wholly aquatic inland water mammals are confined to two orders, Cetacea and Sirenia. In the former, four of the five species of the family Platanistidae (the river dolphins) are confined to river systems and the fifth occurs in estuarine and coastal waters. A number of other cetaceans may enter the lower reaches of river systems but all are predominantly marine. Two of the four living species of Sirenia – the Amazonian manatee *Trichechus inunguis* and the West African manatee *T. senegalensis* – are wholly or very largely confined to freshwaters, and a third, the Caribbean manatee *T. manatus*, is found in both inland and marine waters.

Amongst other groups, a number of species may lead more or less aquatic lives but all these are effectively amphibious, in that at the very least they produce young on land. For most of these, a predominantly aquatic life is evident from direct observations, but for some (chiefly the Muridae) it is inferred from morphological adaptations. In direct contrast to terrestrial systems, where the majority of mammal species are herbivores, a high proportion of these amphibious species (around 50 percent) are carnivores, with only eight true herbivores and most of the aquatic murids believed to be omnivores (though predominantly insectivorous or piscivorous). It is also noteworthy that at least four of these species – the two beavers *Castor* spp., the hippopotamus *Hippopotamus amphibius* and the capybara *Hydrochaeris hydrochaeris* – feed very largely or entirely on land plants. In addition to the species included in Table 7.6, a number of other mammals are largely or wholly confined to wetland habitats (marshes, floodplains and swamps). These include three African antelopes – the Nile lechwe *Kobus megaceros*, red lechwe *Kobus leche* and sitatunga *Tragelaphus spekei* – and the South American marsh deer *Blastocerus dichotomus*.

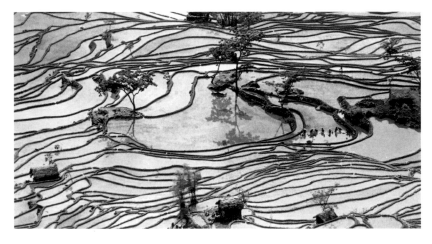

Freshwater is essential to human survival.

HUMAN USE OF AND IMPACT ON INLAND WATERS

Freshwater – as precipitation, groundwater or in inland water ecosystems – is essential for human survival, chiefly because humans must drink and also because it is needed, in far greater quantity, to produce food. It also has a wide range of subsidiary uses – for transport, industrial production, cleaning, waste disposal, generation of hydroelectric power, recreation, esthetic purposes and in the form of inland water ecosystems as sites for the production of food.

Many of these demands conflict with each other, so that for example the use of water for disposal of noxious wastes is incompatible with the provision of safe drinking water. Moreover, while the amount of freshwater available is limited, demands on it continue to grow steadily as the global human population continues to expand. This problem is exacerbated by the fact that freshwater is unevenly distributed around the world, so that it is often not available where and when needed, nor in the appropriate amounts, nor with the necessary quality. The two last are particularly important for the maintenance of freshwater biodiversity. Freshwater systems are therefore under growing pressure, as flow patterns are disrupted and the load of waste substances increases. Inevitably, per capita shares of water for human use are decreasing and water stress is becoming more widespread[19].

Agriculture consumes around 70 percent of all water withdrawn from the world's rivers, lakes and groundwater[20]. In places, more than half the water diverted or pumped for irrigation does not actually reach the crop, and problems of waterlogging and salinization (deposition in soil of salts left by evaporation of pumped groundwater) are increasing. However, irrigated agriculture produces nearly 40 percent of world food and other agricultural commodities on only 17 percent of the total agricultural land area, and is thus disproportionately important to global food security[20].

The principal use of freshwater species, not considering the properties of aquatic systems themselves, is as food. Subsidiary uses include the aquarium trade, materials for medicinal or ornamental use, and as fertilizer. Inland water fishery production has two components: capture fisheries and aquaculture, although as discussed below the distinction between the two is becoming increasingly blurred. For many human communities, particularly in countries less developed industrially, capture fisheries provide a major portion of the diet. Although it appears to be under-reported, inland water production has usually been regarded as far less important than marine fisheries and, with few exceptions where countries have access to both marine and inland aquatic resources, reported yield from inland waters is a small fraction of marine yield. Even in landlocked countries, the recorded inland harvest is often low both in absolute size and in relation to consumption of meat and other agricultural produce.

INLAND WATER FISHERIES
Capture fisheries and aquaculture

Globally, the reported inland water capture fishery for 1999 amounted to 8.2 million metric tons, with 19.8 million metric tons of aquaculture production recorded[21]; over 85 percent of the former and about 98 percent of the latter comprised finfishes, with virtually all the remainder being freshwater crustaceans and mollusks[22,21]. The crustaceans are mainly crayfishes and freshwater shrimps, both exploited for food, and most of the mollusks are bivalve, taken for pearls and for food. These reported totals compare with reported marine capture fisheries of some 84 million metric tons, and marine and brackish water aquaculture animal production of around 13 million metric tons (see Chapter 6).

The reported global inland water capture fishery has increased slowly in the period 1984-99, by nearly 2 percent per year, although this masks considerable regional variation, with declines in some areas (e.g. Europe and North America) and more marked increases elsewhere (notably Asia)[22]. Reported inland aquaculture production has been rising at a higher rate, and was well over twice the reported production of inland capture fisheries in 1999 (Figure 7.1). A major proportion of global inland aquaculture is produced by countries in Asia. China alone reportedly generates more than one quarter of the global total (Table 7.7), and has been responsible for most of the recent increase in this sector. In this particular instance the dividing line between aquaculture and capture fisheries is indistinct; no husbandry is involved beyond release of hatchery stock, and the fishery operates as a capture fishery[22].

However, national statistics do not adequately reflect the actual magnitude, location or importance of inland fisheries. The reported inland capture production is certainly a gross underestimate because much of the catch is made far from recognized landing places where catches are monitored, and is consumed directly by fishers or marketed locally without ever being reported. The evidence suggests that actual capture fisheries catch may be twice or conceivably even three times the reported total, i.e. around 15-23 million metric tons per year[23]. Because a far higher proportion of inland fisheries than marine fisheries harvest is apparently used directly for human consumption (rather than production of oils and meals, often used for livestock feed), and because discards are believed to be negligible, it has been argued by some that the provision of foodfish from inland waters is not that much less than that from recorded marine catch[24].

Inland water capture fisheries, particularly in countries less developed industrially, certainly provide a staple part of the diet for many human communities. This is the case in West Africa generally, locally in East Africa,

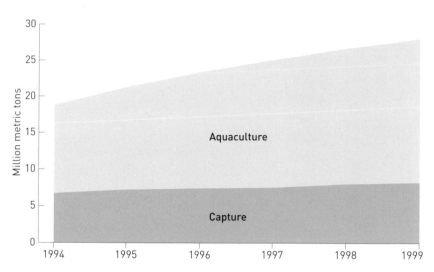

Figure 7.1
Reported global inland fisheries production

Source: FAO[21].

Table 7.7
Major inland fishery countries

Note: The top ten producing countries in 1998.

Source: FAO[21].

	Production (metric tons)	% of world total
China	2 280 000	28.5
India	650 000	8.1
Bangladesh	538 000	6.7
Indonesia	315 000	3.9
Tanzania	300 000	3.7
Russia	271 000	3.4
Egypt	253 000	3.2
Uganda	220 000	2.8
Thailand	191 000	2.4
Brazil	180 000	2.3

Map 7.3
Major areas of diversity of inland water mollusks

This map illustrates the location of areas regarded as globally important for diversity in the bivalve and gastropod mollusks of inland waters, taking into account species richness and local endemism.

Note: For numbered locations see Appendix 6.

Source: Compiled using information and expertise provided by the IUCN/SSC Mollusc Specialist Group; first published in WCMC[1].

Mollusk diversity

■ Important areas

and in parts of Asia and Amazonia. In some landlocked countries inland fisheries are of crucial importance, providing more than 50 percent of animal protein consumed by humans in Zambia[25] and nearly 75 percent in Malawi[26]. In the low-income food-deficit countries fish protein may be particularly important in times of food scarcity.

It is impossible at the global level to carry out any meaningful analysis of the relative contribution of different species or species groups to inland capture fisheries because of the inadequacy of reporting. In FAO (Food and Agriculture Organization of the United Nations) statistics, by far the largest group recorded is 'freshwater fishes not elsewhere included', that is those that are completely unclassified other than being identified as finfishes. These make up just under half of all reported landing

by weight with a further 15 percent consisting of mollusks and crustaceans similarly classified. The majority of the remaining catch is classified into broad species groups (e.g. cyprinids, characins, siluroids), with only three individual fish species having annual reported global landings of more than 100 000 metric tons. These are the Nile perch *Lates niloticus* (ca 330 000 metric tons reported in 1997), Nile tilapia *Oreochromis niloticus* (226 000 metric tons in 1997) and the common carp *Cyprinus carpio* (100 000 metric tons in 1997).

It is, however, evident that the importance of different species of freshwater finfishes varies considerably between different areas. In terms of food security for local subsistence or mixed market/subsistence communities, particularly in the tropics, there is an increasing amount of evidence that the diversity of

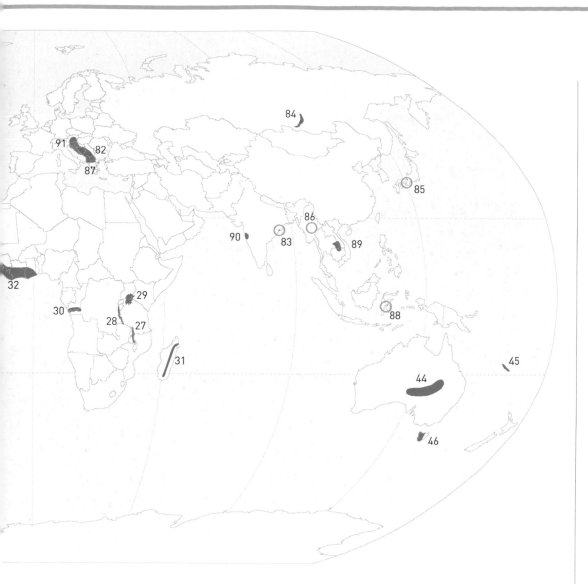

species harvested is in itself a major factor in ensuring a continuous food supply. Many of the species that contribute to these fisheries are often small and would be considered 'trash' fishes in orthodox fisheries, with much lower market value than larger (often non-native) species that might therefore be considered for introduction. However, these small species are easy to preserve and keep under local conditions and moreover are eaten whole, providing a valuable source of calcium and other minerals. Larger species, such as the introduced Nile perch in Lake Victoria, cannot be easily preserved locally and are in any case not eaten whole, leading to a danger of calcium deficiency. Fisheries for such species tend to become industrialized or semi-industrialized, producing fish products for commercial high-value

markets, often for export. While these may improve the balance of payments for the countries concerned, they may ultimately worsen the nutritional status of local people. Additionally, there are some indications that fish populations in mixed species fisheries are more stable over time, less susceptible to 'boom and bust' than those based on a small number of often introduced species.

It is difficult rigorously to assess the condition of inland fish stocks because they appear able to respond rapidly to changing environmental conditions. However, there is a consensus that, regionally, most stocks are fully exploited and in some cases over-exploited. Exploitation has become more efficient because of new technologies, and developing infrastructure has allowed easier access to freshwater resources. Some stocks,

Map 7.4
Major areas of diversity of selected inland water crustacean groups

This map represents a preliminary assessment of areas believed to support high diversity among three of the several major crustacean groups that occur in inland waters, taking into account species richness and local endemism.

Note: For numbered locations see Appendix 6.

Source: Compiled using information and expertise provided by the IUCN/SSC Inland Water Crustacean Specialist Group; first published in WCMC[1].

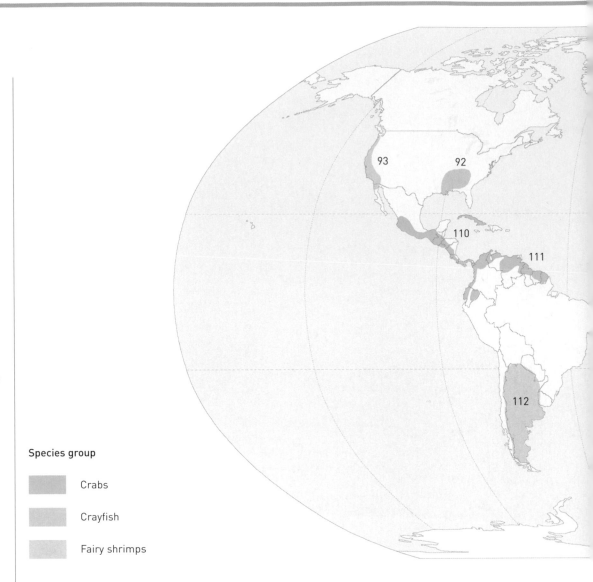

Species group

Crabs

Crayfish

Fairy shrimps

especially in river fisheries, appear to be in decline, but this is seemingly a result mainly of anthropogenic changes to the freshwater environment.

In many parts of the world fishing has high recreational value, as well as being a means of food gathering. Locally, notably in the Amazon basin and in parts of Southeast Asia, capture for the ornamental fish trade may be an important source of income with potential impact on wild populations. Increasingly it is becoming difficult to distinguish between truly wild fish stocks and those that are artificially managed or enhanced in some way.

Other harvested species
Other exploited animal groups in inland waters are far less important globally than finfishes, but may still be highly significant.

Apart from crustaceans and mollusks, mentioned above, these include: frogs (chiefly family Ranidae), exploited for food; crocodilians, hunted mainly for leather; freshwater chelonians, taken for food and to a lesser extent for medicinal purposes, particularly in eastern Asia; waterfowl which are hunted for recreation and for food; fur-bearing mammals, such as beavers *Castor* spp., otters (subfamily Lutrinae) and muskrats (*Ondatra zibethicus* and *Neofiber alleni*), taken for their skins; and manatees (family Trichechidae), taken mostly for food although also used non-consumptively on a small scale for biological control of weeds.

Rice is the principal cultivated wetland plant of global importance to food security. Most of the relatively few plants associated with inland waters that are heavily exploited in

the wild state are also marginal or wetland species. Some species (e.g. *Aponogeton*, in Madagascar) are collected for use as ornamentals; reeds are used as building materials (e.g. thatch); and some are collected for food or as medicines (e.g. *Spirulina* algae). Rhizomes, tubers and seeds (rarely leaves) of aquatic and wetland plants are used as a food source, mainly in less-developed regions where they can be important to food security in times of shortage, but globally they make a relatively minor contribution to human nutrition. Most important are some forms of edible aroid (Araceae), notably some cultivars of *Colocasia* (taro) and the giant swamp taro *Cyrtosperma chamissonis* that grow in flooded conditions and are important food crops in the Caribbean, West Africa and the Pacific islands. Conservation and collection of wild

forms of these is considered a high priority. Sago palms *Metroxylon* spp. in Southeast Asia and the Pacific and watercress *Rorippa nasturtium-aquaticum* in Europe are other examples of cultivated aquatic plants, the wild relatives of which merit conservation. Aquatic plants have been widely used for medicinal purposes, documented for at least two millennia, but such use appears at present to be minor and probably of real significance in few areas. However, interest in ornamental or aquarium water plants is widespread and of some economic importance.

OTHER MAJOR IMPACTS
Physical alteration and destruction of habitat
Destruction of inland water ecosystems is most simply effected by the removal of water. Although humans have always made use of

freshwater systems, the last 200 years (spanning the Industrial Revolution, the growth of cities, the spread of high-input agriculture) have brought about transformations on an unprecedented scale. The global rate of water withdrawal rose steeply at the start of the 20th century, and further after mid-century. Major changes in the distribution of water have resulted mainly from withdrawals for irrigation and secondarily from domestic and industrial use. It has been estimated that humans use 26 percent of the total evapotranspiration from land surfaces and 54 percent of the accessible runoff[27]. Unregulated withdrawal can lead to the wholesale destruction of inland water ecosystems, as has occurred with the Aral Sea in central Asia (Box 7.2). Similarly, many wetlands have been completely destroyed by drainage, often for conversion to agriculture. Other factors can modify or destroy particular habitats within inland water ecosystems. For example, canalization, usually undertaken to improve navigability, generally destroys riparian (shoreline) habitats while flood-control systems drastically alter regimes in floodplains[28].

Box 7.2 Wetland loss in Asian drylands

Aral Sea[29]

Until the mid-20th century the Aral Sea was the world's fourth largest inland water body (after the Caspian Sea, Lake Superior and Lake Victoria). Located within a catchment area of some 1.9 million km² extending over six countries, the Aral is fed by two major rivers, the Amu Darya, rising in the Pamir, and the Syr Darya, rising in the Tien Shan. Starting in the 1960s, excess water withdrawal from these rivers, primarily for cotton irrigation, has severely affected the Aral Sea. Its area had reduced from more than 65 000 km² to about 28 500 km² in 1998, with volume falling by 75 percent, water level falling by around 20 meters (m), and salinity greatly increasing. In consequence, problems with drinking water quality and availability, and with dustborne pollutants, have severely affected the health status of the human population; the commercial fishery has collapsed; waterlogging and salinization have degraded agricultural lands; and the deltaic marshlands of the two feeder rivers have largely been replaced by sandy drylands.

Mesopotamia[30, 31]

Serial satellite images confirm a loss of around 90 percent of the lakes and marshlands in the lower Mesopotamian wetlands during the last three decades. The only significant permanent marshland remaining is in the Al-Hawizeh region. The large number of dams now present on upstream parts of the Tigris-Euphrates system may have contributed to this loss, but it appears to be primarily the result of major hydrological engineering works in southern Iraq, notably the completion of the major outfall drain (or 'third river') which diverts water to the head of the Gulf. This loss has placed further pressure on the Ma'dan (Marsh Arabs), now largely displaced within Iraq or in refugee camps in Iran. Recent information is scarce, but biodiversity in the region will inevitably have been affected, probably including the endemic form of smooth-coated otter (*Lutrogale perspicillata*, an otherwise oriental species, assessed as globally threatened).

Azraq oasis[32]

Groundwater extraction for urban needs in Jordan rose from about 2 million m³ in 1979 to about 25 million m³ in 1993, with an additional 25 million m³ per year used for agricultural irrigation. The important Azraq wetlands natural reserve, formerly extending over some 12 000 hectares, and a vital staging site for bird migrants, now supplies around one quarter of Amman's water needs, and as a consequence has lost most of its marshland and migrant bird populations.

Dams and reservoirs

Dams, particularly large dams, have a major impact on the rivers on which they are built. They affect flow regimes, often dramatically, destroy large areas of existing habitat (while at the same time creating new ones) and can catastrophically disrupt the life cycles of species that migrate up and down rivers[33]. Large dams are unevenly distributed across the world's major catchments, with a particularly high concentration in North America, especially within the contiguous states of the United States, where at least eight catchments have in them more than 100 large dams each. In contrast, small and medium-sized dams, which may cumulatively have as substantial an impact, are concentrated in eastern Asia, particularly China[34]. Dams may be primarily for the generation of hydroelectric power, or to create reservoirs for the storage of water, or both. The size of a dam is not necessarily directly related to the area or volume of the impoundment created or to its downstream impact.

Pollution and water quality

Assessment of anthropogenic changes in water quality is not always easy as such changes are invariably superimposed on natural background variations. Historically, a similar sequence of water quality issues has became apparent in both Europe and North America during rapid socioeconomic development over the past 150 years. Problems of fecal and other organic pollution were evident in the mid-19th century, followed by salinization, metal pollution and eutrophication in the first half of the 20th century, with radioactive waste, nitrates and other organic micropollutants, and acid rain most prominent in recent decades. Newly industrializing countries are likely to face these problems over a much more compressed period, and typically without the capacity to monitor and analyze water quality, or manage water use appropriately[35].

Different kinds of pollutants appear to affect different classes of water system to differing extent[36]. With regard to quality for human use, contamination by pathogens of fecal origin is the major problem in river systems, and eutrophication probably the most widespread problem affecting lake and reservoir waters[36-38].

Acid deposition through precipitation has been recognized as a regional transboundary phenomenon since the 1960s. Industrial emissions of sulfur and nitrogen oxides (SO_2, NO_x), mainly a result of fossil fuel combustion, are the principal source of acid rain. Most evidence of acid rain and its effects relates to North America and Europe, but emission rates are rising steeply in rapidly industrializing countries elsewhere. Acid rain in one country may be a consequence of compounds released into the atmosphere by industry in another country hundreds of kilometers distant. The geology, soil and vegetation of drainage basins strongly influence the acidification process. Acid rain has been shown to decrease species diversity in lakes and streams but has not been implicated in any recorded species extinction or any major species decline. It has not yet been shown to be a significant issue in tropical freshwaters, where global freshwater diversity is concentrated[1].

Sedimentation

Removal or extension of forest cover, or any anthropogenic interference with soils and landcover (e.g. agriculture, urbanization, road construction, mining), modifies the rate of runoff from catchment slopes and also the density of particles carried in the drainage system. All moving waters carry some mass of suspended material, and there is considerable natural variation in this in space and time, but logging can increase sediment load

by up to 100 percent for a short period, and 20-50 percent over the longer term. Sediment reaching lakes will be deposited and in effect enter long-term storage; depending on water velocity, sediment in rivers will settle out on floodplains or other parts of the course, or be carried into the coastal marine environment.

Increased sedimentation can have several effects on aquatic biodiversity: deposition can radically change the physical environment of species restricted to particular conditions of depth, light penetration and velocity; it is a major carrier of heavy metals, organic pollutants, pathogens and nutrients; and it can interfere mechanically with respiration in gill-breathing organisms[1]. The endemic

cichlid fishes of the African Great Lakes rely on complex visual signals in breeding, and reduced water clarity because of sedimentation is suspected to be affecting their breeding success.

Introduced species

Unplanned or poorly planned introduction of non-native species and genetic stocks is a major threat to freshwater biodiversity. Such introductions can have negative or positive effects on fishery production; it is a reasonable assumption that all successful introductions will have an impact on existing population size and community structure, and many changes are likely to be undesirable[39]. The incomplete

Although some factories have introduced cleaner production procedures, industrial emissions are still the principal source of acid rain.

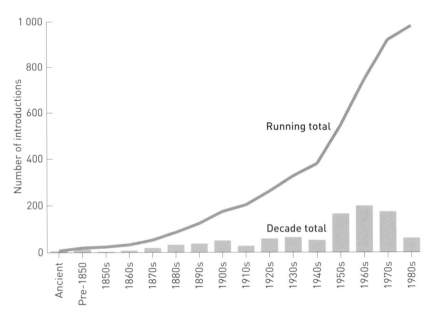

Figure 7.2
Inland water fish introductions

Source: Welcomme[59].

Figure 7.3
Freshwater population trends

Note: A simplified representation of the average population change in a sample of 194 inland water species, see text.

Source: Loh[46].

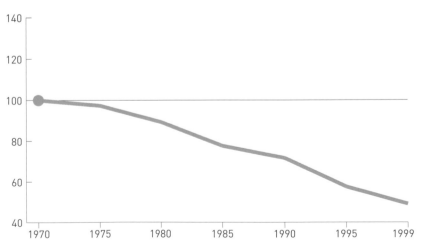

information available suggests that although a significant number of fish introductions took place during the 19th century, the three decades starting with the 1950s were particularly important (Figure 7.2). A classic example of the effect of introduced species is the impact of the Nile perch *Lates niloticus* on the haplochromine cichlids of Lake Victoria discussed further below.

Several species of aquatic plant, in particular free-floating species that are able to spread rapidly by vegetative growth (most notoriously the South American water hyacinth *Eichhornia crassipes*), but also other forms, have dispersed widely over the globe and become major pest species. They block

drainage channels, sluices and hydroelectric installations, impede boat traffic and hinder fishing. In recent decades the question of how best to control or eradicate pest species has been the foremost issue in conservation and management of aquatic plants[1].

THE CURRENT STATUS OF INLAND WATER BIODIVERSITY

As with marine species, assessment of the status of wholly aquatic inland water species is hampered by difficulties of direct observation. However, because these species also in general have far more restricted ranges than marine species, it is easier to infer their status from assessment of habitat condition and from sampling efforts. Amphibious or surface-dwelling species may be relatively easier to monitor. Where such species are of economic importance – as for example with those European and North American waterfowl that play a role in the recreational hunting industry – they may be among the best monitored of all wild species.

Threatened and extinct species

In the few cases where elements of inland water faunas – usually fishes – have been studied in any detail, it has generally been found that more species than suspected are threatened or cannot be re-recorded[40-43]. Among the 20 or so countries where the entire inland water fish fauna has been evaluated, an average of 17 percent of the species are regarded as globally threatened (categories 'critically endangered', 'endangered' or 'vulnerable' in the IUCN (World Conservation Union) threat categorization system[44]) (Table 7.8). A far larger proportion is likely to be in local decline, although not in danger of global extinction. The proportion of inland water chelonians that is believed threatened is even higher: 99 such species were categorized as threatened in 2000, equivalent to about 60 percent of the number of inland water chelonians listed in Table 7.6.

Amongst mammals and birds the proportions are considerably lower, probably because many semi-aquatic species are able to disperse from one inland water body to another relatively easily. Nevertheless, somewhat more species than average for the groups as a whole are regarded as threatened. Table 7.9 shows the category and taxonomic distribution of threatened inland water vertebrates.

Two groups of species, the Lake Victoria cichlid fishes and the Mobile Bay drainage gastropod mollusks, serve as exemplary case studies illustrating the major threats faced by inland water biota worldwide.

Lake Victoria, the largest tropical lake in the world, provides a classic example of the potential negative impacts of species introductions. Until some 30 years ago, when the large top predator, the Nile perch *Lates niloticus*, was introduced, the lake supported an exceptional 'species flock' of more than 300 species of haplochromine cichlid fishes, as well as smaller numbers from other families. The cichlids are of enormous interest in the study of evolutionary biology. Not all the species have yet been formally described; many of these are known among aquarists and others only by informal common names. At least half of the native species are believed to be extinct or so severely depleted that too few individuals exist for the species to be harvested or recorded by scientists.

Predation by the Nile perch is believed to be the major cause of this decline, but important additional factors include increasing pollution and sediment load, excess fishing pressure, and possible competition from introduced tilapiine cichlids. The lake itself has now become depleted of oxygen, and a shrimp tolerant of oxygen-poor waters provides a major food source for the Nile perch. In recent years the Nile perch and one of the introduced tilapiines have formed the basis of a high-yielding fishery and an important national and export trade. However, it is thought unlikely that such high yields will be maintained because of continued overfishing and the suspected instability of the already highly disturbed lake ecosystem[1].

Dam construction is the prime cause of extinction in the gastropod fauna of the Mobile Bay drainage in Alabama, United States. Historically, the freshwater snail fauna of Mobile Bay basin was probably the most diverse in the world, followed by that of the

Mekong River. Nine families and about 118 species were known at the turn of the century to occur in the Mobile Bay drainage. Several genera and many species were endemic, particularly in the Pleuroceridae. Recent surveys suggest at least 38 species are extinct (32 percent); decline in species richness ranges between 33 percent and 84 percent in the main river systems. The richest fauna was in the Coosa River and this system has undergone the greatest decline (from 82 to 30 species). Almost all the snail species presumed extinct were members of the Pleuroceridae and grazed on plants growing on rocks in shallow oxygen-rich riffle and shoal zones. The system has 33 major hydroelectric dams and many smaller impoundments, as well as locks and flood control structures. A combination of siltation behind dams and submergence of shallow water shoals has removed the snails' former habitat. Where habitat remains it has diminished in area and become fragmented[45].

The inland water living planet index
An impression of the overall trend in a large sample of species for which indicators of

Table 7.8
Numbers of threatened freshwater fishes in selected countries

Notes: These are the 20 countries whose fish faunas have been evaluated completely, or nearly so, and which have the greatest number of globally threatened freshwater fish species. The estimates of total fish species present are all approximations.

Source: Total species estimates from UNEP-WCMC database; threatened species data from online Red List http://www.redlist.org (accessed March 2002).

	Total species	Threatened species	% threatened
USA	822	120	15
Mexico	384	82	21
Australia	216	27	13
South Africa	94	24	26
Croatia	64	22	34
Turkey	174	22	13
Greece	98	19	19
Madagascar	41	13	32
Canada	177	12	7
Papua New Guinea	195	11	6
Romania	87	11	13
Italy	45	11	24
Bulgaria	72	11	15
Hungary	79	10	13
Spain	50	10	20
Moldova	82	9	11
Portugal	28	9	32
Sri Lanka	90	9	10
Slovakia	62	9	15
Japan	150	9	6

Table 7.9
Taxonomic distribution and status of threatened inland water vertebrates

Notes: Family selection based on Table 7.6. Numbers refer to species (not subspecies and geographic populations) recorded in the Red List database as occurring in freshwater; only the birds and mammals have been comprehensively assessed for species at risk.

i Emydidae here includes batagurine turtles sometimes treated as a separate family (several of the species are primarily terrestrial not aquatic).

Source: Status categories from 2000 Red List database, www.redlist.org[44] (accessed February 2002).

Phylum and class or order	Family	Common name	'Critically endangered'	'Endangered'	'Vulnerable'
Craniata – fishes					
Petromyzontiformes		Lampreys		1	2
Carchariniformes		Ground sharks	1	1	
Pristiformes		Sawfish	2	3	
Myliobatiformes		Rays		5	2
Acipenseriformes		Sturgeons	6	10	8
Atheriniformes		Silversides	6	5	31
Beloniformes		Needlefishes, sauries, etc.	2	3	8
Characiformes		Characins		1	1
Clupeiformes		Herrings and anchovies		2	3
Cypriniformes		Carp, minnow, loaches	42	37	117
Cyprinodontiformes		Rivulines, killifish, etc.	18	20	26
Gasterostiformes		Sticklebacks	1		
Ophidiiformes		Pearlfishes, etc.			6
Osteoglossiformes		Bonytongues		1	
Perciformes		Perches, etc.	51	25	104
Percopsiformes		Trout-perches, cavefishes	1		3
Salmoniformes		Salmonids	8	8	22
Scorpaeniformes		Gurnards, scorpionfishes, etc.	2		4
Siluriformes		Catfishes	8	7	22
Synbranchiformes		Swamp eels		1	
Syngnathiformes		Pipefishes, seahorses, etc.	1		
Craniata – Amphibia					
Caudata	Cryptobranchidae	Giant salamanders and hellbenders			1
	Proteidae	Mudpuppies and olm			1
Anura	Pipidae	Clawed frogs and pipid toads			1
Craniata – Reptilia					
Chelonia	Carettochelidae	Pig-nosed soft-shelled turtle			1
	Chelidae	Austro-american side-necked turtles	3	4	7
	Chelydridae	Snapping turtles			1
	Dermatemydidae	Central American river turtle		1	
	Emydidae[i]	Pond and river turtles	13	22	16

population change are available can be derived from the WWF living planet index. This method is designed to represent the change in the 'average species' in the sample from one five-year interval to the next, starting in 1970. The inland waters sample represents 194 species of mammals, birds, reptiles and fishes, and the index suggests a significant declining trend over the last three decades of the 20th century (Figure 7.3)[46].

The sample includes a large number of wetland and water margin species in addition to truly aquatic forms. The declining trend in the inland water sample is a little steeper than the equivalent marine index, and substantially steeper than the terrestrial index.

Assessment of the status of inland water ecosystems

Indicators of habitat condition in river catchments

While the living planet index methodology provides an indication of global trends in inland water biodiversity, an alternative approach aims to assess the overall condition of inland water ecosystems. In one approach to a global assessment[1] two high-order indicators of likely habitat condition in different river catchments were combined. First, a wilderness measure for each river catchment was calculated using the wilderness index methodology developed by the Australian Heritage Commission[47, 48] (see Chapter 4 and particularly Map 4.5). This

Phylum and class or order	Family	Common name	'Critically endangered'	'Endangered'	'Vulnerable'
	Kinosternidae	Mud and musk turtles			4
	Pelomedusidae	Side-necked turtles		2	6
	Trionychidae	Soft-shelled turtles	4	5	6
Crocodilia	Alligatoridae	Caimans and alligators	1		
	Crocodylidae	Crocodiles	3	2	3
	Gavialidae	Gharial and false gharial		1	
Craniata – Aves					
Anseriformes	Dendrocygnidae	Whistling-ducks			1
	Anatidae	Ducks, swans and geese	5	7	12
Charadriiformes	Charadriidae	Plovers, etc.	1	1	2
	Laridae	Gulls, terns, skua, auks			1
	Rhynchopidae	Skimmers			1
	Scolopacidae	Curlews, etc.		1	
Ciconiiformes	Ardeidae	Herons, egrets		3	5
	Ciconiidae	Storks		3	2
	Threskiornithidae	Ibis, spoonbill	2	2	
	Phoenicopteridae	Flamingos			2
Gruiformes	Heliornithidae	Limpkin and sungrebes			1
	Rallidae	Rails, gallinules and coots	1	7	6
Pelicaniformes	Pelecanidae	Pelicans and shoebill			1
Passeriformes	Cinclidae	Dippers			1
Podicipediformes	Podicipedidae	Grebes	2		2
Craniata – Mammalia					
Artiodactyla	Hippopotamidae	Hippopotamus			1
Carnivora	Mustelidae	Mustelids, otters		2	3
	Phocidae	Earless seals		1	1
Cetacea	Platanistidae	River dolphins	1	2	1
Insectivora	Soricidae	Shrews	2	1	
	Talpidae	Moles and desmans			2
	Tenrecidae	Tenrecs and otter shrews		4	
Rodentia	Muridae	Mice, voles, etc.	1	5	3
Sirenia	Trichechidae	Manatees			3

Figure 7.4
River basin richness and vulnerability

Notes: Each symbol represents a river basin scored on fish family richness and basin vulnerability; the darker symbols score high on both counts and may be regarded as high priority. See Map 7.5 and Table 7.10.

Source: WCMC[1].

provides a measure of the spatial extent of human impacts in the land area of the catchment as a whole and does not directly reflect the condition of riverine ecosystems. Secondly, a national water resource vulnerability index (WRVI)[49] developed on the basis of three water resource stress indices (reliability, use-to-resource and coping capacity) was resolved at catchment level (by measuring that proportion of each catchment that lies within any given country and weighting this proportion by the national WRVI). This provides an indirect measure of vulnerability for each catchment. By normalizing and combining these two measures, a single value representing vulnerability or stress level for each major river

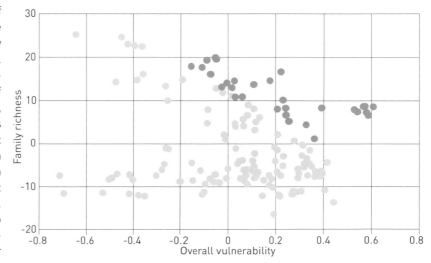

Map 7.5
Priority river basins

A possible scheme for prioritizing river basins at global level. Major catchment basins are assessed for fish diversity (at family level) and for vulnerability (a combined indication of disturbance and potential water stress). Systems with both high diversity and vulnerability are proposed as high priority for investment and management action; those with low diversity and little disturbance are lower priority. See text and source.

Source: WCMC[1].

Table 7.10
Thirty high-priority river basins

Notes: These are 30 river basins that support high biodiversity (assessed as fish family richness) and are most vulnerable to future pressures (have a low wilderness score, and are high on the water resource vulnerability index). See text for further explanation and Figure 7.4. Basins are listed in alphabetical sequence.

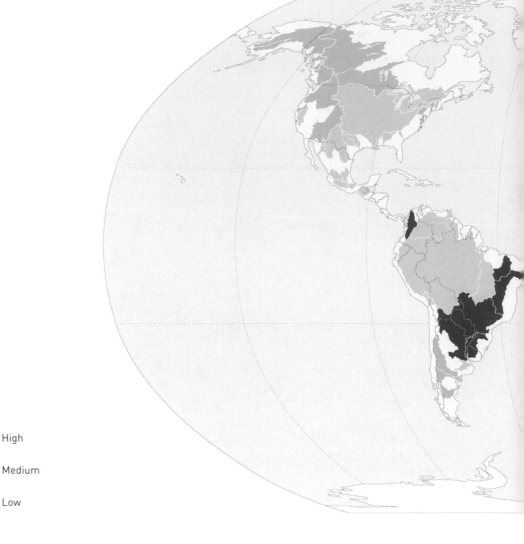

Priority

■ High

▢ Medium

▨ Low

catchment can be calculated. Overall, the pattern mapped agrees well with what might intuitively be expected, with few evident anomalies. Globally, the most stressed catchments are to be found in South Asia (the Indian subcontinent), the Middle East and western and northcentral Europe. The least stressed are those in the northwestern part of North America. Further refinements of this analysis would involve applying water resource vulnerability measures for individual catchments[50] rather than to countries and incorporating measures of direct impacts on inland waters, in particular water quality and the number and kind of dams.

The number of freshwater fish families present in each basin can be calculated from the family density surface (Map 7.1) in order to provide an indication of biodiversity value. Plotting family number against vulnerability allows the basins with high diversity and high vulnerability to be identified, and these can reasonably be regarded as global priorities for management intervention designed to minimize biodiversity loss (Figure 7.4, Table 7.10, Map 7.5).

River basins			
Ca	Irrawaddy	Nile	Senegal
Cauvery	Krishna	Pahang	Sittang
Chao Phraya	Ma	Parana	Song Hong (Red)
Gambia	Magdalena	Parnaiba	Tapti
Ganges-Brahmaputra	Mahanadi	Penner	Tembesi-Hari
	Mekong	Perak	Uruguay
Godavari	Narmada	Salween	Volta
Indus	Niger	Sao Francisco	

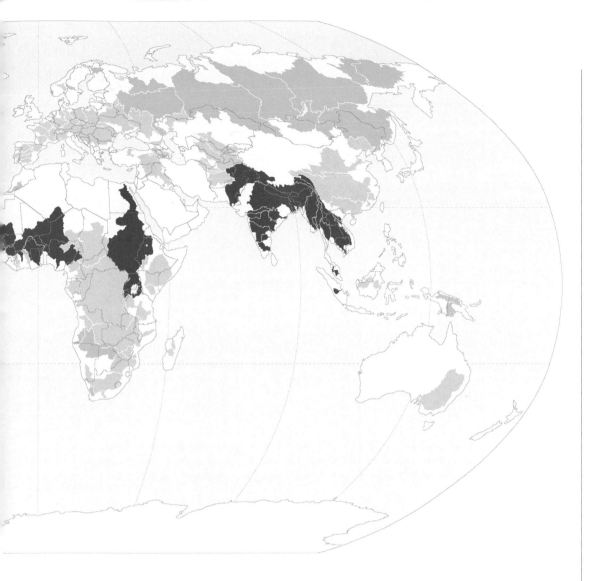

Figure 7.5
Changes in condition of a sample of freshwater lakes between 1950s and 1980s

Note: Number following area name is number of lakes in sample.

Source: Various sources[52, 53].

Habitat condition in lakes

One semi-quantitative study attempted to evaluate the changing condition of freshwater lakes during the last three decades of the 20th century[51]. A baseline was provided by Project Aqua, a project initiated by the Societas Internationalis Limnologiae in 1959, which collated and later published[52] information provided by national and regional specialists in relation to more than 600 water bodies. Many of these lakes were treated in later information sources relating to the 1980s and 1990s[53], and in some 93 cases it was possible to make an assessment that, although imprecise, is likely to be indicative of changing conditions. Each lake was scored according to whether its condition appeared to have deteriorated (or impacts increased), or to have improved, or whether no change (or no new information) was reported. Improvement was reported in a very small number of lakes, but the overwhelming trend was for a deterioration in conditions (see Figure 7.5).

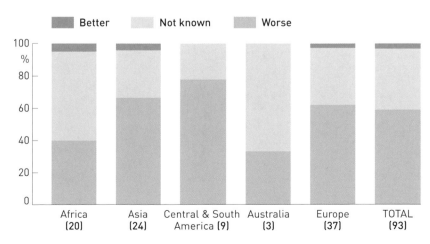

REFERENCES

1 WCMC 1998. *Freshwater biodiversity: A preliminary global assessment.* By Groombridge, B. and Jenkins, M. WCMC–World Conservation Press, Cambridge.

2 Gorthner, A. 1994. What is an ancient lake? Speciation in ancient lakes. In: Martens, K., Goddeeris, B. and Coulter, G. (eds). *Archiv für Hydrobiologie. Ergebnisse der Limnologie* 44: 97-100.

3 Horne, J.A. and Goldman, C.R. 1994. *Limnology.* McGraw-Hill Inc., New York.

4 Guégan, J.-F., Lek, S. and Oberdorff, T. 1998. Energy availability and habitat heterogeneity predict global riverine fish diversity. *Nature* 391: 382-384.

5 Oberdorff, T., Guégan, J.-F. and Hugueny, B. 1995. Global scale patterns of fish species richness in rivers. *Ecography* 18: 345-352.

6 Kottelat, M. and Whitten, T. 1996. *Freshwater biodiversity in Asia with special reference to fish.* World Bank Technical Paper No. 343.

7 Olson, D.M. *et al.* (eds) 1997. *Freshwater biodiversity of Latin America and the Caribbean: A conservation assessment.* Proceedings of a workshop. World Wildlife Fund, Washington DC.

8 Abell, R. *et al.* 1998. *A conservation assessment of the freshwater ecoregions of North America.* Final report submitted to the US EPA, April. World Wildlife Fund-US.

9 Olson, D.M., and Dinerstein, E. 1998. The Global 200: A representation approach to conserving the earth's distinctive ecoregions. *Conservation Biology* 12: 502-515.

10 WWF. Online at http://www.panda.org/resources/programmes/global200/pages/list.htm (accessed March 2002).

11 Watter, G.T. 1992. Unionids, fishes, and the species-area curve. *Journal of Biogeography* 19: 481-490.

12 Tittley, I. 1992. Contribution to chapter 7, Lower plant diversity. In: World Conservation Monitoring Centre. *Global biodiversity.* Groombridge, B. (ed.) Chapman and Hall, London.

13 Goh, T.K. and Hyde, K.D. 1996. Biodiversity of freshwater fungi. *Journal of Industrial Microbiology* 17: 97-100.

14 Shearer, C.A. 1993. The freshwater ascomycetes. *Nova Hedwigia* 56: 1-33.

15 Shearer, C.A. Freshwater ascomycetes and their anamorphs. Available online at http://fm5web.life.uiuc.edu:23523/ascomycete/ (accessed March 2002).

16 Sculthorpe, C.D. 1967. *The biology of aquatic vascular plants.* Edward Arnold, London.

17 Palmer, M. *et al.* 1997. Biodiversity and ecosystem processes in freshwater sediments. *Ambio* 26(8): 571-577.

18 Nelson, J.S. 1994. *Fishes of the world.* 3rd edition. John Wiley and Sons, Inc., New York.

19 UN (CSD) 1997. *Comprehensive assessment of the freshwater resources of the world.* Economic and Social Council. Fifth Session, 5-25 April. E/CN.17/1997/9. Available online at http://www.un.org/esa/sustdev/csd.htm (accessed April 2002).

20 FAO 1996. *Food production: The critical role of water.* Technical background document. World Food Summit. Available online in pdf at http://www.fao.org (accessed April 2002).

21 FAO 2000. *The state of world fisheries and aquaculture 2000.* Fisheries Department, Food and Agriculture Organization of the United Nations, Rome. Available online at http://www.fao.org/DOCREP/003/X8002E/X8002E00.htm (accessed April 2002).

22 FAO 1999. *Review of the state of world fishery resources: Inland fisheries.* FAO Fisheries Circular No. 942. FIRI/C942.

23 Coates, D. 1995. Inland capture fisheries and enhancement: Status, constraints and prospects for food security. Paper presented at international conference on sustainable contribution of fisheries to food security. Kyoto, Japan, 4-9 December 1995. Food and Agriculture Organization of the United Nations, Rome. KC/FI/95/Tech/3.

24 Borgström, R. 1994. Freshwater ecology and fisheries. In: Balakrishnam, M., Borgström, R. and Bie, S.W. *Tropical ecosystems,* pp. 41-69. International Science Publishers, Oxford and IBH, New Dehli.

25 Scudder, T. and Conelly, T. 1985. *Management systems for riverine fisheries.* FAO Fisheries Technical Paper No. 263. Food and Agriculture Organization of the United Nations, Rome.

26 Munthali, S.M. 1997. Dwindling food-fish species and fishers' preference: Problems of conserving Lake Malawi's biodiversity. *Biodiversity and Conservation* 6: 253-261.

27 Postel, S.L., Daily, G.C. and Ehrlich, P.R. 1995. Human appropriation of renewable fresh water. *Science* 271: 785-788.

28 L'Vovich, M.I. *et al.* 1990. Use and transformation of terrestrial water systems. In: Turner II, B.L. *The earth as transformed by human action. Global and regional changes in the biosphere over the past 300 years*, pp. 235-252. Cambridge University Press with Clark University.

29 FAO AQUASTAT. Online at http://www.fao.org/landandwater/aglw/aquastat/ regions/fussr/main8.htm (accessed March 2002).

30 Evans, M.I. 2001. The ecosystem. Draft paper for conference, Iraqi Marshlands: Prospects, London, 21 May 2001. Available online at http://www.amarappeal.com/documents/ Draft_Report.pdf (accessed April 2002).

31 UNEP 2001. UN study sounds alarm about the disappearance of the Mesopotamian marshlands. 18 May. Available online at http://www.grida.no/inf/news/news01/ news42.htm, and also see http://www.grid.unep.ch/./activities/sustainable/ tigris/marshlands/marshlands.pdf (accessed April 2002).

32 Fariz, G.H. and Hatough-Bouran, A. 1998. Population dynamics in arid regions: The experience of the Azraq Oasis Conservation Project (Case study: Jordan). In: de Sherbinin, A. and Dompka, V. (eds). *Water and population dynamics: Case studies and policy implications.* Available online at http://www.aaas.org/international/ehn/waterpop/ front.htm (accessed April 2002).

33 Dynesius, M. and Nilsson, C. 1994. Fragmentation and flow regulation of river systems in the northern third of the world. *Science* 266: 735-762.

34 Avakyan, A.B. and Iakovleva, V.B. 1998. Status of global reservoirs: The position in the late twentieth century. *Lakes and Reservoirs: Research and Management* 3: 45-52.

35 Meybeck, M. and Helmer, R. 1989. The quality of rivers: From pristine stage to global pollution. *Palaeogeography, palaeoclimatology, palaeoecology* 75: 283-309.

36 Chapman, D.V. 1992. *Water quality assessment.* Chapman and Hall, London.

37 UNEP 1991. United Nations Environment Programme. *Freshwater pollution.* UNEP/GEMS Environment Library No. 6.

38 UNEP 1995. United Nations Environment Programme. *Water quality of world river basins.* UNEP Environment Library No. 14.

39 Moyle, P.B. 1996. Effects of invading species on freshwater and estuarine ecosystems. In: Sandlund, O.T. *et al.* (eds). *Proceedings of the Norway/UN Conference on Alien Species, Trondheim, 1-5 July 1996*, pp. 86-92. Directorate for Nature Management and Norwegian Institute for Nature Research. Trondheim.

40 Moyle, P.B. and Leidy, R.A. 1992. Loss of biodiversity in aquatic ecosystems: Evidence from fish faunas. In: Fielder, P.L. (eds). *Conservation biology, the theory and practice of nature conservation, preservation and management*, pp. 129-169. Chapman and Hall, New York and London.

41 Stiassny, M. 1996. An overview of freshwater biodiversity with some lessons learned from African fishes. *Fisheries* 21(9): 7-13.

42 Reinthal, P.N. and Stiassny, M.L.J. 1991. The freshwater fishes of Madagascar: A study of an endangered fauna with recommendations for a conservation strategy. *Conservation Biology* 5(2): 231-243.

43 Kirchhofer, A. and Hefti, D. (eds) 1996. *Conservation of endangered freshwater fish in Europe.* Advances in Life Sciences. Birkhäuser Verlag, Basel.

44 Hilton-Taylor, C. (compiler) 2000. *2000 IUCN Red List of threatened species*. IUCN-the World Conservation Union, Gland and Cambridge. Also available online at http://www.redlist.org/ (accessed April 2002).

45 Bogan, A.E., Pierson, J.M. and Hartfield, P. 1995. Decline in the freshwater gastropod fauna in the Mobile Bay basin. In: LaRoe, E.T. *et al.* (eds). *Our living resources: A report to the nation on the distribution, abundance, and health of U.S. plants, animals, and ecosystems*, pp. 240-252. US Department of the Interior, National Biological Service, Washington DC.

46 Loh, J. (ed.) 2000. *Living planet report 2000*. WWF – World Wide Fund for Nature, Gland.

47 Lesslie, R., *in litt.*, 30 May 1998.

48 Lesslie, R. and Maslen, M., 1995. *National wilderness inventory, handbook and procedures, content and usage*. 2nd edition. Commonwealth Government Printer, Canberra.

49 Raskin, P. *et al.* 1997. *Water futures: Assessment of long-range patterns and problems*. Background Report No. 3 of Comprehensive Assessment of the Freshwater Resources of the World. Stockholm Environment Institute.

50 Alcamo, J., Henrichs, T. and Rösch, T. 2000. *World water in 2025*. Global modeling and scenario analysis for the World Commission on Water for the 21st Century. Kassel World Water Series. Report No. 2. University of Kassel, Centre for Environmental Systems Research.

51 Loh, J. *et al.* 1998. *Living planet report 1998*. WWF–World Wide Fund for Nature, Gland.

52 Luther, H. and Rzóska, J. 1971. *Project Aqua: A source book of inland waters proposed for conservation*. IBP Handbook No. 21. IUCN Occasional Paper No. 2. International Biological Programme. Blackwell Scientific Publications, Oxford and Edinburgh.

53 For example, Scott, D.A. (ed.) 1989. *A directory of Asian wetlands*. IUCN, Gland and Cambridge; Scott, D.A. and Carbonell, M. 1986. *A directory of neotropical wetlands*. IUCN, Gland and Cambridge; International Lake Environment Committee (ILEC), World lakes database, available online at http://www.ilec.or.jp/database/database.html (accessed March 2002).

54 Anon 1978. Centre for Natural Resources, Energy and Transport of the Department of Economic and Social Affairs, United Nations. Register of International Rivers. *Water Supply and Management* 2: 1-58. (Special issue). Pergamon Press.

55 Shiklomanov, I.A. 1998. *World water resources at the beginning of the XXIst century*. Report to UNESCO (Division of Water Sciences). Available online in summary form at http://webworld.unesco.org/water/ihp/db/shiklomanov/summary/html/summary.html (accessed March 2002).

56 Martens, K., Goddeeris, B. and Coulter, G. (eds) 1994. Speciation in ancient lakes. *Archiv für Hydrobiologie. Ergebnisse der Limnologie* 44.

57 Hutchinson, G.E. 1993. *A treatise on limnology. Vol. IV. The zoobenthos*. John Wiley and Sons, Inc., New York.

58 Eschmeyer, W.N. *et al.* 1998. *A catalog of the species of fishes*. Vols 1-3. California Academy of Sciences, San Francisco. Available online in searchable format at http://www.calacademy.org/research/ichthyology/catalog/fishcatsearch.html (accessed January 2001).

59 Welcomme, R.L. (compiler) 1988. *International introductions of inland aquatic species*. FAO Fisheries Technical Paper No. 294. Food and Agriculture Organization, Rome.

60 Berra, T.M. 2001. *Freshwater fish distribution*. Academic Press, San Diego and London.

61 Johnson, T.C. *et al.* 1996. Late Pleistocene desiccation of Lake Victoria and rapid evolution of cichlid fishes. *Science* 273: 1091-1093.

62 Snoeks, J. 2000. How well known is the ichthyofauna of the large East African Lakes? *Advances in Ecological Research* 31: 17-38.

8 Global biodiversity: responding to change

CONCERN OVER THE RATE OF CHANGE IN THE BIOSPHERE, particularly in populations of larger animals and in the naturalness of landscapes, dates primarily from the turn of the 20th century. Although some early measures were taken, prompted by pioneering conservation organizations, concerted international effort did not develop until mid-century. Since this period, actions have tended to focus on conservation of individual species or of large areas of habitat, as national parks and other protected areas.

During the 1970s several international agreements aimed at conserving wild species and habitats were agreed and entered into force. During the 1980s the word 'biodiversity' was coined, and a new paradigm formulated, aiming to integrate biodiversity conservation with sustainable human development. In 1992 the pivotal Earth Summit was held in Rio, culminating in agreement on the Convention on Biological Diversity, and Agenda 21 – a plan of action for sustainable human development.

After a decade of planning and implementation, the status of some species and the condition of some ecosystems have improved. In some areas, this has been achieved by restoration of degraded ecosystems, an approach likely to be of increasing importance in the future. However, given future trends in human population growth and development, it seems probable that pressures on biodiversity will continue to intensify in coming decades. In many parts of the world, the most significant challenge to conservation will be to minimize losses of biodiversity while improving the livelihood of human populations, particularly those experiencing severe poverty.

BIODIVERSITY CHANGE

Change is a dominant theme in the biosphere; species have diversified and become extinct throughout the history of life, and habitats too have expanded and declined, along with changes in community composition, climate and landforms. However, the recent rate of change in global biodiversity appears higher than that prevailing over most periods of geological time. Earlier chapters have indicated how in well-assessed groups of organisms, such as birds and mammals, a significant proportion of species now appears to be threatened with global extinction. Countless other species exist in reduced numbers and as fragmented populations, many of which are threatened with extinction at national or more local scale. The rate of extinction over the past few centuries, so far as this can be reliably estimated, appears to be much higher than the average background rate estimated from the fossil record. Whatever proximate causes may be implicated, the increasing numbers and material aspirations of the human species, the increasing burden of waste and continuing inequities in the distribution of wealth and resources, together appear to drive most contemporary biodiversity loss.

Despite abundant evidence of change in biodiversity, often involving radical modification of landcover or water bodies, much of the relevant information on the status of species or populations is qualitative or anecdotal in nature. At the level of individual species, a qualitative assessment of trends in numbers or range may be adequate evidence of the need for management intervention, or a guide to its effectiveness, but it remains difficult to develop a comprehensive and quantitative view of global species trends. The relatively few large-scale population monitoring programs that exist have tended to concentrate on marine fishery stocks, on particular groups (e.g. farmland birds) or the larger threatened species of animals. The resulting series of data are not always directly comparable, and often not amenable to presentation in a manner appropriate for assisting global and regional planning. Much recent discussion[1] has focused on the design of biodiversity indicators that could serve this purpose, but few operational systems yet exist.

This chapter first provides an overview of current approaches to conservation, focusing

Concern about the decline of species became the focus of sustained international attention only in the last 30 years.

on species, areas and ecosystems, and the particular role of protected areas, as well as the potential for ecological restoration. The international dimension is then considered, with specific reference to multilateral environmental agreements and conventions relating to biodiversity. Finally, changes in global biodiversity that might be anticipated in future are explored through description of recently developed scenarios. Such changes highlight the challenges facing current and future efforts aimed at the conservation and sustainable use of global biodiversity.

RESPONSES TO BIODIVERSITY CHANGE

Although concern about the decline of species and loss of undisturbed habitats arose within developed industrialized societies around the start of the 20th century, they became subject to sustained international attention only during the 1970s. In general, major sectors that used biodiversity, such as forestry and fisheries, failed to incorporate consideration of biodiversity into their planning and regulations. The issue was widely perceived as marginal to other concerns, and conservation of nature was commonly seen as an impediment to human development.

The past three decades have been marked by the emergence of concerted responses to the crisis in biodiversity at all levels from the local to the global. Civil society, largely in the form of a hugely diverse and increasingly sophisticated network of non-governmental organizations (NGOs), has undoubtedly been the driving force for most of this change (see Box 8.1). Response to pressure from civil society can be seen at both governmental and intergovernmental levels.

Much of the progress that has been made in recent decades can be attributed to the vision articulated in the World Conservation Strategy by the International Union for Conservation of Nature and Natural Resources, the United Nations Environment Programme and the World Wildlife Fund in 1980[2]. This document helped set the conservation agenda during the period leading up to the UN Conference on Environment and Development (the Earth Summit) in Rio de Janeiro in 1992. It emphasized conservation for development, and embraced the notion of the sustainable use of natural resources as a means of achieving this. Maintenance of biological diversity as part of a functioning biosphere was presented as a fundamental prerequisite of sustainable human development, not an impediment to it. On the ground, this was manifested in increasing numbers of integrated conservation and development projects in developing countries. The Earth Summit was undoubtedly the major environmental milestone of the 1990s, and from it emerged the Convention on Biological Diversity, which entered into force in December 1993.

At the start of the 21st century concerns about loss of species and habitats have intensified. Additional issues have emerged, such as climate change, with its inevitable effects on coastal, montane and other ecosystems, and the ability of humans directly to modify gene structure and expression in living organisms, with debate over the risks involved when genetically modified organisms are released into the biosphere.

Protecting species

Maintaining biodiversity requires that viable populations of diverse organisms are maintained in viable ecosystems. This may involve activities carried out on site or off site. While the latter (*ex situ*) has an important role, as with the seed banks and germplasm collections for agricultural plants, it can only have very limited application. The former (*in situ*) is essential for the vast majority of organisms, and ecosystem conservation self-evidently depends entirely on maintaining environments in which communities of organisms can interact and evolve.

Because the vast majority of the world's biodiversity exists within the territorial boundaries established by nations, most conservation action is carried out within the policy and legal systems established by national governments (or in a few instances, by regional or provincial governments). A wide range of national measures exists, varying to some extent from country to country depending on the social, political and economic environment. Despite this variety, most such measures involve regulation of the taking, possession and trade in a set of species, typically named and listed in legislation. The protection of wild fauna has generally been given much more attention than the protection of wild flora, but most modern examples of species-specific legislation cover both flora and fauna, as exemplified by the US Endangered Species Act, passed in 1973.

The species approach has been supported at the international level by the IUCN (World Conservation Union) Red Data Book program of activities and information tools. The focus is on documenting and disseminating information on species at risk of extinction, and on

Box 8.1 Pioneering NGOs in biodiversity conservation

Some of the earliest established non-governmental organizations remain influential today. The Sierra Club was founded in the United States in 1892, with a focus on maintaining wilderness areas in North America. The Society for the Preservation of the Wild Fauna of the Empire – now Fauna and Flora International (FFI) – was founded in 1903 in order to safeguard southern Africa's declining large mammal populations and is now worldwide in scope. Initiated in 1962, Operation Oryx was a landmark project of FFI, and averted extinction of the Arabian oryx by means of a captive breeding and reintroduction program. The International Council for Bird Preservation – now BirdLife International (BLI) – was founded in 1922. It has played a lead role in advancing standards for conservation information and has promoted grassroots involvement in conservation through its worldwide membership structure.

The International Union for the Protection of Nature – now IUCN-The World Conservation Union – was created in Switzerland in 1948. Sir Peter Scott became the chair of two of IUCN's key commissions: on protected areas and on species survival (SSC), and later of FFI, and in 1966 he initiated the Red Data Book approach to documenting species at risk. The World Wildlife Fund (WWF) – now known as WWF–World Wide Fund for Nature outside North America – was designed originally to generate public contributions to support IUCN's work, and was launched in 1961 with a campaign on black rhino. These two NGOs have been driving forces in global policy development.

detailing the management steps – often including preservation of important areas or reduction in exploitation levels – needed to reduce that risk. In some cases, targets and indicators of progress may also be defined.

In practice, conservation attention has tended to focus on species that are large, charismatic and possibly also ecologically important or highly threatened, or both. The tiger *Panthera tigris*, Arabian oryx *Oryx leucoryx*, white rhinoceros *Ceratotherium simum* and the blue whale *Balaenoptera musculus* are familiar mammalian examples. Considerable success has been achieved with the species named above and the many others (mainly animals) subject to similar conservation efforts. However, the species approach could never be extended to cover more than a minute fraction of the approximately 300 000 larger organisms (plants and vertebrates) in the biosphere, let alone the other several million that probably exist. In fact, the primary benefit may be that large organisms, and terrestrial vertebrates in particular, generally require large areas of suitable habitat, and if such areas can be managed to minimize risk, other species may be safeguarded.

Protecting areas

The regulation of access to and use of particular areas has always been seen as complementary to the focus on individual species. Protected areas are in many ways the most important form of legislative measure for the conservation of biodiversity. Whereas the initial purpose of many such areas was to protect spectacular scenery and provide recreational facilities, the concept evolved to encompass habitats of threatened species and ecosystems rich in biodiversity.

By the beginning of the 20th century many countries had either already established protected areas or were contemplating doing so. The concept, however, was slow to be put into practice and it was not until the 1940s that protected areas were being established in any significant number. The rate at which land was being incorporated into the system did not increase markedly until the early 1960s. The first World Parks Congress held in Seattle, in the United States, in 1962 was an important stimulus for the increase. This meeting signified the emergence of the modern protected area network with over 80 percent of the world's protected areas being established since then[3]. The cumulative and periodic growth of world protected areas are plotted in Figure 8.1 (the creation of the Greenland National Park in 1974, covering some 97 million hectares, and the Great Barrier Reef Marine Park in the 1980s, extending over around 34 million hectares, have had marked individual effects on the global totals). The location of protected areas in the IUCN/WCPA (World Commission on Protected Areas) categories I-VI greater than 100 000 hectares in extent is shown in Map 8.1.

Many protected areas, particularly those where secure funding allows management to be maintained, are effective in conserving species, habitats and landscapes of value. An extensive questionnaire study involving 93 protected areas in the tropics, in 22 countries, showed that areas where basic management activities are in place tend to avoid wholesale land clearance (but often still suffer some degree of disturbance, from hunting or logging, for example)[4]. An integrated measure of their effectiveness is not yet available, and it may be that firm management within park boundaries can have the effect of increasing pressure on outside land that is less controlled; on the other hand, many protected areas are badly under-resourced, reducing their effectiveness.

Wise management of areas outside the protected areas network, by means of planning controls and voluntary agreements and by incorporating conservation principles in landuse planning, also plays an essential role in conservation of biodiversity. Indeed, in many countries, management of landuse outside the national network of protected areas – in agricultural landscapes, for example – will play as important a role in the maintenance of national biodiversity as will the network itself.

Maintaining ecosystems

Increasingly in recent years a more holistic approach has emerged in which ecosystems themselves have become the focus of conservation efforts. Many would argue that the primary goal of conservation action is the long-term maintenance of ecosystem processes at the global scale and over foreseeable human generations. This is essentially equivalent to achieving sustainability in human development and use of biodiversity. The Convention on Biological

Figure 8.1
Development of the global network of protected areas

Source: UNEP-WCMC database, maintainted in collaboration with IUCN World Commission on Protected Areas.

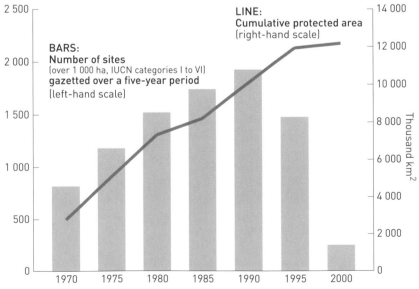

Diversity adopted 'the ecosystem approach' as the guiding framework for actions taken in pursuit of its goals and, in elaborating on the meaning of this term, stressed the need to conserve ecosystem structure and function in order to maintain ecosystem services. However, making an assessment of the organization, vigor and resilience of ecosystems – the key components of 'ecosystem health' – is fraught with immense practical difficulties. Much conservation activity therefore is quite properly based on a strong form of the precautionary principle (Box 8.2), and focuses on maintaining so far as possible the prominent elements of ecosystems, i.e. species and populations and their physical environment, in anticipation that the system will thus be perpetuated.

Defining priority areas for protection

Management action aimed at biosphere conservation demands financial resources, but these are limited, while human numbers and impacts continue to increase. This implies that choices must be made between possible actions, whether by design or by default.

Rational and informed decision-making should seek to increase the efficiency with which conservation funds are used. Several studies have been undertaken that aim to identify and sometimes to rank areas of high biodiversity value. The approach is generally based on the premise that 'more is better'. That is, an area with greater biodiversity value is more worth conserving than an area with lower value. At its simplest biodiversity may be equated with species number overall or per unit area, but other selection criteria are possible. Higher priority may be accorded to an area with populations of threatened species, or one rich in endemic species (especially if there are endemics in several different groups), or in species of commercial or cultural importance; or one which is particularly representative of an ecosystem (perhaps one that is widely degraded elsewhere). While the results of such assessments are usually of considerable biological interest, they may also have direct application as a basis for choice between different courses of action, on the grounds that greater

benefit will accrue from ensuring the integrity of areas of higher biodiversity value than areas of lesser value.

One early area-based approach identified some 12 'megadiversity' countries, which between them include a large proportion of global biodiversity in selected major groups[5].

Box 8.2 The precautionary principle

A concise definition of the precautionary principle is provided by Principle 15 of the Rio Declaration, made at the 1992 Earth Summit:

'In order to protect the environment, the precautionary approach shall be widely applied by States according to their capabilities. Where there are threats of serious or irreversible damage, lack of full scientific certainty shall not be used as a reason for postponing cost-effective measures to prevent environmental degradation.'

Another early study delineated 18 endemic-rich botanical 'hotspots', which between them included around 20 percent of the world's known plant species in less than 1 percent of the land surface, and were also undergoing rapid habitat conversion[6,7]. A strength of the former approach is that it is focused at country level, and it is at this administrative level that most conservation action is undertaken; a weakness is that by evaluating species richness alone it was not able to address uniqueness, and adjacent countries rich in species are likely to have many species in common. A strength of the latter approach is that it focused on areas rich in restricted-range endemic species and such areas by definition make a large contribution to global biodiversity.

The country-based approach has been extended in a study[8] using a database of estimates of richness and endemism in land vertebrates and vascular plants for all countries of the world (Appendix 5). Indices of overall diversity (weighting richness and endemism equally) and diversity adjusted for country area have been produced. Within the relatively wide margin of error associated with species inventory (see Chapter 2), these indices can yield a useful view of variation in diversity in geopolitical terms or, if the data are treated as geographic samples, a view of general global variation in diversity (see Map

Map 8.1
World protected areas

An overview of the world's surface nominally subject to protection and appropriate management. The location of protected areas in IUCN/WCPA categories I-VI greater than 100 000 hectares in area is shown, represented by a point symbol. Protected areas greater than 1 million hectares in extent are represented by polygons instead of point symbols whenever boundary data are available.

Source: UNEP-WCMC database (data extracted March 2002), maintained in collaboration with IUCN World Commission on Protected Areas.

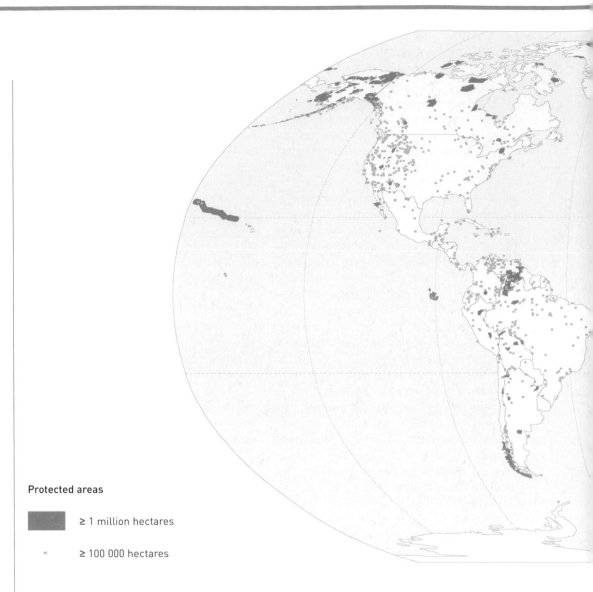

Protected areas

≥ 1 million hectares

≥ 100 000 hectares

5.4). The hotspots approach has been further developed by Conservation International in a recent synthesis, and adopted as the basis for its conservation planning and action[9]. Twenty-five terrestrial hotspot regions were identified, combining high species endemism, particularly among plants, and high rates of habitat loss, with at least 70 percent of original natural vegetation having been lost. Coral reef areas of the world have now also been evaluated in this way (see Chapter 6 and Map 6.1), and freshwater ecosystems are currently being evaluated. The term 'hotspot' is now often applied loosely to any area that has a concentration of diversity, whether by a measure of simple richness, or endemism or number of threatened species, and whether undergoing habitat conversion or not.

Of studies based on biogeographic rather than geopolitical areas, the Centres of Plant Diversity project[10] remains one of the largest. It relied heavily on extensive consultation among botanists to identify several hundred important sites worldwide, defined semi-quantitatively on the basis of a general combination of richness and endemism (Map 8.2). A broadly similar but less structured approach, also based on expert knowledge, has been taken to identify 43 areas of special importance for amphibian diversity, again using species richness and endemism as criteria[11] (Map 8.3). A preliminary selection of areas of importance for inland water biodiversity has also been made on the basis of expert knowledge of richness and endemism among fishes, mollusks and crustaceans (see Chapter 7, and Maps 7.2-7.4).

However, the most systematic and com-

plete global level assessment to date has involved bird species. Birds are by far the best-known major group of organisms on the planet, with a relative wealth of distribution and population data available, and global analyses by BirdLife International and its partners have set a standard yet to be matched for other taxonomic groups. The distributions of all restricted-range bird species (defined as those in which the area encompassing all distribution records is less than 50 000 km²), amounting to 25 percent of all birds, have been mapped in digital format. The co-occurrence of restricted-range species defines a set of 218 endemic bird areas (EBAs)[12]; these are shown, ranked in three categories according to biodiversity importance, in Map 8.4. 'Importance' here takes account of the number of restricted-range

species in the EBA and the number of EBAs in which they are present, taxonomic uniqueness and EBA area.

There are a number of limitations to area-based methods for establishing priorities among possible conservation actions. First, there is no single unequivocal way of comparing value in different categories: how many vulnerable species is a single critically endangered species worth? Or how many endemic beetles is a single endemic bird worth? Second, information is always incomplete, in that it never covers all taxa at all sites of interest. Resolving the former requires more or less arbitrary assigning of value. Attempts to resolve the latter entail the search for indicators, that is groups of species or other variables that can act as surrogates for wider measures of biodiversity.

Map 8.2
Centers of plant diversity

This map shows the location of the sites and areas identified as important centers of plant diversity at regional and global levels, using expert knowledge, and mixed criteria emphasizing species richness and endemism. See the three-volume source cited below for further details and extended documentation.

Source: WWF and IUCN[10].

Plant diversity

Areas and centers

The search for biological indicators of this kind has generated a great deal of research and discussion. Findings to date have generally been equivocal, and depend in part on spatial scale. In general, though, it seems that at coarse scales there may be quite close agreement between different taxa so that, for example, many EBAs (which may be up to 600 000 km^2 in extent) also hold significant numbers of other restricted-range species[12]. At this scale, birds may serve as indicators of high biodiversity value more generally. In other instances, and perhaps more generally at finer resolution, the relationship appears to break down. Studies in areas as disparate as North America, South Africa[13], Cameroon[14] and the British Isles[15, 16] indicate that areas important for rare species in different groups often do not coincide (and these may be negatively correlated with areas of high species richness). Also, richness levels in any one group do not necessarily provide an indication of species richness in others.

A slightly different approach that avoids some of the limitations faced by methods based on species-specific information is to identify areas on the basis of their biological communities, and on biogeographic criteria, with reduced emphasis on richness or endemism in particular taxa, and none on current habitat condition. The conservation organization WWF's 'ecoregions' system, mentioned in Chapters 5-7 above, is the principal example of this approach, and currently covers much of the world's marine and inland waters, together with the entire land surface. It has been adopted as the basis for conservation planning and action by WWF.

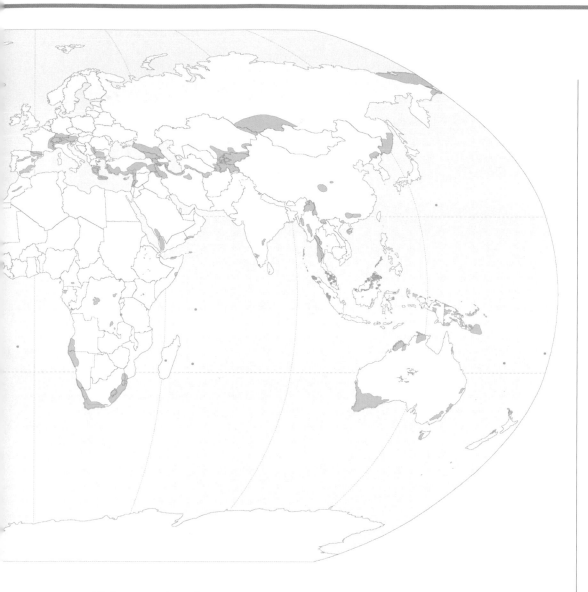

Whilst a number of different groups and organizations have explored the use of spatial biodiversity information in suggesting conservation priorities, these generally remain disconnected and to some extent duplicative[17]. Recent developments suggest that a more integrated approach is feasible. Firstly, the conceptual and methodological framework of a more systematic approach to priority setting has been elaborated[18] (see Box 8.3). Secondly, sustained effort to map species distribution in a globally consistent manner is beginning to allow robust analysis of large-scale multi-taxon patterns of biodiversity[19, 20]. Combining a systematic approach with a geographically extensive set of spatial biodiversity data has the potential to allow priorities to be defined in a way that is flexible, transparent, and robust.

Box 8.3 Systematic conservation planning

Key stages in systematic planning for conservation by means of adding to or modifying protected area systems:

1. Compile data on the biodiversity of the planning region.
2. Identify conservation goals for the planning region.
3. Review existing conservation areas.
4. Select additional conservation areas.
5. Implement conservation actions.
6. Maintain the required values of conservation areas.

Source: Margules and Pressey[18].

Map 8.3
Major areas of amphibian diversity

The location of areas identified on the basis of expert knowledge as globally important for amphibian diversity, using data on species richness and endemism. Areas are here shown classified in five categories according to species richness. See source cited below for further details.

Source: Adapted from Duellman[11].

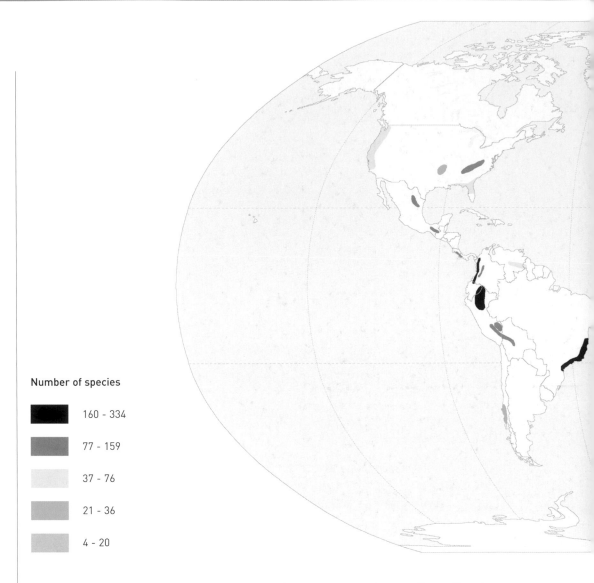

Number of species

- 160 - 334
- 77 - 159
- 37 - 76
- 21 - 36
- 4 - 20

It should, however, be emphasized that many priority-setting exercises are likely to remain only theoretical. This is because the opportunities to establish entirely new terrestrial protected areas, or redesign existing protected area networks in the light of priorities identified, are in general extremely limited. The designation and management of protected areas, as so much other conservation action, is generally driven and constrained far more by socio-political and economic considerations than it is by conservation biology.

Management of inland waters
To a greater extent than is typical for terrestrial ecosystems, inland waters are often subject to management and use decisions made within many different sectors, including agriculture and forestry, navigation, public utility supply (power, water and waste disposal) and recreation. Most inland waters or their catchments also typically intersect several subnational administration units (counties, provinces, etc.) Although it has been recognized for some time that the catchment basin is the fundamental unit within which management must be formulated, it has proved difficult to reconcile the many different interests concerned and coordinate action. This is particularly problematic because sources of adverse impacts on such systems often originate far distant from where they are felt: thus pollutants and other inputs that enter the top reaches of a river system may have an impact in all downstream parts of the system as far as the river mouth and beyond. Similarly, water abstraction may have little effect in upper reaches, but be a serious constraint downstream.

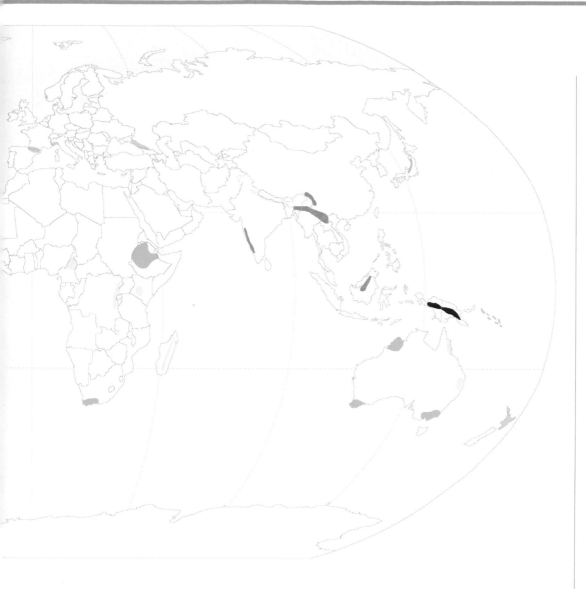

Biodiversity maintenance and conservation of inland water capture fisheries (other than some lucrative sports fisheries) have not ranked highly among these competing interests, so that it has been difficult to impose catchment-wide regulations or remedial measures for their benefit. This problem is exacerbated by the fact that it is often difficult to pinpoint a distant source of a problem or to unequivocally demonstrate that actions in one place are having an adverse effect somewhere else (e.g. to convince farmers that application of large doses of nitrogenous fertilizer on upstream agricultural land is causing deleterious eutrophication of estuarine wetlands).

Although they cannot solve catchment-wide problems, inland water protected areas may play a valuable role in safeguarding particular sites or populations of species from immediate threats. Protection may be most effective where sites are relatively small and thus manageable, and have a relatively low level of allochthonous inputs. Wetlands, with their often abundant and highly conspicuous avifauna, have in general received most attention in this regard.

Notable wetland protected areas include the Moremi game reserve in the Okavango delta (Botswana), Camargue national reserve (France), Keoladeo (Bharatpur) national park (India), Doñana national park (Spain) and Everglades national park (United States). Inland water ecosystems are unusual in that an international convention is dedicated specifically to them: the Convention on Wetlands of International Importance especially as Waterfowl Habitat (the Ramsar Convention, see below).

Map 8.4
Endemic bird areas

More than one quarter
(2 561) of the world's bird
species, including more
than 70% of the threatened
birds, have a range
restricted to less than
50 000 km[2]. Virtually all
occur within the 218
endemic bird areas (EBAs)
defined by BirdLife
International[12].
The world's EBAs are
shown on this map
categorized 1, 2 or 3
according to increasing
biodiversity importance
(based on the number of
restricted range species,
whether shared between
EBAs, taxonomic
uniqueness and EBA size).

Source: Data provided by BirdLife
International, and see Stattersfield[12].

Category

■	3
■	2
■	1

Transboundary inland waters

Waters that delineate or cross international boundaries present a special class of management issue. Such waters and the living resources they contain are shared by one or more countries, and require positive international collaboration for effective use and management.

Available water in any given country within an international basin (or other administrative unit within a basin more generally) can be divided into endogenous, i.e. locally generated runoff available in national aquifers and surface water systems, and exogenous, i.e. remotely generated runoff imported in flow from upstream. Some countries (e.g. Canada and Norway) have an abundance of water from endogenous sources; others (e.g. Egypt and Iraq) have a small endogenous supply but large exogenous volumes (others have small supplies from both sources). Use of exogenous water carries an increasing risk because of dependence on sufficient supply from upstream countries.

There are well over 200 major international rivers and a host of smaller ones[21]. As demands on inland water resources continue to grow through the 21st century, as they undoubtedly will, management of these and the biological resources they contain will also grow ever more challenging.

Management of marine ecosystems

Management of the terrestrial environment is typically carried out alongside more or less severe anthropogenic disturbance. Although the particular nature of the marine biosphere has to some extent buffered it from the

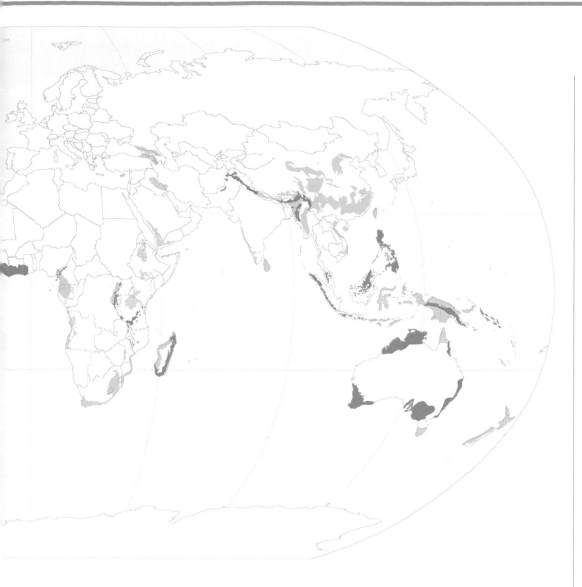

impacts of humans, it also imposes its own set of difficulties and constraints for rational management. Firstly, because almost all of it is generally out of sight, impacts are not immediately apparent, so that extreme deterioration may take place before anyone is aware of the fact. There may also be less incentive to take action than with terrestrial ecosystems where such deterioration has a direct impact on people. Secondly, with some exceptions (such as communal property resources on reefs and other inshore areas in parts of the South Pacific), living marine resources have been widely considered open-access resources, particularly those outside territorial waters (usually up to 12 nautical miles (nm) from shore). There is thus, quite simply, an incentive for any given individual to exploit a resource as fast and as intensively

as possible before someone else does. Thirdly, the ability of water to transport large amounts of dissolved and suspended materials, including living organisms, means that it is extremely difficult to manage limited areas of marine habitat in isolation.

With the introduction of the exclusive economic zone (EEZ) under the United Nations Convention on the Law of the Sea (UNCLOS), which allows nations control over resources in an area up to 200 nm offshore, a far greater proportion of the world's seas now come within the control of individual nations. At present 99 percent of world fisheries catch is taken within EEZs. Although this should theoretically allow more rational management of marine resources, and more effective enforcement of management measures, progress in both has in practice been limited

Map 8.5
Marine protected areas

The map shows the location of protected areas in IUCN categories I-VI that are entirely or in part marine, with map symbols graded according to protected area size (including any land present).

Note: For presentation purposes it has been necessary to use symbols that in most cases at this map scale greatly exceed the size of the protected area represented, giving the impression that much more of the world's coastal waters are protected than is in fact the case.

Source: UNEP-WCMC database (data extracted March 2002), maintained in collaboration with IUCN World Commission on Protected Areas.

Protected area (km²)

▬ > 5 000

● > 1 000

● > 100

· < 100

to date, as evidenced by the increasing proportion of world's fisheries that are over-exploited. This is because fisheries management regimes are frequently subject to political pressure, so that in many countries quotas are habitually set higher than those recommended by fisheries biologists, and also because the active enforcement of regulations is difficult and expensive. Many countries lack the resources or the political will, or both, to enforce such regulations adequately.

It is also increasingly apparent that individual marine resources cannot be effectively managed in isolation from each other. Complex interactions between populations of different organisms, when combined with perturbations in the environment and variations in human impact (e.g. changes in fisheries technology or fishing effort) create responses that may be far from intuitively predictable. Recognizing that the understanding of such responses will require modeling and management of large-scale ecosystem processes, a number of large marine ecosystem (LME) units have been identified[22], based on the world's coastal and continental shelf waters, which are regarded as central to such analysis. Over 95 percent of the usable annual global biomass yield of fishes and other living marine resources is produced within 64 identified LMEs, nearly all of which lie within and immediately adjacent to the boundaries of EEZs of coastal nations. Many LMEs include the coastal waters of more than one state. In these cases, it will be effectively impossible for individual nations to assess whether their use of marine resources

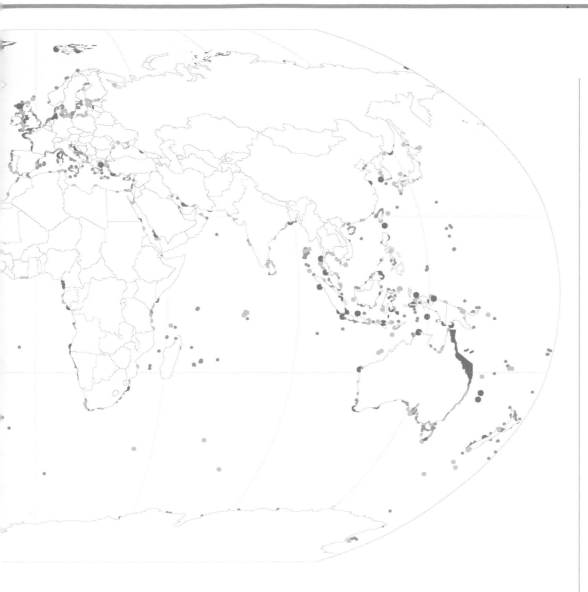

is sustainable in isolation from neighboring nations. Coordination between states in monitoring and resource management will thus become increasingly necessary as the pressures placed on these areas increase.

A critical need in monitoring marine ecosystems is the development of consistent long-term databases for understanding between-year changes and multi-year trends in biomass yields. For example, marked alterations in fish abundances were observed during the late 1960s when there was intense fishing within the northeast US continental shelf LME. The biomass of economically important finfish species (e.g. cod, haddock, flounders) declined by approximately 50 percent, and this was followed by increases in the biomass of lower-valued small elasmobranchs (dogfish and skates). Management of

marine fisheries will need to take these kinds of species dominance shifts into account in the development of strategies for long-term, economic sustainability of the fisheries[23]. Monitoring the changing states of LMEs has received considerable attention, with several now being assessed and managed from a more holistic ecosystem perspective[22, 24].

Marine protected areas
The long-term management of LMEs is highly complex and there is an urgent need for smaller-scale and more immediate approaches. As with terrestrial ecosystems, the establishment of protected areas in marine ecosystems has been viewed as a major contribution to maintenance of biodiversity. A 1995 review[24] identified just over 1 300 marine protected areas in existence at that time,

Map 8.6
International protected area agreements

The map shows the location of protected areas managed under the Ramsar Convention on Wetlands, under the World Heritage Convention or as a biosphere reserve within the UNESCO Man and the Biosphere Programme.

Source: UNEP-WCMC database (data extracted March 2002), maintained in collaboration with IUCN World Commission on Protected Areas.

World protected areas

⚬ Ramsar site

· World Heritage site

■ MAB biosphere reserve

ranging in size from 1 hectare to 34.4 million hectares (the Great Barrier Reef Marine Park) (Map 8.5). Effective management and control of marine protected areas is problematic, particularly if, as is often the case, they are in areas of intensive and potentially conflicting resource use. As noted above, marine ecosystems are also in general more difficult to protect than terrestrial ones from allochthonous inputs (i.e. those originating elsewhere). Although a no-catch regime can be effective in small marine reserves, in general it has been found that large, carefully zoned, multiple-use areas are more practical and effective than small reserves. Sanctuaries or strict reserves may still be required for critical habitat areas such as nutrient sources, areas of high biological diversity, nesting sites of threatened species or to protect breeding stocks of important fishes[25, 26].

Reversing change: restoration and reintroduction

Increasing recognition of widespread environmental degradation has led to a growth of interest in both the science and practice of ecological restoration. The main aim of such restoration is to reestablish the key characteristics of an ecosystem, such as composition, structure and function, which were present before degradation took place. It has been suggested that ecological restoration is a crucial complement to the establishment of protected areas for safeguarding biodiversity[27], and it is widely expected that restoration will become a central activity in environmental management in the future. Such efforts are being supported by development of national and international policies. For example, the UN Convention on Biological

Diversity, Article 8f, states that parties should 'rehabilitate and restore degraded ecosystems and promote the recovery of threatened species, through the development and implementation of plans or other management strategies'.

A large number of restoration projects have now been initiated in different parts of the world, focusing on a variety of different ecosystem types, including grasslands, wetlands and forests. Although a number of national governments are active in ecological restoration, sometimes on a very large scale (notably in North America), most projects are being undertaken by NGOs, often as grassroots or community-based initiatives. For example, together with a variety of local partners, the Forests for Life program of WWF/IUCN is implementing restoration

programs in areas such as the Lower Mekong, New Caledonia, the Mediterranean, India and the Carpathians. WWF/IUCN is increasingly recognizing the critical importance of developing plans for restoration at the landscape scale, and the need to provide benefits to local communities as well as to biodiversity.

Experience of restoration projects to date has highlighted how difficult such ecological rehabilitation can be in practice. Although many degraded ecosystems display an ability to recover through natural processes if the causes of degradation are removed, in many areas the extent of degradation has been so severe that greater management intervention is required for restoration to be effective. For example, severely deforested areas may require large-scale tree planting in order for forest ecosystems to reestablish on a particular site. Restoration projects may also be difficult to manage or monitor, as it is often hard to define with precision what the structure, composition or function of a given ecosystem was prior to degradation, particularly in areas where degradation occurred a long time ago. Another key challenge to

Restoration projects have been initiated in different parts of the world, focusing on a variety of ecosystem types, including grasslands, wetlands and forests.

restoration projects is the high cost involved: for example, a plan to restore the Florida Everglades has recently been launched, at a total cost of US$7.8 billion[28].

Efforts at restoring degraded ecosystems can be complemented by programs focusing on the reintroduction or reestablishment of species that have become extinct within a particular area. For such reintroductions to be successful, thorough knowledge of a species and its habitat requirements are needed, in addition to a clear understanding of the original causes of extinction. In some cases, such as large vertebrate predators, there may be considerable public antipathy to reintroduction being attempted. However, there have been some notable examples of successful reintroductions, such as the Arabian oryx (*Oryx leucoryx*) to Oman[29], the white-tailed eagle (*Haliaeetus albicilla*) to Scotland, and the Mexican gray wolf (*Canis lupus baileyi*) and California condor (*Gymnogyps californianus*) to parts of the United States. Such examples provide important lessons for successful reintroductions and, together with habitat

restoration projects, illustrate how positive action can contribute to reversing the trends of biodiversity loss.

THE INTERNATIONAL DIMENSION
National boundaries do not enclose all the world's biological diversity: the high seas, the deep seabed and Antarctica all contain natural resources, some of great interest or economic importance. Management of biodiversity in such areas can, by definition, only be achieved by means of international measures. Similarly, areas or communities of particular interest may be crossed by national boundaries, and in such cases international cooperation is essential for conservation measures to be planned and implemented effectively.

More generally, it is important to develop policy and planning at the global level to place national efforts in a broader context. In a hypothetical example, an individual country might devote more effort to conserving species that are rare or peripheral at national level, but widespread elsewhere, than to more common, nationally endemic species. Although the former may be regarded as national priorities, the latter may be more important to global biodiversity. Regardless of differences between national and global priorities, an international forum is needed to develop conservation science, to provide exposure for diverse opinions and an opportunity for NGOs and other bodies to comment on policy, and to formalize agreements that can guide the way individual countries manage their environments. Such opportunities are offered by the international agreements that have recently been developed.

International agreements
A multilateral treaty is an international agreement concluded between three or more states and governed by international law (see Box 8.4). Existing international treaties that deal entirely or in part with biological diversity have evolved in an uncoordinated manner. Despite this, and the consequent gaps and duplications in overall coverage, a handful of such treaties whose text was agreed during the 1970s have come to exert a

powerful influence on the conservation and management of elements of biodiversity. Among the most notable are the 1971 Convention on Wetlands of International Importance especially as Waterfowl Habitat (Ramsar); the 1972 Convention Concerning the Protection of the World Cultural and Natural Heritage (World Heritage); the 1973 Convention on International Trade in Endangered Species of Wild Fauna and Flora (CITES); and the 1979 Convention on the Conservation of Migratory Species of Wild Animals (CMS). The 1982 UN Convention on the Law of the Sea (UNCLOS), which entered into force in 1994, has strong potential for enhancing marine and coastal conservation.

The names of these major treaties indicate their sectoral focus and, even if the many important regional and species-related treaties are also considered, it is clear that the total obligations explicit in existing treaties fall short of the demands of an adequately comprehensive system. The Convention on Biological Diversity (CBD), agreed at the 1992 Earth Summit in Rio, attempted to meet many of these demands. It was the first treaty planned to concentrate specifically on the conservation and use of global biodiversity. Its text establishes the conservation and use of biological resources as a matter of common interest to all. It has as its objectives the conservation of biological diversity, the sustainable use of biological resources and the equitable sharing of benefits arising from the use of genetic resources, recognizing that a careful balance must be maintained between these if biological resources are to be used wisely. It not only acknowledges the control of individual states over their biological resources, but it also states their responsibility for protecting them and using them sustainably.

The United Nations Framework Convention on Climate Change (UNFCCC) was also agreed at the Earth Summit and is relevant to biodiversity management. The CBD, the UNFCCC and the United Nations Convention to Combat Desertification (UNCCD) (which arose from the Summit but was not agreed until 1994) are sometimes termed the 'Rio conventions'.

Those with a particular focus on biodiversity (CBD, CITES, CMS, Ramsar, World Heritage) are informally termed 'the biodiversity-related conventions'. They each impose more or less rigorous reporting requirements on parties to them, and also generate a significant demand for information from their parties and others. Meeting these demands can place a substantial burden on governments, particularly those with limited resources, and work is proceeding on harmonizing information management among the treaties.

In addition to the Rio conventions, the Rio Declaration (a set of guiding principles), and a comprehensive plan of action – Agenda 21 – were also agreed and adopted by more than 178 governments at the Earth Summit. Agenda 21 is designed to support sustainable

---

Box 8.4 Negotiating a multilateral treaty

Text

Negotiation of the text of the treaty can require many years and numerous meetings. It is concluded by the adoption of the text, typically when all the participating states reach agreement over the wording. Adoption of a treaty does not by itself create any obligations.

Consent

A treaty does not come into force until two or more states consent to be bound by it. The expression of such consent is usually by signature, notification, acceptance, approval or accession or by other means where so agreed. Signature followed by ratification is the most frequent means of expressing consent. *Signature* refers to the signature of the diplomats negotiating the treaty and is often synonymous with the adoption of the treaty. *Ratification* is the need for approval of the treaty by the head of state or the legislature. *Accession* is the normal way that states which did not participate in the negotiations become parties to the treaty and has the same effect as signature and ratification combined.

Entry into force

This final stage usually occurs when all the negotiating states have expressed their consent to be bound by the treaty; the date may be delayed to provide time for parties to adapt themselves to its requirements. When a large number of states participate in the drafting of a large multilateral treaty, it often enters into force once a specified number have ratified.

---

human development and stewardship of the environment, and to be implemented at a range of scales – globally, nationally and locally – by organizations within the United Nations system, governments and other major groups.

Table 8.1
Major global conventions relevant to biodiversity maintenance

Notes: Conventions are listed in order of entry into force. 'Year' is date of agreement, 'Entry' is year in which agreement entered into force, 'Parties' is number of party states as indicated at each agreement website in March 2002.

Convention	Scope
Convention on Wetlands of International Importance especially as Waterfowl Habitat (Convention on Wetlands or Ramsar Convention) 	All aspects of wetland conservation and wise use. Parties are required to list at least one wetland of international importance for special management and protection.
Year — 1971 / Entry — 1975 / Parties — 131	
Convention Concerning the Protection of the World Cultural and Natural Heritage (World Heritage Convention) 	To define and conserve the world's heritage, by drawing up a list of sites whose outstanding values should be preserved for all humanity, and to ensure their protection through a closer cooperation among nations. Sites may be of importance as cultural heritage or natural heritage or both.
Year — 1972 / Entry — 1975 / Parties — 167	
Convention on International Trade in Endangered Species of Wild Fauna and Flora (CITES) 	Aims to prevent species being threatened with extinction because of international trade. Parties act by banning commercial international trade in an agreed list of endangered species (Appendix-I listed species) and by regulating and monitoring trade in others that might become endangered or whose trade needs to be regulated to ensure control over trade in Appendix-I species (Appendix-II listed species).
Year — 1973 / Entry — 1975 / Parties — 154	
Convention on the Conservation of Migratory Species of Wild Animals (CMS or Bonn Convention) 	Aims to protect migratory species and their habitats. Parties cooperate in research relating to migratory species and provide immediate protection for species listed in Appendix I of the convention. For those species listed in Appendix II, parties are required to endeavor to conclude 'range state' agreements on their conservation and management, a number of which have been concluded.
Year — 1979 / Entry — 1983 / Parties — 79	
Convention on Biological Diversity (CBD) 	The major international agreement on biodiversity, the CBD sets out a framework within which parties undertake to carry out national and international measures aiming to: conserve biodiversity, make sustainable the use of its components, and share equitably the benefits derived from the use of genetic resources. The Conference of the Parties has so far met five times. The Cartagena Protocol on Biosafety was adopted in 2000 but is not yet in force.
Year — 1992 / Entry — 1993 / Parties — 183	

Convention	Scope
United Nations Framework Convention on Climate Change (UNFCCC)	Aims to stabilize greenhouse gas concentrations in the atmosphere at safe levels. Parties are required to make an inventory of their sources and sinks of greenhouse gases and to formulate policies and measures to mitigate and/or adapt to the effect of climate change. Developed country parties were required to reduce their emissions of greenhouse gases to their 1990 level by the year 2000. The Kyoto Protocol establishes further reduction commitments for developed country parties.

Year	Entry	Parties
1992	1994	179

Convention	Scope
United Nations Convention on the Law of the Sea (UNCLOS)	Contains a comprehensive codification of the principles and rules relating to the seas. UNCLOS establishes rights and obligations relating to navigation, the conservation and use of marine resources, and the protection of the marine environment

Year	Entry	Parties
1982	1994	138

Convention	Scope
United Nations Convention to Combat Desertification in those countries experiencing serious drought and/or desertification, particularly in Africa (UNCCD – Desertification Convention)	Aims to ensure improved management of dryland ecosystems and use of development aid. National action programs (NAPs) will address the underlying causes of desertification and drought and seek to identify preventative or remedial measures. Subregional and regional action programs (SRAPs, RAPs) will be developed, particularly when transboundary resources such as lakes and rivers are involved.

Year	Entry	Parties
1994	1996	179

Convention	Scope
Agreement for the Implementation of the Provisions of the UN Convention on the Law of the Sea relating to the Conservation and Management of Straddling Fish Stocks and Highly Migratory Fish Stocks (Straddling Fish Stocks Agreement)	The objective is to ensure the long-term conservation and sustainable use of straddling and highly migratory fish stocks. Emphasizes the precautionary approach, protection of marine biodiversity and the sustainable use of fisheries resources.

Year	Entry	Parties
1995	2001	30

International protected area systems

Two international conventions and one international program include provision for designation of sites internationally important for biodiversity conservation. These are the World Heritage Convention, the Ramsar (Wetlands) Convention, and the UNESCO Man and the Biosphere (MAB) Programme. The location of protected areas managed under these agreements is shown in Map 8.6.

Ramsar Convention

The Convention on Wetlands of International Importance especially as Waterfowl Habitat was signed in Ramsar (Iran) in 1971, and provides a framework for international cooperation for the conservation of wetland habitats. It places general obligations on contracting party states relating to the conservation of wetlands throughout their territories, with special obligations pertaining to those wetlands which have been added to the List of Wetlands of International Importance. Each state party is obliged to list at least one site. Wetlands are defined by the convention as: areas of marsh, fen, peatland or water, whether natural or artificial, permanent or temporary, with water that is static or flowing, fresh, brackish or salt, including areas of marine waters, the depth of which at low tide does not exceed 6 meters. There are currently (March 2002) 131 contracting parties to the convention; 1 148 wetlands have been designated for inclusion in the List of Wetlands of International Importance, covering more than 96 million hectares.

World Heritage Convention

The Convention Concerning the Protection of the World Cultural and Natural Heritage was adopted in Paris in 1972, and provides for the designation of areas of 'outstanding universal value' as World Heritage sites, with the principal aim of fostering international cooperation in safeguarding these important

The Ramsar Convention provides a framework for international cooperation for the conservation of wetland habitats.

areas. Sites must be nominated by the signatory nation responsible and are evaluated for their world heritage quality before being listed by the international World Heritage Committee. Of the 721 sites distributed among 124 countries that are currently listed, 144 cover natural heritage and 23 sites are mixed cultural and natural.

Article 2 of the World Heritage Convention considers as natural heritage: natural features consisting of physical and biological formations or groups of such formations which are of outstanding universal value from the esthetic or scientific point of view; geological or physiographical formations and precisely delineated areas which constitute the habitat of threatened species of animals and plants of outstanding universal value from the point of view of science or conservation; and natural sites or precisely delineated areas of outstanding universal value from the point of view of science, conservation or natural beauty. Criteria for inclusion in the list are published by the United Nations Educational, Scientific and Cultural Organization (UNESCO).

Biosphere reserves

The establishment of biosphere reserves is not covered by a specific convention, but is part of an international scientific program, the UNESCO Man and the Biosphere (MAB) Programme. The objectives of the network of biosphere reserves, and the characteristics which biosphere reserves might display, are identified in the Action Plan for Biosphere Reserves. There are currently 411 biosphere reserves, spread over 94 countries.

Biosphere reserves differ from the preceding types of site in that they are not exclusively designated to protect unique areas or important wetlands, but for a range of objectives which include research, monitoring, training and demonstration, as well as conservation. In most cases the human component is vital to the functioning of the biosphere reserve, which does not necessarily hold for either World Heritage or Ramsar sites. Some biosphere reserves coincide spatially with Ramsar or World Heritage sites.

POSSIBLE FUTURES

Previous chapters have introduced some of the impacts on the biosphere of human expansion and development, and this chapter has outlined some of the general ways in which humans have attempted to manage the extent or severity of these impacts on biodiversity. These approaches reflect current conditions of climate and human development and the present policy environment, but is there a way to incorporate possible future conditions within the planning process?

Scenarios

Scenarios are increasingly being used to inform policy development and implementation, by illustrating the possible outcome of current trends, and by highlighting the implications of different policy decisions. The development of scenarios has become an important tool by which scientists communicate the results of their research to decision-makers, as well as constituting a significant research endeavor in its own right. An example is provided by the Intergovernmental Panel on Climate Change (IPCC), which provides scientific, technical and socioeconomic advice to the world community on the specific issue of climate change. Climate change scenarios have been developed by the IPCC through workshops and meetings involving experts in modeling, climate impact assessment and emissions scenarios. These scenarios have formed the basis of the decisions made by parties to the UN Framework Convention on Climate Change, which provides the overall policy framework for addressing the climate change issue.

Few attempts have been made to develop scenarios of biodiversity change. One recent approach[30] involved consultation between ecological specialists from a number of different regions, who developed global scenarios for ten terrestrial biomes for the year 2100, based on global scenarios of changes in environment and landuse. Five key determinants of changes in biodiversity were identified: landuse change, atmospheric carbon dioxide concentration, nitrogen deposition and acid rain, climate and the introduction of exotic species into an ecosystem. The

expected changes in these drivers were then considered for different ecosystem types (biomes), and the relative impact of the different drivers on biodiversity was also estimated. Three scenarios of future biodiversity were developed for each biome, based on various assumptions about the interactions between the drivers affecting biodiversity change.

The results of this investigation suggest that when all biomes are considered together, landuse change is the driver expected to have the largest impact on biodiversity at the global scale, with climate change ranking second in importance. Considering the impact of all drivers together, Mediterranean ecosystems appear to be most at risk, with grassland and savannah ecosystems also being at relatively high risk.

To further illustrate possible future trends in global biodiversity, examples of two alternative approaches to scenario development are described below. Both these approaches were developed as contributions to the Global Environment Outlook (GEO) produced by the United Nations Environment Programme (UNEP), in collaboration with a wide range of partners. The latest report of the GEO process (GEO-3) provides further details of these scenarios, and the methods employed in their development[31].

RIVM IMAGE scenario evaluation

For GEO-3, UNEP developed scenarios through a highly participative process involving GEO collaborating centers and other partners throughout the world, allowing particular issues identified at regional scale to be incorporated. The scenarios involved development of narratives or storylines, respectively termed *Markets First*, *Policy First*, *Security First* and *Sustainability First*. These each describe possible futures based on different interpretations of prevailing global driving forces.

A number of models and analytical tools were used to develop the scenarios, to help quantify the dynamics described qualitatively in the narratives, and to compare impacts across regions. These included PoleStar, developed at SEI-Boston; IMAGE, developed at RIVM in the Netherlands; WaterGAP,

Figure 8.2
Possible future scenarios
from GEO 3, evaluated
with RIVM IMAGE:
pressures on remaining
natural areas in 2002 (top)
and changes in pressure
between 2002 and 2032
under the *Markets First*
scenario (bottom)

Notes: The pressures
selected are those believed to
have a major influence on
biodiversity and for which
data are available. These
include habitat loss,
population density, primary
energy use, temperature
change, and restoration time
for agricultural land and
deforested areas. The lower
map shows the relative
increase or decrease in
pressure between 2002 and
2032 (difference divided by
the pressure in year 2002).
No change means less than
10% change over the
scenario period; small
increase or decrease means
between 10 and 50% change;
substantial increase or
decrease means 50 to 100%
change; and strong increase
means more than doubling.

Source: RIVM IMAGE; Brink[34]; Brink[35];
Brink *et al.*[36].

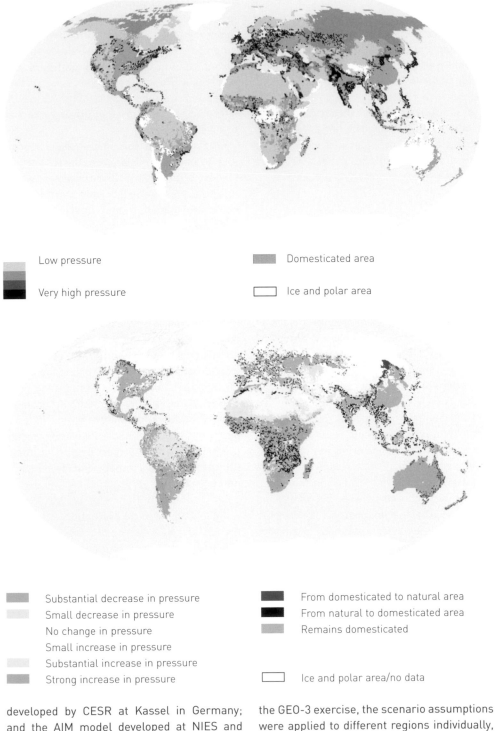

Low pressure

Very high pressure

Domesticated area

Ice and polar area

Substantial decrease in pressure
Small decrease in pressure
No change in pressure
Small increase in pressure
Substantial increase in pressure
Strong increase in pressure

From domesticated to natural area
From natural to domesticated area
Remains domesticated

Ice and polar area/no data

developed by CESR at Kassel in Germany;
and the AIM model developed at NIES and
Kyoto University in Japan. PoleStar offers a
simple modeling approach to explore how
assumptions affect social and environmental
performance, and incorporates scenarios
developed by the Global Scenario Group[32]. For

the GEO-3 exercise, the scenario assumptions
were applied to different regions individually,
and were then used to initiate and inform the
scenario design process. PoleStar was also
used to test the plausibility of storylines
identified through the participatory process in
most of the regions.

The GEO-3 scenarios look 30 years ahead. *Markets First* is a world in which market-driven developments converge on the values and expectations that prevail in industrialized countries. *Policy First* is a world in which concerted action on environment and social issues occurs through incremental policy adjustments. *Security First* is a world of fragmentation, where inequality and conflict prevail, brought about by socioeconomic and environmental stresses. *Sustainability First* is a world in which a new development paradigm emerges in response to the challenge of sustainability, supported by new values and institutions.

In order to assess the impact on biodiversity of such diverging trends, IMAGE was used to assess changes in the natural capital index. This index is designed as a measure of biodiversity in natural ecosystems and agricultural land, and is calculated as the product of remaining area and its quality. The current remaining natural area is taken as a percentage of the total land area. Changes over time are caused by the conversion of natural ecosystems into agricultural and built-up areas, and vice versa. Ideally, ecosystem quality should incorporate measures of the abundance of characteristic species relative to a low-impacted baseline state, but such data are not generally available at global scale, and ecosystem quality is therefore approximated by means of four pressure factors that are assumed to have a major influence on biodiversity and for which data are available. For each pressure factor, a preliminary range is defined from a review of relevant literature, from no effect to complete deterioration of the habitat if the maximum value is exceeded. The four selected pressure factors are population density (min-max: 10-150 persons per km²); primary energy use (0.05-100 petajoules per km²); rate of temperature change (0.2-2.0°C in a 20-year period); and a restoration time for abandoned agricultural land and deforested zones in reconversion towards natural, low-impacted ecosystems (min-max: 100-0 years after conversion or clear-cutting). The proxy for ecosystem quality is a reversed function of these pressures, calculated as a percentage of the low-impacted baseline state. The higher the pressure, the lower the quality. Finally, the

percentages for habitat area and quality are multiplied, resulting in a pressure-based natural capital index. The calculations have been carried out on a detailed latitude-longitude grid, before aggregation to sub-regions and regions.

The maps (see Figure 8.2) illustrate the combined effect of decreasing area and quality of habitat. The *Markets First* scenario suggests a strong decrease of habitat quality and quantity in most regions, but in some regions agricultural land is taken out of production and may therefore revegetate naturally through the process of succession. However, in biodiversity terms this reconverted land must be considered of low quality during initial decades.

GLOBIO

GLOBIO (Global methodology for mapping human impacts on the biosphere) was launched by UNEP in 2001 as a simple global model to help visualize the growing impacts of infrastructural development on biodiversity and ecosystem resources[33]. GLOBIO provides an assessment of the probability of such impacts occurring by defining buffer zones around infrastructure. The extent of impact zones varies with the type of human activity and density of infrastructure, region, vegetation type, climate and sensitivity of species and ecosystems, and is based on extensive literature surveys of scientific studies assessing environmental impacts resulting from human development.

By linking impacts in different ecosystems and regions with satellite imagery (AVHRR data from 1992-93 on 1 km² resolution in the Global Land Cover Characterization database version 2.0), available resources and infrastructure, overviews of the cumulative impacts of continuous development can be derived. Scenarios for impacts on biodiversity are based upon data describing existing infrastructure, historic growth rates of infrastructure, availability of petroleum and mineral reserves, vegetation cover, population density, distance to coast and projected development. The outcome is a simple overview of the current and possible future cumulative impacts of infrastructural development on biodiversity, assuming continued

Figure 8.3
Possible future scenarios, GLOBIO: the pressures on the environment in 2002 (top) and in 2032 under the *Markets First* scenario (bottom)

Notes: GLOBIO provides projections of human impacts based on the development of infrastructure. With an economy dominated by market forces, biodiversity will be affected by infrastructural development on over 72% of the land area by 2032, according to GLOBIO projections.

Source: GLOBIO[33], GRID Arendal, UNEP.

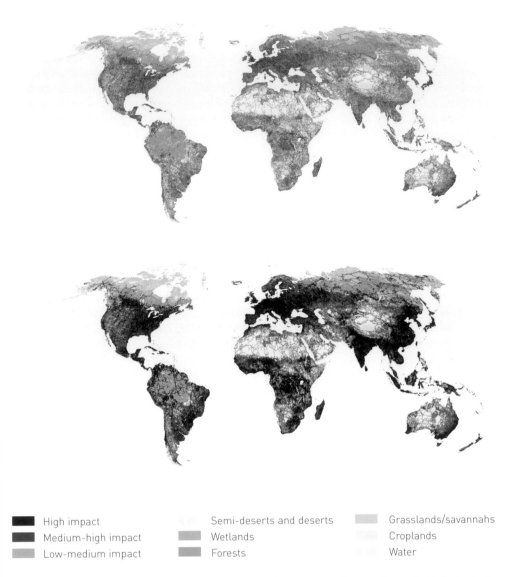

■ High impact	Semi-deserts and deserts	Grasslands/savannahs
■ Medium-high impact	Wetlands	Croplands
■ Low-medium impact	Forests	Water

growth in demand for natural resources and the associated infrastructure development.

GLOBIO defined four different degrees of human impact:

- Human populations and urban areas (high impact). This includes all areas that are within 0.5 km of a road, or within a few hundred meters of any development, including urban areas.
- Converted land (medium-high impact). This includes areas converted to agricultural lands, plantations, and with a high density of infrastructure (i.e. generally within 0.3-2 km of any infrastructure).
- Areas under conversion and fragmentation (medium-moderate impact). This includes areas within 2-10 km of infrastructure.
- Relatively intact (low impact). These are areas that still retain their natural state to a large extent, i.e. more than 10 km from any infrastructure.

GLOBIO scenarios are based on two assumptions: that stable gross domestic product (GDP) or an increase in GDP will require further development in infrastructure, and that impacts of such developments on biodiversity will decrease with distance from the infrastructure. The scenario illustrated here (*Markets First*, Figure 8.3) assumes a 3-4 percent annual increase in economic growth.

Because of differences in history, population density, coastal distance, landcover, availability of mineral and petroleum resources (included as part of the model) and current economic capacity, the rate of change is not the same in all parts of the world. The implication of this scenario is that if further expansion is not controlled, increasingly severe losses of biodiversity will occur as a result of continuing human development. GLOBIO projections suggest that increased industrial exploration for oil, gas and minerals will accelerate road construction, draining of rivers for hydro-power and irrigation, and increased immigration, logging, and conversion of land to plantations and farmland, resulting in heavy losses of biodiversity in many areas. During the last 150 years, humans have directly impacted and altered close to 47 percent of the global land area. Under the scenario presented here, biodiversity will be threatened in almost 72 percent of the land area by 2032. Losses of biodiversity are likely to be particularly severe in Southeast Asia, the Congo basin and parts of the Amazon. As much as 48 percent of these areas will become converted to agricultural land, plantations and urban areas, compared with 22 percent today, suggesting wide depletions of biodiversity.

Conclusions

In considering the results of such scenario-building exercises, it is important to recognize that the future is very uncertain. None of the approaches described above attempts to provide accurate predictions of what is likely to happen in the future; rather, the aim is to illustrate what could occur if current trends continue. Although differing in the methodologies adopted and the results obtained, each of these approaches suggests that human activities will continue to have a major impact on biodiversity in coming decades. It is clear that those areas characterized by the highest numbers of species, such as tropical forests and Mediterranean ecosystems, are at particularly high risk of suffering major losses of biodiversity. In contrast, northern temperate areas appear less likely to experience such severe impacts, although significant losses of biodiversity may still be anticipated. The challenge for future efforts at scenario development will be to identify those policy decisions that could make the greatest contribution to the conservation of biodiversity while securing the economic benefits of sustainable development. With widespread declines in biodiversity likely to continue, given current trends in economic development and population growth, it seems probable that such decisions will at best contribute to slowing the rate of biodiversity loss. Significantly, none of the scenario approaches suggests much scope for the ecological recovery or rehabilitation of ecosystems that have already been degraded, except in localized areas.

REFERENCES

1 UNEP 1997. Recommendations for a core set of indicators of biological diversity. Background paper prepared by the Liaison Group on Indicators of Biological Diversity. UNEP/CBD/SBSTTA/3/Inf.13. Available online from the CBD Secretariat website, http://www.biodiv.org/ (accessed April 2002).

2 IUCN, UNEP, WWF 1980. *World conservation strategy: Living resource conservation for sustainable development*. IUCN-the World Conservation Union, Gland.

3 IUCN 1998. *1997 United Nations list of protected areas*. Prepared by UNEP-WCMC and WCPA. IUCN-the World Conservation Union, Gland and Cambridge.

4 Bruner, A.G. *et al.* 2001. Effectiveness of parks in protecting tropical biodiversity. *Science* 291: 125-128.

5 Mittermeier, R. 1988. Primate diversity and the tropical forest. Case studies from Brazil and Madagascar and the importance of megadiversity countries. In: Wilson, E.O. (ed.) *Biodiversity*, pp. 145-154. National Academy Press, Washington DC.

6 Myers, N. 1988. Threatened biotas: 'Hotspots' in tropical forests. *Environmentalist* 8: 1-20.

7   Myers, N. 1990. The biodiversity challenge: Expanded hot-spots analysis. *Environmentalist* 10: 243-256.

8   WCMC 1998. Development of a national biodiversity index. Version 2. Unpublished report for the World Bank. World Conservation Monitoring Centre.

9   Mittermeier, R.A., Myers, N. and Mittermeier, C.G. 1999. *Hotspots: Earth's biologically richest and most endangered terrestrial ecoregions*. Conservation International, Washington DC, distributed by University of Chicago Press. Available online at http://www.conservation.org/xp/CIWEB/strategies/hotspots/hotspots.xml (accessed March 2002).

10 WWF and IUCN 1994. *Centres of plant diversity. A guide and strategy for their conservation*. 3 vols. IUCN Publications Unit, Cambridge.

11 Duellman, W.E. (ed.) 1999. *Patterns of distribution of amphibians: A global perspective*. Johns Hopkins University Press, Baltimore.

12 Stattersfield, A.J. *et al.* 1998. *Endemic bird areas of the world*. BirdLife International, Cambridge.

13 Van Jaarsveld, A.S. *et al.* 1997. Biodiversity assessment and conservation strategies. *Science* 279: 2106-2108.

14 Prendergast, J.R. *et al.* 1998. Biodiversity inventories, indicator taxa and effects of habitat modification in tropical forest. *Nature* 391: 72-76.

15 Prendergast, J.R. *et al.* 1993. Rare species, the coincidence of biodiversity hotspots and conservation strategies. *Nature* 365: 335-337.

16 Prendergast, J.R. and Eversham, B. 1997. Species richness covariance in higher taxa: Empirical tests of the biodiversity indicator concept. *Ecography* 20: 210-216.

17 Mace, G. *et al.* 2000. It's time to work together and stop duplicating conservation efforts. *Nature* 405: 393.

18 Margules, C.R. and Pressey, R.L. 2000. Systematic conservation planning. *Nature* 405: 243-253.

19 Da Fonseca, G.A.B. *et al.* 2000. Following Africa's lead in setting priorities. *Nature* 405: 393-394.

20 Brooks, T. *et al.* 2001. Towards a blueprint for conservation in Africa. *BioScience* 51: 613-624.

21 Wolf, A.T. *et al.* 1999. International river basins of the world. *International Journal of Water Resources Development* 15: 387-427.

22 Large marine ecosystems of the world. Available online at http://www.edc.uri.edu/lme/default.htm (accessed March 2002).

23 Murowski, S.A. 1996. Can we manage our multispecies fisheries? In: *The northeast shelf ecosystem: Assessment, sustainability, and management*, pp. 491-510. Blackwell Science Ltd, Oxford.

24 Sherman, K. and Busch, D.A. 1995. Assessment and monitoring of large marine ecosystems. In: Rapport D.J., Guadet, C.L. and Calow, P. (eds). *Evaluating and monitoring the health of large-scale ecosystems*. Springer-Verlag, Berlin, published in cooperation with NATO Scientific Affairs Division. NATO Advanced Science Institutes Series 1: *Global Environmental Change* 28: 385-430.

25 Salm, R.V. and Clark, J.R. 1984. *Marine and coastal protected areas: A guide for planners and managers*. IUCN-the World Conservation Union, Gland.

26 Kelleher, G., Bleakley, C. and Wells, S. (eds) 1995. *A global representative system of marine protected areas*, Vols 1-4. IBRD/World Bank, Washington DC.

27 Dobson, A.P., Bradshaw, A.D. and Baker, A.J.M. 1997. Hopes for the future: Restoration ecology and conservation biology. *Science* 277: 515-522.

28 WRI 2000. *World resources 2000-2001: People and ecosystems. The fraying web of life*. World Resources Institute, Washington DC.

29 Spalton, A. 1990. Recent developments in the reintroduction of the Arabian oryx (*Oryx leucoryx*) to Oman. *Species* 15: 27-29.

30 Sala, O.E. *et al.* 2000. Global biodiversity scenarios for the year 2100. *Science* 287: 1770-1774.

31 UNEP 2002. *Global environment outlook*. Report of GEO-3. United Nations Environment Programme, Nairobi.

32 Raskin, P. *et al.* 1998. *Bending the curve: Toward global sustainability.* Stockholm Environment Institute, Stockholm.

33 GLOBIO. Online at http://www.globio.info (accessed April 2002).

34 Brink, ten B. 2000. Biodiversity indicators for the OECD Environmental Outlook and Strategy, a feasibility study. RIVM report 402001014, Bilthoven.

35 Brink, ten B. 2001. The state of agro-biodiversity in the Netherlands. Integrating habitat and species indicators. Paper for the OECD workshop on agri-biodiversity indicators, 5-8 October, Zurich, Switzerland.

36 Brink, ten B. *et al.* 2000. Technical report on Biodiversity in Europe: An integrated economic and environmental assessment. Prepared by RIVM, EFTEC, NTUA and IIASA in association with TME and TNO under contract with the Environment Directorate-General of the European Commission. RIVM report 481505019, Bilthoven.

APPENDIX 1:
THE PHYLA OF LIVING ORGANISMS

The phyla are listed in alphabetical sequence within each higher group, as are the 'kingdoms' within the Eukarya. The symbols associated with each phylum name indicate whether the species occur in marine, inland water or terrestrial habitats. Where more than one symbol is shown, this does not mean that species are equally distributed between them. In some cases, the text notes the principal habitat. For parasitic forms the symbol refers to the host habitat.

Where possible, an estimate has been given of the number of described extant species in each eukaryote phylum. Because of the vagaries of taxonomy and lack of a consolidated catalogue of such species (see Chapter 2), even this number is subject to considerable uncertainty in many cases; '?' indicates no estimate available.

An attempt has also been made to gauge whether the given number represents a low, medium or high proportion of the possible total number of species in each phylum (i.e. including currently undescribed species). In rough terms, a low proportion is taken to indicate that the total

diversity of the phylum may be an order of magnitude (or more) higher than the number of currently described species. A high proportion indicates that well over half the total number of species has almost certainly already been described. Generally, these estimates reflect the likelihood that microscopic, aquatic (especially marine) and parasitic groups are less thoroughly sampled than most macroscopic terrestrial forms. Because applying the species concept to prokaryotes is so problematic, no figures for species diversity are given for Archaea or Bacteria: it can be safely assumed that in almost all these phyla the proportion of existing diversity characterized to date is low.

Source: Margulis, L. and Schwartz, K.V. 1998. *Five kingdoms. An illustrated guide to the phyla of life on earth.* 3rd edition. W.H. Freeman and Company, New York.

▼ **Marine**
◆ **Freshwater**
● **Terrestrial**

ARCHAEA

Crenarchaeota ▼ ◆	Bacteria adapted to hot, acidic sulfur-rich environments often found in hot springs and around submarine vents. *Pyrolobus* grows at temperatures of 113°C. *Sulfolobus* tolerates temperatures up to 90°C and may die if the temperature drops below 55°C; it also tolerates highly acid conditions (pH of less than 1, or stronger than concentrated sulfuric acid).
Thermoacidophils	
Size Microscopic	
Nutrition Chemoautotrophs or heterotrophs	
Mode of life Free-living	

Euryarchaeota ▼ ◆ ●
Methanogens and halophils

Size Microscopic
Nutrition Methanogens are chemoautotrophs halophils are photosynthetic
Mode of life Free-living or symbiotic, inhabiting the intestines of animals
Euryarchaeota share similarities in ribosomal RNA sequence but consist of two very different groups. Methanogens cannot tolerate oxygen (are obligate anaerobes) and free-living forms tend to occur in

swamps, bogs and estuary sediments; many others live in the guts of herbivorous animals, from termites to cows. They are chemoautotrophs that obtain energy by reducing CO_2 (carbon dioxide) and oxidizing H_2 (hydrogen) to produce CH_4 (methane) and H_2O (water). They are responsible for production of most natural gas and for liberation of organic carbon from sediments into the atmosphere where it can be reused, involving around 2 billion tons of methane annually. Halophils are aerobes that live in extremely salty or highly alkaline environments, such as soda lakes, where they may be visible as a pink scum.

BACTERIA

Actinobacteria
Actinomycetes, actinomycota and their relatives

◆ ●

Size	Microscopic
Nutrition	Heterotrophic
Mode of life	Some free-living; some symbionts

A large and diverse group of heterotrophic unicellular rod-shaped bacteria (coryneforms), and filamentous, multicelled bacteria (actinomycetes) originally regarded as fungi. Some form pathogenic lesions on skin, while others are found in leaf litter; some of the latter can break down cellulose. *Frankia* is a nitrogen-fixing symbiont in plants. *Streptomyces* produces streptomycin and other antibiotics.

Aphragmabacteria

? ? ●

Size	Microscopic
Nutrition	Heterotrophic
Mode of life	All symbionts; some parasitic

Very small bacteria, lacking a cellwall, widespread in insect, plant and vertebrate tissues. Normally benign, but pathogenic in some conditions, and responsible for some forms of pneumonia and tick-borne diseases (e.g. *Ehrlichia*). Eight named genera to date.

Chlorobia
Anoxygenic green sulfur bacteria

▼ ◆ ●

Size	Microscopic
Nutrition	Photosynthetic
Mode of life	Mainly free-living; some symbiotic with other bacteria

Phototrophic obligate anaerobes, inhabiting sunlit sulfide-rich habitats, particularly anaerobic muds. Some are tolerant of extremely high or low temperatures and salinities. Most use hydrogen sulfide or sodium sulfide in photosynthesis, instead of water, releasing sulfur instead of oxygen. Others form symbiotic associations with oxygen-respiring heterotrophic bacteria.

Chloroflexa
Green nonsulphur phototrophs

◆ ●

Size	Microscopic
Nutrition	Photosynthetic
Mode of life	Free-living

Anaerobic filament-forming bacteria known from sulfur-rich habitats such as hot springs. Whilst these forms are typically photosynthetic, *Chloroflexus* can also grow heterotrophically in the dark. Three genera named to date.

Cyanobacteria
Blue-green bacteria and chloroxybacteria

▼ ◆ ●

Size	Microscopic but relatively large
Nutrition	Photosynthetic
Mode of life	Free-living

Photosynthesizing bacteria, present in a great variety of habitats. Until recently called 'blue-green algae' and considered to be plants. These bacteria dominated the landscape in the Proterozoic eon between 2 600 and 545 million years ago. *Prochlorococcus* occurs at the base of the photic zone throughout the world's oceans and may be one of the commonest bacteria. Many fix atmospheric nitrogen. Form reef-like stromatolites in some shallow-water marine environments. Around 1 000 named genera.

Deinococci
Heat- or radiation-resistant bacteria

◆

Size	Microscopic
Nutrition	Heterotrophic
Mode of life	Free-living

Spherical, heterotrophic, obligate or facultative aerobic bacteria highly resistant to heat (*Thermus*) or radiation (*Deinococcus*). Most metabolize sugars. *Thermus aquaticus*, isolated from hot springs in Yellowstone National Park, United States, is the source of Taq polymerase used in the polymerase chain reaction technique.

BACTERIA

Endospora
Endospore-forming and related bacteria

Size Microscopic
Nutrition Heterotrophic
Mode of life Many symbionts and parasites

A very large, important and varied group of heterotrophic bacteria, some obligate anaerobes, others facultative or obligate aerobes. Most form endospores (propagules within the parent cell resistant to heat and desiccation). Some can break down lignin and cellulose; others are fermenters, breaking down sugars to produce compounds such as lactic acid and ethanol. Some, such as *Streptococcus*, are associated with infections.

Pirellulae ◆
Proteinaceous-walled bacteria and their relatives

Size Microscopic
Nutrition Heterotrophic
Mode of life Mostly aquatic in freshwaters; some symbionts, some parasitic

Diverse bacteria with proteinaceous cell walls; mostly obligate aerobic heterotrophs living in freshwaters. *Chlamydia* is parasitic, inhabiting animal cells and with apparently no independent means of producing energy. *C. psittaci* causes psittacosis; *C. trachomatis* causes trachoma blindness.

Proteobacteria ▼ ◆ ●
Purple bacteria

Size Microscopic
Nutrition Include virtually all nutritional modes known
Mode of life Major parasites and symbionts; some free-living aquatic forms

An enormous and extremely varied group of bacteria including many disease-causing forms (e.g. *Salmonella*, a cause of food poisoning, and *Neisseria*, which causes gonorrhea) and symbionts

such as *Escherichia coli*. Proteobacteria show a great range of physical structure and metabolic activity; the group includes heterotrophs, chemotrophs, chemoheterotrophs, chemolithoautotrophs, photoautotrophs, photoheterotrophs, methylotrophs, hydrogen-oxidizers and sulfide-oxidizers. Many are facultative aerobes, respiring oxygen when this is available but able to survive by respiring, for example, nitrogen (N_2) or sulfate (SO_4^{2-}) when not. Responsible for a significant proportion of atmospheric nitrogen fixation.

Saprospirae ▼ ◆ ●
Fermenting gliders

Size Microscopic
Nutrition Heterotrophic
Mode of life Free-living, or symbiotic in animals

One group includes anaerobic fermenters restricted to anoxic environments. Some free-living; some (e.g. *Bacteroides*) inhabit intestinal tract of vertebrates, including humans, in enormous numbers. A second group includes aerobic forms inhabiting decaying vegetation and other organic-rich environments.

Spirochaetae ▼ ◆ ●
Spirochaetes

Size Microscopic
Nutrition Heterotrophic
Mode of life Free-living or symbiotic; some parasitic

Spiral-shaped bacteria occurring in marine and freshwater habitats, including deep muddy sediments, and in animals, where many are major parasites or symbionts. Some respire gaseous oxygen, others are poisoned by it. *Treponema pallidum* causes syphilis and yaws, and *Leptospira* causes leptospirosis. Twelve genera named to date.

Thermotogae ▼ ◆ ●
Thermophilic fermenters

Size Microscopic
Nutrition Heterotrophic
Mode of life Free-living

Recently discovered obligate anaerobic bacteria known from submarine hot vents, terrestrial hot springs and subterranean oil reservoirs. Highly heat tolerant, living at temperatures of 50°C to 80°C. Ferment sugar and other organic compounds.

EUKARYA: ANIMALIA

Acanthocephala ▼ ◆ ●
Thorny-headed worms
No. species described   Over 1 000
Proportion of group known   Low/moderate
Size   Between 1 mm and 1 m in length
Nutrition   Heterotrophic
Mode of life Parasitic worms that lack a free-living stage

Adult individuals anchor themselves to the gut wall of vertebrates. Infection generally occurs after an intermediate invertebrate host is ingested. Thorny-headed worms appear to alter host behavior so as to increase probability of host being ingested by predator, and so transfer parasite to further host. Humans are seldom parasitized.

Annelida ▼ ◆ ●
Annelids
No. species described   ca 16 000
Proportion of group known   Low/moderate
Size   From 0.5 mm to 3 m
Nutrition   Heterotrophic
Mode of life Segmented worms, mostly free-living in soils and sediments; some parasitic

A large phylum including polychaetes (9 000 species), oligochaetes (6 000) and leeches (500). Most are active predators and scavengers. Polychaetes include free-living and tube-dwelling marine species, mainly benthic but some pelagic. Oligochaetes occur in freshwater, estuaries and deep sea, but are most numerous on land where earthworms are very important to soil structure. Leeches are mainly free-living predators of vertebrates and invertebrates in freshwaters or water film on land; formerly more widely used for medicinal purposes (*Hirudo medicinalis*).

Brachiopoda ▼
Lampshells
No. species described   ca 350
Proportion of group known   Low/moderate
Size   From 2 mm to 10 cm
Nutrition   Heterotrophic
Mode of life Benthic, mainly sessile, marine animals

Cosmopolitan. Present between the intertidal zone and 4 000 m depth. Usually occur cemented to surface by stalk (pedicle); some species free-living on or in marine sediment. Unlike mollusks, brachiopods have a lophophore, a specialized surface for gas exchange and food collection, and have dorso-ventral symmetry instead of lateral symmetry. Previously very diverse, especially during the Paleozoic era (see Chapter 3). Some 30 000 extinct species have been described, and *Lingula*, with fossils from 400 million years ago, may be the oldest genus with living species.

Bryozoa ▼ ◆
Ectoprocts
No. species described   ca 4 000
Proportion of group known   Low
Size   Individuals mainly microscopic; colonies to 0.5 m diameter
Nutrition   Heterotrophic
Mode of life Sessile, colony-forming filter-feeders, mainly marine

Marine bryozoans mainly intertidal, but also on seafloor to considerable depths. About 50 freshwater species known, with jelly-like colonies on plant surfaces in slow streams. Large colonies (0.5 m in diameter) derived from the asexual budding of zooids (less than 1 mm in length) may contain several million individuals. Marine forms contribute to reef diversity.

Cephalochordata ▼
Lancelets
No. species described   23
Proportion of group known   Moderate/high
Size   From 5 cm to 15 cm
Nutrition   Heterotrophic
Mode of life Free-living, filter-feeding marine animals

Lancelets occur in estuary sediments and shallow sandy seafloors, and live with the head protruding in order to screen out small plankton and organic materials. They make up a small phylum of chordates with a cartilaginous rod dorsal to the gut (notochord), a dorsal hollow nerve cord, and persistent gill slits in the pharynx, but without an internal bony skeleton or cerebral ganglion. They are the closest living relatives of vertebrate animals. Used as human food in some areas.

EUKARYA: ANIMALIA

Chaetognatha ▼
Arrow worms
No. species described ca 70
Proportion of group known Low
Size From 0.5 to 15 cm
Nutrition Heterotrophic
Mode of life Worm-like, planktonic marine predators

Common plankton in open seas, especially abundant in warm seas down to 200 m. Detect prey, mainly copepods, by vibration sensors, and can inject neurotoxins. Important to marine fisheries as a source of food for fishes.

Chelicerata ▼ ◆ ●
Chelicerates
No. species described ca 75 000
Proportion of group known Low
Size Macroscopic; the largest species of Pycnogonida have a leg span of almost 80 cm
Nutrition Heterotrophic
Mode of life Generally free-living and in most habitats
A very large and diverse arthropod phylum characterized by claws (chelicerae) on the anterior pair of appendages, and sharing other features (e.g.

segmented bodies, chitinous exoskeleton, jointed appendages) with insects and crustaceans. Most chelicerates are in Arachnida (more than 75 000 species); others are horseshoe crabs (Merostomata) and sea spiders (Pycnogonida). Arachnids are ubiquitous on land, with a few freshwater species; the group includes ticks and mites, some of considerable importance as vectors of disease in humans and livestock. Merostomata include *Limulus*, superficially unchanged since the Silurian (see Chapter 3). Pycnogonids range from the shallows to the deep ocean (6 800 m) and from pole to pole.

Cnidaria ▼ ◆
Cnidarians, hydras
No. species described ca 9 000
Proportion of group known Moderate
Size Mainly macroscopic
Nutrition Heterotrophic, mostly carnivorous; reef-building corals contain photosynthetic symbionts
Mode of life Aquatic, almost all marine; colonial and solitary, free-swimming and sedentary forms known

A diverse phylum of radially symmetrical animals, including sea anemones, jellyfishes and corals. Specialized stinging cells called cnidoblasts are diagnostic. Largest individuals (e.g. lion's mane *Cyanea*) may have tentacles many meters long. Reef-building corals are of major importance in clear-water coastal shallows in tropics and subtropics. Coral reefs often highly diverse and of great economic value. Photosynthetic symbionts (dinomastigotes) of reef corals occur in polyp tissue at density up to 5 million/cm^2; these require sunlight and limit reef growth to upper part of the photic zone. Non-reef coral without symbionts range down to 3 000 m.

Craniata ▼ ◆ ●
Craniates or vertebrates
No. species described ca 52 500
Proportion of group known High
Size From about 1 cm to 35 m
Nutrition Heterotrophic
Mode of life Free-living species present in most habitat types
This group contains the vertebrates; a very large and very diverse phylum of chordates, all of which, unlike the acraniate chordates (urochordates, cephalochordates), have a brain enclosed within a skull (cranium). The majority have a bony internal skeleton. A small group of mainly marine species (lampreys and hagfish) lack jaws and are grouped in

Agnatha in contrast to other vertebrates, all of which possess jaws (Gnathostomata). About half of all described vertebrate species are fishes: the Chondrichthyes (sharks and rays), Osteichthyes (bony fishes), the lungfishes and coelacanths. These make up around 25 000 species in total. The tetrapods (the four-limbed non-fish vertebrates) include amphibians, reptiles, birds and mammals. Although not so versatile as bacteria, vertebrates between them extend from the air above the highest mountain to abyssal ocean depths, from sand desert to tropical forest, and from hot springs to polar ice and subzero waters. The vertebrates include the most familiar animals, and, with mollusks and crustaceans, most of those of direct nutritional importance to humans.

EUKARYA: ANIMALIA

Crustacea   ▼ ◆ ●
Crustaceans
No. species described   ca 40 000
Proportion of group known   Low/moderate
Size   Mostly macroscopic
Nutrition   Heterotrophic
Mode of life   Mostly free-living in aquatic and humid terrestrial habitats; some parasitic
A very large and very diverse arthropod phylum distinguished by having two pairs of antennae. Species occur in virtually all habitats. Includes crabs, crayfish, prawns, barnacles, copepods, brine shrimp, water fleas, woodlice, etc. Size ranges between 0.25 mm (*Alonella*) and 2.8 m (*Macrocheira* claw span). They are the predominant arthropods in most freshwaters and widespread in all marine habitats, from pelagic waters to ocean depths at 5 000 m, and in moist terrestrial situations. Numerous parasitic and commensal forms exist; pentastomids sometimes parasitize humans. Crustaceans such as krill (*Euphausia superba*) form key components of the marine food web. There are numerous important crustacean fisheries.

Ctenophora   ▼
Comb jellies
No. species described   ca 100
Proportion of group known   Low
Size   Typically around 1 cm; largest up to 2 m in length
Nutrition   Heterotrophic
Mode of life   Free-swimming marine organisms
A small phylum of translucent, soft-bodied predators. Widespread in marine waters and possibly the most abundant planktonic animals between 400 and 700 m depth. Their fragility makes them difficult to collect and study.

Echinodermata   ▼
Echinoderms
No. species described   ca 7 000
Proportion of group known   Moderate
Size   A few centimeters to near 2 m
Nutrition   Heterotrophic
Mode of life   Mostly free-living, benthic marine species
Invertebrates with five-part radial symmetry, an internal calcium carbonate skeleton, and a water vascular system. Includes starfish, sea urchins, sea cucumbers and others. Mostly benthic in intertidal or subtidal habitats. Sea lilies extend to 10 000 m, and sea cucumbers in places make up nearly the entire animal biomass at these abyssal depths. Viviparous forms exist. Some, such as dried sea cucumbers (trepang), used as human food.

Echiura   ▼
Spoon worms
No. species described   ca 140
Proportion of group known   Low/moderate
Size   From a few millimeters to 40 cm
Nutrition   Heterotrophic
Mode of life   Exclusively free-living marine organisms
Echiurans live in U-shaped burrows in marine sediments, rock crevices and mangrove, with some forms extending to abyssal depths around 10 000 m. Echiurans have a flexible proboscis which may extend to 1.5 m from the bulbous, unsegmented body. Cilia move food items down the proboscis to the mouth.

Entoprocta   ▼
Entoprocts
No. species described   ca 150
Proportion of group known   Low
Size   Very small macroscopic animals, up to 1 cm
Nutrition   Heterotrophic
Mode of life   Mostly sessile, colonial marine organisms
Widely distributed in shallow coastal waters. One freshwater species known. Colonies permanently attached by stalks, horizontal stolons and basal discs to solid substrate, algae or other animals. Often form conspicuous mat-like growth on seaweed and rocks. Filter-feeders, consuming diatoms, desmids, other plankton and detritus. *Loxosomella* is free-living, and moves by somersaulting basal disc over tentacles.

EUKARYA: ANIMALIA

Gastrotricha ▼ ◆
Gastrotrichs
No. species described   ca 400
Proportion of group known   Low
Size   Average length 0.5 mm
Nutrition   Heterotrophic
Mode of life Free-living, worm-like animals, mainly marine

The ventral side is ciliated and often glued to substrate; exposed surfaces bear bristles or scales. Most occur in subtidal or intertidal sediments where form part of the marine meiobenthos; freshwater forms most abundant in small still waters. Important scavengers of dead bacteria and plankton.

Gnathostomulids ▼
Jaw worms
No. species described   ca 80
Proportion of group known   Low
Size   Average length around 1.5 mm
Nutrition   Heterotrophic
Mode of life Free-living marine worms

A small phylum of translucent, benthic, worms capable of surviving in sediments very low in oxygen and high in hydrogen sulfide. Graze on bacteria, protoctists and fungi in marine sediments. Have been found at several hundred meters depth. Population densities may exceed 6 000 per liter of sediment, outnumbering nematodes.

Hemichordata ▼
Acorn worms
No. species described   ca 90
Proportion of group known   Low/moderate
Size   Adults between 2.5 and 250 cm
Nutrition   Heterotrophic
Mode of life Sedentary, benthic marine species

A small phylum of soft-bodied, benthic marine worm-like animals. Adults mostly sedentary and live burrowed in soft sediment of shallow seas, or in secreted tubes. Sexual and asexual reproduction occurs; colonies may be formed by budding. Hemichordates were previously classified as chordates, and resemble them in having ciliated gill slits in pharynx.

Kinorhyncha ▼
Kinorhynchs
No. species described   ca 150
Proportion of group known   Low/moderate
Size   Up to 1 mm in length
Nutrition   Heterotrophic
Mode of life Free-living marine animals

Kinorhynchs are cosmopolitan in muddy bottom habitats, including estuaries and the intertidal zone, and to a depth of approximately 5 000 m. Some species are commensal with hydrozoans, bryozoans and sponges.

Loricifera ▼
Loriciferans
No. species described   ca 100
Proportion of group known   Low
Size   Microscopic
Nutrition   Heterotrophic
Mode of life Benthic marine species

Widespread, probably cosmopolitan part of interstitial fauna. Life history incompletely known. Adults are sedentary on sand or gravel, sometimes ectoparasites; the larvae are believed to be free-living and mobile. Protective plates cover the abdomen, into which the neck and head with mouth cone can be retracted.

Mandibulata ▼ ◆ ●
Mandibulates
No. species described   ca 950 000
Proportion of group known   Low
Size   From near microscopic to many centimeters
Nutrition   Heterotrophic
Mode of life Most are free-living terrestrial species
An exceptionally large and diverse arthropod phylum distinguished by a pair of crushing mandibles. Jointed chitinous exoskeleton and a single pair of

antennae. Includes insects (Hexapoda), centipedes and millipedes (Myriapoda), symphyla and pauropods. The largest group of animals, Hexapoda, contains some 950 000 described species and may number in millions. Insects range up to 30 cm length (giant stick insect *Pharnacia serratipes*). Social insects such as termites, ants, some bees and wasps can form large colonies. Many insects are important crop pests or disease vectors; others are beneficial because of crop pollination or pest control.

EUKARYA: ANIMALIA

Mollusca   ▼ ◆ ●
Mollusks
No. species described   At least 70 000 and possibly more than 100 000
Proportion of group known   Moderate
Size   Range from near microscopic to several meters
Nutrition   Heterotrophic
Mode of life   Mainly free-living species present in most habitat types; some parasitic
A large and highly diverse phylum, with species occurring in benthic and pelagic marine waters, in freshwaters of all kinds, and on land, from forests to

deserts. Includes snails, slugs, mussels, chitons, octopods, squid and others. Most species are free-living, although some are parasitic or commensals. Many, e.g. bivalves, are sedentary as adults. Size reaches maximum, approaching 20 m, in the giant squid *Architeuthis*. Many important as food source. Some forms act as intermediate hosts to parasites (e.g. *Schistosoma*) that cause serious human disease; other species can cause significant damage to crops and constructions (e.g. *Dreissena*). Venom of some marine gastropods is of medical interest. Monoplacophorans, the most primitive mollusks, were first seen alive in the 1970s but abundant in Paleozoic (see Chapter 3).

Nematoda   ▼ ◆ ●
Nematodes
No. species described   ca 20 000
Proportion of group known   Low
Size   From 0.1 mm to 9 m in length
Nutrition   Heterotrophic
Mode of life   Free-living or parasitic worm-like animals

Possibly the most abundant animals living on Earth, found in virtually all habitats and in many other organisms. Free-living forms are key to decomposition and nutrient cycling. Many species are important parasites of plants and animals, including humans (e.g. filariasis). Nematodes have provided important research animals in genetics and cell differentiation.

Nematomorpha   ▼ ◆ ●
Nematomorphs
No. species described   ca 240
Proportion of group known   Low
Size   Ranging from 10 to 70 cm in length
Nutrition   Heterotrophic
Mode of life   Adults are free-living and usually aquatic; all are endoparasitic at some stage

A small phylum of leathery, unsegmented, worm-like animals. Occur widely in aquatic or moist terrestrial habitats. Eggs hatch into minute motile larvae which enter host and metamorphose into immature worms. These burst out, killing the host, when near water or during rain. Hosts include annelids and arthropods. Rarely found in humans, where appear non-pathogenic.

Nemertina   ▼ ◆ ●
Ribbon worms
No. species described   ca 900
Proportion of group known   Low/moderate
Size   Macroscopic, from 0.5 mm to 30 meters in length, mostly small
Nutrition   Heterotrophic
Mode of life   Mostly free-living, predatory marine worms

Characterized by the slender anterior proboscis, used for predation, defense and locomotion. Abundant in the intertidal zone; some forms are pelagic. Freshwater and terrestrial forms are known. Some species are symbionts or parasites: recorded from echinoderms, nemertines, annelids, mollusks and flatworms. May affect host reproduction. *Malacobdella* is a filter-feeder, inhabiting the mantle cavity of clams.

Onychophora   ●
Velvet worms
No. species described   ca 100
Proportion of group known   Low
Size   From 1 to 20 cm
Nutrition   Heterotrophic
Mode of life   A small phylum of free-living, terrestrial, worm-like animals

Require high humidity levels to counter water loss through thin chitinous cuticle. Many occur in forest; some in caves. All carnivorous. Walk slowly on 14-43 pairs of stumpy legs. Two geographic groups exist: one mainly warm northern hemisphere; the other southern hemisphere. Many tropical species are viviparous.

EUKARYA: ANIMALIA

Orthonectida ▼
Orthonectida
No. species described   ca 20
Proportion of group known   Low
Size   Microscopic
Nutrition   Heterotrophic
Mode of life Worm-like internal parasites or
             symbionts of marine invertebrates

Recorded from echinoderms, nemertines, annelids, mollusks and flatworms. Less benign than rhombozoans; may affect host reproduction.

Phoronida ▼
Phoronids
No. species described   14
Proportion of group known   Low/moderate
Size   From 1 mm to 50 cm
Nutrition   Heterotrophic
Mode of life Sedentary filter-feeding marine worms

Most inhabit leathery chitinous tubes encrusted with sand or shell fragments; some burrow in mollusk shells or rock. From coastal shallows to 400 m depth. Filter-feed on plankton and detritus. Cosmopolitan but not abundant; half the known species occur on Pacific coast of North America.

Placozoa ▼
Trichoplax
No. species described   1
Proportion of group known   High
Size   Up to 1 mm
Nutrition   Heterotrophic
Mode of life Very small marine animal

Discovered in a seawater aquarium in 1883, and since reported in shallow marine water and marine research stations. *Trichoplax adhaerens* is the least complex of all living animals, consisting of a few thousand cells but no distinct tissues; little is known of its life history.

Platyhelminthes ▼ ◆ ●
Flatworms
No. species described   ca 20 000
Proportion of group known   Low/moderate
Size   Often a few millimeters; tapeworms to
             30 m in length
Nutrition   Heterotrophic
Mode of life Free-living or symbionts, many
             parasitic; found in freshwater, marine
             and terrestrial environments

Flatworms, flukes and tapeworms. Four classes. Free-living soil flatworms most abundant in tropics; aquatic forms mainly temperate. Some can survive in environments low in oxygen by oxidizing hydrogen sulfide. Parasitic forms include flukes such as *Schistosoma*, the cause of schistosomiasis, and tapeworms, obligate parasites of vertebrate gut. Many flatworm parasites have complex life cycle with infective larvae and intermediate hosts.

Pogonophora ▼
Beard worms
No. species described   More than 120
Proportion of group known   Low
Size   From 10 cm to 2 m in length
Nutrition   Heterotrophic
Mode of life Sessile, benthic marine worms
Pogonophora live in fixed upright chitin tubes secreted in sediments, shell or decaying wood on the ocean floor. Most abundant in cold, deep waters,

shallow polar seas or (the vestimentiferans) around hot submarine vents with a high hydrogen sulfide and methane content. Greatest diversity in the western Pacific. Adult pogonophorans have no gut and probably absorb nutrients directly from tentacles. Vent-living forms derive nutrients and energy from the oxidation of hydrogen sulfide through symbiotic chemoautotrophic bacteria, which can occur at densities of 1 billion per gram of body tissue.

Porifera ▼ ◆
Sponges
No. species described   5 000–10 000
Proportion of group known   Moderate
Size   Macroscopic, some to 2 m in height
Nutrition   Heterotrophic
Mode of life Sedentary aquatic animals

The vast majority of sponges are marine and about 100 species are freshwater. Filter-feeders (one Mediterranean form passively captures crustaceans and digests them externally). Many include photosynthetic symbionts, e.g. cyanobacteria, and brown, red or green algae. Sponges have simple structure with no tissues or organs and are generally supported by calcareous or siliceous spicules or fibrous, proteinaceous matrix.

EUKARYA: ANIMALIA

Priapulida ▼
Priapulids
No. species described 17
Proportion of group known Low
Size Range between 0.5 mm and 30 cm
Nutrition Heterotrophic
Mode of life Exclusively marine, free-living, worm like animals

Found in sand or mud, from intertidal pools to abyssal depths, and from tropical waters to the Antarctic. Approximately half of the described species are part of the marine meiobenthos (i.e. small bottom- or sediment-living species between about 0.5 and 1 mm).

Rhombozoa ▼
Rhombozoans
No. species described ca 70
Proportion of group known Low
Size Up to 5 mm
Nutrition Heterotrophic
Mode of life Worm-like internal parasites or symbionts of benthic cephalopod mollusks

Mostly found in the kidneys of squid and octopus in temperate waters.

Rotifera ▼ ◆ ●
Rotifers
No. species described ca 2 000
Proportion of group known Low
Size Mostly microscopic, some to 2 mm
Nutrition Heterotrophic
Mode of life Mostly free-swimming in freshwaters

Mainly freshwater, also in moist habitats on land; about 50 species occur in benthic and pelagic marine habitats. Rotifers are the most abundant and cosmopolitan of the freshwater zooplankton. Mostly free-living. Many live on other invertebrate organisms; many are endoparasites of invertebrates. Most free-living rotifers reproduce parthenogenetically.

Sipuncula ▼
Peanut worms
No. species described ca 150
Proportion of group known Low
Size A few millimeters to 0.5 m in length
Nutrition Heterotrophic
Mode of life Exclusively marine, worm-like, burrowing or crevice-dwelling organisms

Benthic species, mainly in shallow, warm marine habitats; most abundant on rocky shores, but also present in polar regions and down to 7 000 m in the abyssal ocean. Ingest diatoms and other protoctists or organic debris. Used locally as human food in the Indo-Pacific and China.

Tardigrada ▼ ◆ ●
Water bears
No. species described ca 750
Proportion of group known Low
Size Mainly microscopic
Nutrition Heterotrophic
Mode of life Free-living animals, mostly in moist terrestrial and freshwater habitats

Widely distributed from pole to pole. All are aquatic; land species live in the water film on mosses, forest litter and other habitats. A few marine species. Move on four pairs of stumpy legs. Mainly ingest liquid food obtained by piercing protoctists, animals or plants. The Mesotardigrada (genus *Thermozodium*) inhabit hot springs. Can survive extreme desiccation with low metabolism or in encysted form, and when dormant (cryptobiotic) may be tolerant of exceptionally high or low temperatures, approaching absolute zero.

EUKARYA: ANIMALIA

Urochordata ▼	Adults may be either benthic and sedentary (class

Urochordata   ▼
Sea squirts
No. species described   ca 1 400
Proportion of group known   Low/moderate
Size   From 1 mm to 2 cm
Nutrition   Heterotrophic
Mode of life Small, marine, filter-feeding animals

Adults may be either benthic and sedentary (class Ascidiacea, tunicates) or pelagic and free-swimming (class Larvacea); Thaliacea or salps also free-swimming. All are ciliary filter-feeders. Urochordata have a dorsal hollow nerve cord, a cartilaginous rod dorsal to the gut (notochord) and gill slits in the pharynx at some stage, these being features of chordates.

EUKARYA: FUNGI

Ascomycota   ▼ ◆ ●
Ascomycotes
No. species described   ca 30 000
Proportion of group known   Low
Size   Mainly microscopic but reproductive body up to several centimeters
Nutrition   Heterotrophic
Mode of life Mainly terrestrial, in soil and leaf litter, or on and in other organisms
A large diverse phylum distinguished from other fungi by the microscopic, sac-like, spore-producing reproductive structure (ascus). Includes yeasts, morels, truffles, blue-green molds and lichens. Thread-like hyphae form network (mycelium)

through substrate. Many are free-living; many are parasitic. Several hundred forms occur in freshwaters. More than 10 000 are the heterotrophic components of lichens, which are joint organisms formed by ascomycotes with either photosynthetic green algae or cyanobacteria. As with other fungi, ascomycotes fulfil a key ecological role in breaking down organic material and transferring inorganic nutrients and water from soil to plants. Some are important in food preparation (e.g. baker's yeast). Many cause important diseases of plants and animals, including humans, while others are sources of key medicinal substances such as penicillin and similar antibiotics.

Basidiomycota   ? ◆ ●
Basidomycotes
No. species described   ca 22 250
Proportion of group known   Low/moderate
Size   Generally microscopic or somewhat larger, but hyphae extend considerable distance, and fruiting bodies to 10 cm or more
Nutrition   Heterotrophic
Mode of life Mainly terrestrial, in soil and leaf litter, or on trees and other organisms
A large phylum including typical mushrooms, puffballs, stinkhorns, and rusts and smuts (some of which cause economically important plant diseases).

All are characterized by microscopic, club-shaped, spore-producing reproductive structures, typically borne in great numbers on a basidiocarp – the familiar mushroom. Largest known mushroom specimen grew to 146 cm wide and 54 cm high. Many basiomycotes form mycorrhizas with trees and shrubs; as with zygomycotes, the fungi move basic nutrients from soil to plant, and plant carbohydrates move into the fungus. Many ascomycotes fulfil a key ecological role in breaking down organic material, and transferring inorganic nutrients and water from soil to plants. Several species are valued wild food items. A few freshwater forms are known.

Zygomycota   ? ◆ ●
Zygomycotes
No. species described   ca 1 100
Proportion of group known   Low
Size   Generally microscopic or a little larger, but hyphae may extend considerable distance through soil
Nutrition   Heterotrophic
Mode of life Mainly terrestrial, in soil and leaf litter

Most are saprobic (saprophytic), secreting digestive enzymes into organic material and absorbing nutrients released. Many are parasites on protoctists, small animals, plants or other fungi. Zygomycotes include about 100 species that form mycorrhizal associations, in which fungal partner contacts or enters plant roots and assists inflow of nutrients and absorbs organic plant substances in exchange. Most vascular plants probably have such relationships with fungi and these may be critical in nutrient-poor soils. A few occur in freshwaters.

EUKARYA: PLANTAE

Anthocerophyta
Horned liverworts
No. species described   ca 100
Proportion of group known   Moderate
Size   Low-growing plants
Nutrition   Photosynthetic
Mode of life   Terrestrial, in moist habitats

A small group of non-vascular plants of moist habitats, typically on woodland floor or water margins. Present worldwide in temperate and tropical regions. Among first colonists of bare substrates, including rocks. Some species have associated nitrogen-fixing cyanobacteria.

Anthophyta
Flowering plants, angiosperms
No. species described   ca 270 000
Proportion of group known   High
Size   From less than 1 mm in length (*Wolffia angusta*) to more than 100 m (*Eucalyptus regnans*)
Nutrition   Photosynthetic
Mode of life   Flowering plants occurring in most habitat types
An extremely diverse, geographically cosmopolitan, phylum of vascular seed plants, distinguished by flowers, and fruits that enclose the fertilized seeds. The great majority of species are terrestrial, in virtually all habitat types. Many occur in or around

lakes, rivers and wetlands, and seagrasses occur subtidally in shallow marine waters. Two main groups are distinguished, according to whether the germinating seed has one or two seed leaves: Monocotyledones and Dicotyledones. Monocots include palms, lilies and the economically vital grasses; most monocots are herbaceous and woody forms lack special tissue that secondarily adds width to the trunk. Dicots form the larger group. Success of anthophytes appears linked to coevolution with animals, in particular with specialized modes of pollination and seed dispersal. All major food and medicinal plants, and hardwood timber trees, are found in this phylum.

Bryophyta
Mosses
No. species described   ca 10 000
Proportion of group known   Moderate/high
Size   Low-growing plants
Nutrition   Photosynthetic
Mode of life   Terrestrial, mainly in moist habitats and wetlands
A large phylum of non-vascular plants (i.e. lacking specialized xylem and phloem transport tissue)

consisting of mosses and the genus *Takakia*. Often conspicuous in cold or cool temperate habitats, particularly tundra, where mosses are the dominant plants, and also in heathland, bogs, woodland, waterlogged areas and freshwater margins. Most diverse in moist tropical habitats. Many mosses well adapted to withstand desiccation; some occur in warm arid regions. Peat moss (*Sphagnum*) contributes to the development of new soils.

Coniferophyta
Conifers
No. species described   630
Proportion of group known   High
Size   Shrubs or large trees, up to 100 m in height
Nutrition   Photosynthetic
Mode of life   Terrestrial
Conifers are cone-bearing gymnospermous vascular plants, with needle-shape leaves, and are mostly

evergreen trees. They form extensive forests at high latitudes in the northern hemisphere, and also occur more locally, often on arid mountains; also common in the tropics and in temperate southern forests, where *Araucaria* is widespread. In *Sequoiadendron* and *Sequoia*, conifers include the largest living plants. Many species in mountainous and northern areas have characteristic symbiotic mycorrhizal fungi. Conifers provide timber, paper pulp and ornamental plants, and some have food or medicinal value.

Cycadophyta
Cycads
No. species described   145
Proportion of group known   High
Size   From shrubs to small trees of 18 m in height
Nutrition   Photosynthetic
Mode of life   Terrestrial
A small phylum of seed-bearing, often palm-like, vascular plants restricted to the tropics and

subtropics, where are present in a range of habitats, from moist forest to deserts and coastal mangroves. Diverse in the Cretaceous. Cycads are gymnosperms, i.e. seeds do not become enclosed in a fruit. Many are insect pollinated, often by beetles. All species have symbiotic, nitrogen-fixing cyanobacteria. Cycads provide a variety of materials, including thatch, food, medicines and ornamental plants. Cycad starch for bread requires special treatment to destroy potentially fatal toxins.

EUKARYA: PLANTAE

Filicinophyta ◆ ●
Ferns
No. species described ca 12 000
Proportion of group known High
Size From a few centimeters to 25 m
Nutrition Photosynthetic
Mode of life Terrestrial; a few in freshwater
A diverse phylum of vascular plants, the most species-rich group of plants lacking seeds, widespread from cold temperate areas to the tropics.

Mainly in moist areas, such as forest floor and stream margins; species diversity highest in tropics, where many forms are epiphytic and some species grow as trees to 25 m in height. The aquatic *Azolla*, a very small floating fern, has symbiotic, nitrogen-fixing cyanobacteria. Several food, medicinal and other products are derived from ferns. Ferns, especially tree ferns, were diverse and very abundant in Devonian and Carboniferous times.

Ginkgophyta ●
Ginkgo
No. species described 1
Proportion of group known High
Size To 30 m in height
Nutrition Photosynthetic
Mode of life Terrestrial

A vascular seed-bearing tree, characterized by a fan-shaped leaf with bifurcating veins and a fleshy exposed ovule, now restricted as a wild species to steep forest in southern China. A wide diversity of ginkgophytes, of which *Ginkgo biloba* is the only survivor, existed during the Mesozoic. Now widely planted for ornamental purposes. It is a gymnosperm, i.e. seed not enclosed in a fruit. Leaf extract used as a traditional food and medicine in East Asia.

Gnetophyta ●
Gnetophytes
No. species described ca 70
Proportion of group known High
Size Small trees, shrubs or vines
Nutrition Photosynthetic
Mode of life Terrestrial

A small phylum of vascular seed plants, distinguished from other gymnosperms by having vessels for water transport similar to those of flowering plants. The three living genera differ greatly from each other. Some plants of the genus *Gnetum* in tropical moist forest grow to 7 m in height: some *Welwitschia*, a unique, low-growing, cone-bearing plant of southwest African deserts, may be 2 000 years old.

Hepatophyta ◆ ●
Liverworts
No. species described ca 6 000
Proportion of group known Moderate
Size Low-growing plants
Nutrition Photosynthetic
Mode of life Terrestrial, in moist habitats

Non-vascular plants typically found in moist habitats growing on woodland floor, shaded stream banks, waterfalls or rocks; often epiphytic and often occur with mosses (Bryophyta). Widespread in cold temperate regions, present in Antarctica, but species diversity highest in tropics. Often among first plants to colonize burned or newly exposed substrates.

Lycophyta ◆ ●
Club mosses
No. species described ca 1 000
Proportion of group known Moderate
Size Mainly low-growing herbaceous plants
Nutrition Photosynthetic
Mode of life Terrestrial, in moist and dry habitats

Small seedless evergreen vascular plants found in temperate and tropical habitats, typically on forest floor in temperate regions although most tropical species are epiphytic. A few occur in arid areas. Lycophytes were prominent in Paleozoic plant communities before evolution of flowering plants; although all living species are small, trees up to 40 m in height were dominant in Carboniferous coal forests. Some similarity to mosses and conifers but unrelated to either.

EUKARYA: PLANTAE

Psilophyta   ●
Whisk fern
No. species described   10
Proportion of group known   Moderate/high
Size   Small herbaceous plants
Nutrition   Photosynthetic and symbiotic with fungi
Mode of life Terrestrial

A very small group of vascular plants, the only ones lacking both roots and leaves. Similar to earliest simple, leafless land plants of late Silurian and Devonian times, 400 million years ago, and conceivably direct descendants of them. Present as epiphytes or ground-living species with a restricted range in subtropics and temperate areas. These plants have a mycorrhizal association (also seen in earliest fossil forms) with fungal hyphae that increase the flow of soil nutrients to the non-photosynthetic plant cells.

Sphenophyta   ●
Horsetails
No. species described   15
Proportion of group known   High
Size   Herbaceous plants
Nutrition   Photosynthetic
Mode of life Terrestrial

A small phylum of seedless vascular plants, with jointed ridged stems and tiny scale-like leaves. Found in moist or disturbed areas, including urban areas and roadsides; more typically in moist woods and wetland margins, and also in salt flats. Historically consumed as food in Europe and North America; poisonous to livestock. As with lycophytes, sphenophytes were diverse and abundant in Devonian and Carboniferous forests, with tree-like forms up to 15 m high.

EUKARYA: PROTOCTISTA

Actinopoda   ▼ ◆ ●
Radiolarians
No. species described   ca 4 000
Proportion of group known   Low
Size   Microscopic
Nutrition   Heterotrophic but most hold symbiotic photosynthetic haptomonads
Mode of life Mostly marine, although the Heliozoa is mainly freshwater

Relatively large, generally unicellular protoctists with radial symmetry. Some form large colonies in which many individuals are embedded in a jelly-like matrix. Some occur in open ocean waters, some are benthic. Many have siliceous skeletons with spines or oars used for swimming. Most acantharians include photosynthetic grass-green haptomonad or yellow or green algae symbionts.

Apicomplexa   ? ? ●
Sporozoa
No. species described   ca 5 000
Proportion of group known   Low
Size   Single-celled
Nutrition   Heterotrophic
Mode of life Symbiotic with or parasitic on animals

Many of these spore-forming protoctists are bloodstream parasites with complex life cycles. Coccidians are the best-known group because infection often causes serious or fatal intestinal tract infection. Many, e.g. *Eimeria*, infect livestock; *Isospora hominis* is the only direct coccidian parasite of humans. *Plasmodium* causes malaria, probably at present the single most important infectious disease affecting humans.

Archaeoprotista   ▼ ◆ ●
Amitochondriates
No. species described   ?
Proportion of group known   Low?
Size   Single-celled
Nutrition   Mostly heterotrophic
Mode of life Free-living in aquatic habitats, and symbionts, often parasitic, in animals

Anaerobic and lacking mitochondria. Many forms are parasitic or symbiotic in the intestines of animals, e.g. wood-eating termites and cockroaches. *Giardia* causes giardiasis in humans.

EUKARYA: PROTOCTISTA

Chlorophyta ▼ ◆
Green algae
No. species described ca 16 000
Proportion of group known Low
Size Range from single-celled to macroscopic green seaweeds
Nutrition Photosynthetic
Mode of life Diverse, mostly marine and freshwater algae; a few symbiotic with other organisms

Chlorophytes include unicellular and complex multicellular species as well as forms with many nuclei sharing the same cytoplasm. Major primary producers, they are estimated to fix over 1 billion tons of atmospheric carbon annually. Symbiotic forms include *Platymonas* in the flatworm *Convoluta roscoffensis*. Some forms are resistant to at least periodic desiccation. Some early form of chlorophyta almost certainly gave rise to plants.

Chrysomonada ▼ ◆
Chrysophyta
No. species described ?
Proportion of group known Low?
Size Most single-celled; some form large branching colonies
Nutrition Photosynthetic
Mode of life Free-living, mainly in freshwaters

A large and diverse group of algae with golden-yellow pigments. The silicoflagellates are a component of marine plankton and extract silica from sea water to form shells.

Chytridiomycota ◆ ●
Chytridiomycota
No. species described ca 1 000
Proportion of group known Low
Size Microscopic
Nutrition Heterotrophic
Mode of life Decomposers or parasites in freshwater or moist soils

Feed by extending threadlike hyphae into living hosts or dead material. Simplest forms grow entirely within the cells of their hosts. Some are associated with plant diseases, e.g. *Physoderma zea-maydis* causes brown-spot in maize. Cell walls of chitin; some with cellulose also. Chytrids may be ancestral to fungi.

Ciliophora ▼ ◆
Ciliates
No. species described ca 10 000
Proportion of group known Low
Size Mostly microscopic and single-celled
Nutrition Mostly heterotrophic
Mode of life Mostly free-living; some symbionts or parasites

Although most are unicells, a few multicellular forms resembling slime molds exist. Ciliates feed on bacteria or absorb nutrients from the surrounding medium. Entodiniomorphs live as symbionts in the stomachs of ruminants; the parasite *Balantidium* sometimes causes disease in humans. The free-living *Paramecium* and *Stentor* are well studied and much used in research and education.

Cryptomonada ▼ ◆
Cryptophyta
No. species described ?
Proportion of group known Low?
Size Single-celled
Nutrition Some heterotrophic; others photosynthetic
Mode of life Mostly free-living

Cosmopolitan in moist areas. Most cryptomonads are flattened, elliptical, free-swimming cells in freshwater. Marine species may form blooms on beaches; others are intestinal parasites. Heterotrophs ingest bacteria and protoctists. Some of the photosynthetic forms possess yellow and red pigments in addition to chlorophyll, and some also contain blue-red phycocyanin pigments. Some form colonies of non-mobile cells embedded in a gel-like matrix.

Diatoms ▼ ◆ ●

No. species described ca 10 000
Proportion of group known Low
Size Single-celled; some colonial
Nutrition Mostly photosynthetic; some saprophytes
Mode of life Mostly free-living

Widely distributed in the photic zone of marine and inland waters worldwide. Some occur in moist soils. Diatoms have distinctive paired tests or shells of organic material impregnated with silica extracted from surrounding water. Important basal components of marine and freshwater food webs.

EUKARYA: PROTOCTISTA

Dinomastigota ▼ ◆
Dinoflagellates
No. species described ca 4 000
Proportion of group known Low
Size Single-celled, up to 2 mm, occasionally colonial
Nutrition Some are heterotrophic; others photosynthetic
Mode of life Mostly free-living marine plankton

Typically planktonic; some symbiotic with or live on marine animals or seaweed, some occur in freshwaters. Many adopt very different forms at different life stages. *Gymnodinium microadriaticum* is the most common intracellular photosynthesizing symbiont in corals. Some produce powerful toxins and are an important cause of fish mortality (e.g. *Pfeisteria piscicida*) and may form toxic 'red tides' (e.g. *Gonyaulax tamarensis*). Ciguatera poisoning in humans is caused by accumulations of dinoflagellate toxins in fishes and marine invertebrates. Many species (e.g. *Noctiluca*) are bioluminescent.

Discomitochondria ▼ ◆ ●
Flagellates, zoomastigotes
No. species described ?
Proportion of group known Low?
Size Mostly unicellular
Nutrition Generally heterotrophic; most euglenids are photosynthetic
Mode of life Mainly free-living in a wide range of aquatic and terrestrial habitats

All formerly regarded as protozoan animals, and in medical literature are commonly termed flagellates. Most feed on bacteria or absorb nutrients directly from surroundings; some, i.e. euglenids, are usually photosynthetic. Some are symbiotic or parasitic, the latter including organisms (*Trypanosoma*) responsible for sleeping sickness and Chagas disease.

Eustigmatophyta ▼ ◆
Green eyespot algae
No. species described ?
Proportion of group known Low?
Size Single-celled
Nutrition Photosynthetic
Mode of life Free-living algae, mostly freshwater

Planktonic algae with yellowish-green pigments, typically at the base of freshwater food webs. A few multicellular forms are known. Nine genera described.

Gamophyta ◆
Conjugating green algae
No. species described Several thousand
Proportion of group known Low
Size Multicellular forms are macroscopic
Nutrition Photosynthetic
Mode of life Freshwater algae

Multicellular filament-forming or unicellular green algae found in freshwaters. Many contribute to algal blooms and pond scum. Filamentous forms include *Spirogyra*. Desmids consist of paired cells joined at a narrow bridge through which their cytoplasm is continuous.

Granuloreticulosa ▼ ◆
Foraminifera and reticulomyxids
No. species described ca 4 000
Proportion of group known Low
Size Mostly microscopic, but some several centimeters in diameter
Nutrition Heterotrophic, some with photosynthetic symbionts
Mode of life Mostly benthic, but some are free-swimming planktonic organisms; nearly all marine

Foraminifera have multipored shells (tests) composed of organic matter reinforced with minerals (sand or calcium carbonate). Important in marine food webs. Many marine sediments are composed largely of foraminifera, and fossil species, about 40 000 of which are known, are important in stratigraphy. Some of latter, e.g. *Nummulites*, can be up to 10 cm in diameter. Reticulomyxids lack shells and form soft reticulate masses.

**EUKARYA:
PROTOCTISTA**

Haplospora ▼

No. species described   33
Proportion of group known   Low
Size   Single-celled
Nutrition   Heterotrophic
Mode of life   Unicellular symbionts living in the tissues of marine animals

Life history incompletely known, but characterized by production of spores into water or host tissue. Many are benign symbionts, and exist in multinucleate plasmodium form, but several are parasitic and damage host tissues. Host animals include mollusks, nematodes, trematodes and polychaetes. Often found as parasites of parasites, e.g. within trematode parasites of oysters. Formerly regarded as sporozoans.

Haptomonada ▼ ◆
Prymnesiophytes
No. species described
Proportion of group known   Low
Size   Mostly single-celled
Nutrition   Photosynthetic
Mode of life   Aquatic, with free-living and resting stages

Most are marine; some occur in freshwaters. Two distinct life stages: a motile, golden-colored alga and a resting, coccolithophorid stage, covered in distinctive calcareous plates (coccoliths). Coccolithophorids are important in calcium carbonate sediments and in stratigraphic studies. Some are endosymbionts of radiolaria (Actinopoda).

Hyphochytriomycota ◆ ●
Hyphochytrids
No. species described   23
Proportion of group known   Low
Size   Microscopic
Nutrition   Heterotrophic
Mode of life   Present in freshwaters and soil moisture, saprophytic or parasitic

Feed by extending threadlike hyphae into host tissue, typically algae or fungi, or into organic remains, e.g. insect or plant debris, where digestive enzymes are released and nutrients absorbed. Formerly regarded as fungi.

Labyrinthulata ▼
Slime nets and thraustochytrids
No. species described   ?
Proportion of group known   Low?
Size   Colonies up to a few centimeters long
Nutrition   Heterotrophic
Mode of life   Colonial marine protoctists

Slime nets consist of a complex colonial network of cells that move and grow within an extracellular slime matrix of their own making. *Labyrinthula* grows on eel grass (*Zostera*) where possibly pathogenic. Eight genera described.

Microspora ? ? ●
Microsporans
No. species described   ca 800
Proportion of group known   Low
Size   Single-celled
Nutrition   Heterotrophic
Mode of life   Intracellular parasites of animals

Anaerobic and lacking mitochondria. Frequently orm large single-cell tumors in host animals, some highly pathogenic, some harmless. *Nosema* causes pebrine, a disease of silkworm larvae.

Myxomycota ? ? ●
Plasmodial slime molds
No. species described   ca 500
Proportion of group known   Low
Size   Microscopic cells but macroscopic, up to several centimeters, in plasmodial form
Nutrition   Heterotrophic
Mode of life   Free-living organisms in damp terrestrial habitats

Similar in some respects to cellular slime molds (Rhizopoda). Myxomycotes have a sexual stage, and the plasmodium that develops from the zygote is multinucleate. Fruiting stage develops in drier conditions. Feed by enveloping bacteria and protoctists growing on decaying vegetation. Key organisms in studies of cell motility.

**EUKARYA:
PROTOCTISTA**

Myxospora   ▼ ◆
Myxosporidians
No. species described   ca 1 100
Proportion of group known   Low
Size   Infected tissue may have growths of
several centimeters in diameter
Nutrition   Heterotrophic
Mode of life   Multicellular symbionts, mostly
parasites of fishes but also of marine
and freshwater invertebrates

Myxosporidians penetrate the host integument
and travel to the intestine where amoeboid forms
carried to target organs are released. Many form
large plasmodial masses attached to internal
organs. Hosts include sipunculans and freshwater
oligochaete worms. Most appear benign, including
the fish symbionts, but some are important
pathogens, e.g. *Myxostoma cerebralis* causes
twist disease of salmon. Formerly regarded
as sporozoans.

Oomycota   ◆ ●
Oomycetes
No. species described   Several hundred
Proportion of group known   Low
Size   Microscopic
Nutrition   Heterotrophic symbionts
Mode of life   Mostly in freshwaters or soil; some
parasitic on land plants

Feed by extending threadlike hyphae into host tissue
where release digestive enzymes and absorb
nutrients. Familiarly known as water molds, white
rusts and downy mildews. Many oomyctes are very
important crop pests, e.g. *Phytophthora infestans*
causes potato blight; *Saprolegnia parasitica* attacks
freshwater and aquarium fishes. Formerly regarded
as fungi.

Paramyxa   ▼

No. species described   6
Proportion of group known   Low
Size   Microscopic
Nutrition   Heterotrophic
Mode of life   Obligate symbionts living within the
cells of marine invertebrates

Characterized by production of multicelled spores
within host tissue. Live within annelids, crustaceans,
mollusks and probably other groups of marine
invertebrates. Formerly regarded as sporozoans.

Phaeophyta   ▼ ◆
Brown algae
No. species described   ca 900
Proportion of group known   Moderate/high
Size   Macroscopic plant-like organisms,
mostly a few centimeters; sometimes
much larger
Nutrition   Photosynthetic
Mode of life   Most live anchored to the substrate on
rocky coasts

Most widespread in temperate regions, where they
usually dominate the intertidal zone. Generally fixed
but some, e.g. *Sargassum*, form large floating mats
far out to sea. The largest protoctists: Pacific giant
kelp (*Macrocystis pyrifera*) sometimes reach 65 m in
length. Brown algae are major primary producers in
inshore environments and also provide habitat for a
large number of macroscopic marine organisms.

Plasmodiomorpha   ? ? ●

No. species described   29
Proportion of group known   Low
Size   Microscopic
Nutrition   Heterotrophic
Mode of life   Obligate intracellular symbionts, mainly
of terrestrial plants; some parasitic

Zoospores occur in soil and infect the host; a
plasmodium with many cell nuclei but no dividing
walls develops within the host cell. Most species do
not appear to harm their hosts but *Plasmodiophora
brassicae* causes clubroot disease of brassicas and
Spongospora subterranea powdery scab of potatoes.

**EUKARYA:
PROTOCTISTA**

Rhizopoda   ▼ ◆ ●
Amastigote amoebas and cellular slime molds
No. species described   ca 200
Proportion of group known   Low
Size   Single-celled or multicellular
Nutrition   Heterotrophic
Mode of life   Mainly benthic in aquatic habitats, or in water film on land; some amoebas are parasitic

Often abundant in soil, where cyst-forming types highly resistant to desiccation. *Entamoeba histolytica* is responsible for some forms of amoebic dysentery in humans. Some amoebas construct a coating (test) from detritus and these have a fossil record from Paleozoic times; some fossil acritarchs (see Chapter 3) may represent testate amoebas. Cellular slime molds typically exist amid decaying vegetation, on logs or bark, and feed by enveloping bacteria and protoctists. The reproductive form of slime molds is an aggregation of cells each formerly having independent existence. Key experimental organisms in studies of cell communication and differentiation.

Rhodophyta   ▼ ◆ ●
Red algae
No. species described   ca 4 000
Proportion of group known   Moderate/high
Size   Macroscopic plant-like organisms, up to 1 meter in size
Nutrition   Nearly all photosynthetic; a few are symbionts on other red algae
Mode of life   Virtually all are marine; a few species are freshwater or terrestrial

Red algae occur attached to substrate on beaches and rocky shores worldwide. Most abundant in tropics. Many forms become encrusted with calcium carbonate; calcified red algae have a fossil record from the early Paleozoic. Agar jelly is extracted from red algae, and other extracts are used in food manufacture. Along with the Phaeophyta (brown algae) the largest and most complex protoctists.

Xanthophyta   ▼ ◆ ●
Yellow-green algae
No. species described   ca 600
Proportion of group known   Low
Size   Single-celled or colonial
Nutrition   Photosynthetic
Mode of life   Free-living, mostly freshwater algae

Free-swimming unicells, or highly structured multicellular or multinucleated organisms, with gold-yellow xanthin pigments. Often form scum in pond water and margins. Typically form pectin-rich cellulosic cell walls; cysts often rich in iron or silica.

Xenophyophora   ▼
Xenophyophores
No. species described   42
Proportion of group known   Low
Size   Sometimes several centimeters in diameter
Nutrition   Heterotrophic
Mode of life   Benthic marine forms

Little-known bottom-living marine protoctists from deep sea and abyssal regions. Make shells (tests) from detritus (e.g. foraminiferan shells, sponge spicules). Xenophyophores are the most abundant macroscopic organisms in some deep-sea communities, with several individuals per square meter. Some acritarch fossils (see Chapter 3) may have been xenophyophores.

Zoomastigota   ▼ ◆ ●
Zoomastigotes
No. species described   ?
Proportion of group known   Low?
Size   Single-celled, some colonial
Nutrition   Heterotrophic
Mode of life   Some are free-living in marine and freshwater environments; others are symbionts in the intestines of vertebrates

Many feed by ingesting bacteria. Parasitic forms occur in the intestine of aquatic vertebrates, e.g. the opalinids, found in frogs and toads. One group of colonial forms, the choanomastigotes, may be ancestral to the sponges.

APPENDIX 2:
IMPORTANT FOOD CROPS

FOOD CROPS OF MAJOR SIGNIFICANCE

These species and groups of species are those that, according to national food supply data maintained by the Food and Agriculture Organization of the United Nations (FAO)[2], provide 5 percent or more of the per capita supply of calories, protein or fat in at least ten countries.

Source: List of crops from Prescott-Allen and Prescott-Allen[1]; other information from Mabberley[3], Smartt and Simmonds[4], Smith et al.[5], Vaughan and Geissler[6]; conservation status from Walter and Gillett[7].

CEREALS

Millets
Echinochloa frumentacea, Eleusine coracana, Panicum miliaceum, Pennisetum glaucum, Setaria italica (Gramineae)

Main uses: *Ec. frumentacea:* Quickest growing of all millets, available in six weeks. Used for human consumption in India and East Asia and for animal fodder in the United States. *El. coracana:* Important staple in East and Central Africa and India. Wild cereal is harvested during times of famine. In Africa it is the preferred cereal for brewing beer. Seed heads may be stored for ten years. *Pa. miliaceum:* Cereal cultivated for human consumption, mainly in northern China, Russia, Mongolia and Korea, and as animal feed elsewhere. In Europe it is grown mainly as bird seed. Millets are generally tolerant of poor soils, low rainfall and high temperatures, and are quick maturing. *Pe. glaucum:* Most widely grown of the millets. The main cereal of the Sahel and northwest India. Heat and drought resistant. May contribute to incidence of goitre. *S. italica:* A once important cereal that has declined in popularity, but still grown on a relatively large scale in India and China, mainly for home consumption. Used also for animal fodder and bird seed. Early maturing and stores well.

Origins: *Echinochloa:* Different strains are thought to have partially different origins. Approximately 35 spp. exist in the genus, distributed in warm areas. *Eleusine:* Eastern and southern Africa highlands. Nine spp. in the genus in Africa and South America. *Panicum:* Unknown in wild state. The closest relative *P. miliaceum* var. *ruderale* is native to central China. At least 500 spp. in the genus in tropical to warm temperate areas. *Pennisetum:* Cultigen originated in West Africa from *P. violaceum.* Total of 130 spp. in the genus, found in tropical and warm areas. *Setaria:* Native to temperate Eurasia. Approximately 150 spp. in the genus in tropical and warm areas.

Related species: *Echinochloa: E. pyramidalis* (tropical and southern Africa and Madagascar), used as fodder and locally as flour; *E. turnerana* channel millet (Australia) is a promising forage and grain crop. Several other spp. are weeds. *Panicum: P. hemiotum* (pifine grass, North America) and *P. texanum* (Colorado grass, North America) used as fodder; *P. maximum* (Guinea grass, Africa, naturalized America) is cultivated as a forage crop; *P. sonorum* (Mexico) a minor grain; *P. sumatrense* (little millet, Malaysia) a minor grain. *Pennisetum:* Used as fodder, lawn grasses, some grains. *P. hohenackeri* (moya grass, East Africa to India) is suggested for papermaking; *P. clandestinum* (Kikuyu grass, tropical Africa) pasture grass, erosion control, lawns; *P. purpureum* (elephant or Napier grass, Africa) fodder and paper. *P. violaceum* (Africa) harvested during famines. *Setaria: S. glauca* (yellow foxtail) cattle fodder; *S. pallidifusca* is a cereal in Burkina Faso; *S. palmifolia* (India) shoots are eaten in Java; *S. pumila* is cultivated as cereal; *S. sphacelata* (South Africa) is an important silage crop.

Genetic base: *Echinochloa:* One sp. listed as threatened in 1997. *Eleusine:* Five races of cultivated finger millet recognized from Africa and India. Excellent prospects for improvement. Significant annual yield increases in India, mainly due to the incorporation of African germplasm. *Panicum:* 16 spp. listed as threatened in 1997. *Pennisetum: P. violaceum,* the wild progenitor, is an aggressive colonizer and may be found in large populations around villages in West Africa. The cultivated crop is relatively undeveloped. Open-pollinated cultivars are popular in Africa and India. Five spp. listed as threatened in 1997. *Setaria:* Largely a crop of traditional agriculture systems. Two spp. listed as threatened in 1997.

Breeding: *El. coracana:* Wild spp. in Africa cross with domesticated finger millet to produce fertile hybrids which can be obnoxious weeds. *Pe. glaucum:* Genetic exchange with related wild forms in same geographical area is possible. *S. italica:* Hybridizes easily with wild relative *S. italica* var. *viridis* to produce fertile offspring.

Germplasm collections: 90 500 general millet accessions, 45-60 percent of landraces and 2-10 percent of wild spp. represented.

Barley
Hordeum vulgare (Gramineae)

Main uses: Early maturing grain with high yield potential; can be grown where other crops fail, e.g. above Arctic circle, at high altitude and in desert and saline areas. Most important as animal feed, also for brewing beer and human food. Main producers in Europe, North Africa, Near East, Russia, China, India, Canada, United States.

Origins: Southwestern Asia. Approximately 20 spp. in the genus, distributed in the north temperate region.

Related species: *H. distichon* (2-rowed barley) is possibly *H. vulgare* x *H. spontaneum.*

Genetic base: Landraces have been almost completely replaced by pure line cultivars and the change in genetic structure of barley populations has been profound. Important contribution of Ethiopian barleys highlights need to broaden genetic base. Two spp. were listed as threatened in 1997.

Breeding: Fertile hybrids between wild and cultivated forms occur naturally where ranges overlap. Crosses possible with other spp. in the genus but not utilized in barley cultivars. Ethiopian barleys have been important as genetic source of disease resistance and improved nutritional value.

Rice
Oryza glaberrima, O. sativa (Gramineae)

Main uses: Highest world production of all grains. Primary source of calories and protein in humid and subhumid tropics. The grain is relatively low in protein; brown rice a source of some B vitamins. Rice bran is used in animal feeds and industrial processes. Can grow in flood-prone areas. Main producers are China, India, Indonesia, Bangladesh, Viet Nam. Only 4 percent of world production is exported. Main exporters Thailand, Viet Nam, Pakistan, United States.

Origins: Two cultigens appear to have been domesticated independently. The origin of *O. sativa* is uncertain, possibly derived in several centers from *O. rufipogon* (selected weed in Colocasia fields). Archeological evidence suggests origin in China or Southeast Asia. Center of diversity of *O. glaberrima* is the swampy area of the Upper Niger. Approximately 18 spp. in the genus, distributed in the tropics.

Genetic base: Following agricultural intensification many populations of wild relatives have disappeared or intergraded with domesticated rice. Reduced genetic base has also led to repeated pest epidemics. Great genetic diversity exists in *O. sativa* cultivars; much less in *O. glaberrima.* Rapid spread of improved rice varieties has displaced tens of thousands of landraces, many now extinct. *O. glaberrima* is rapidly being replaced by *O. sativa.* Genetic erosion is reported in China, Philippines, Malaysia, Thailand and Kenya. Three spp. were listed as threatened in 1997.

Breeding: *O. sativa* has formed numerous hybrids with wild spp. *O. nivara* and *O. rufipogon.* Genes improving tolerance to diseases or adverse conditions have been derived from African rices and wild relatives.

CEREALS

Rye
Secale cereale (Gramineae)

Main uses: Cereal crop used as animal feed and for human consumption. Eaten mainly as rye bread and crispbread. Higher in minerals and fiber than wheat bread. Previously more popular as a bread flour; now largely replaced by wheat. Still important in cooler parts of northern and central Europe and Russia, cultivated up to the Arctic circle and up to 4 000 m altitude. Tolerates poor soils. Also used in brewing industry and young plants produce animal fodder. Main producers are Russia and Europe.

Origins: Probably originated from weedy *Secale* types in eastern Turkey and Armenia. *S. montanum* is possible ancestor. Total of three spp. in the genus in Eurasia.

Genetic base: A number of weed ryes, found associated with agriculture throughout the Near East, are now considered to be subspecies of *S. cereale*. A complex of subspecies of *S. montanum* extends from Morocco east to Iraq. Five spp. were listed as threatened in 1997.

Sorghum
Sorghum bicolor (Gramineae)

Main uses: Staple cereal in semi-arid tropics. Mostly grown in developing countries, especially for domestic consumption by small farmers in Africa and India. Used in brewing beer and as animal fodder. Grain stores well. Main producers are United States, India, China, Nigeria, Sudan.

Origins: Developed primarily from the wild *S. arundinaceum* in Africa. Total of 24 spp. in the genus in warm areas of the Old World and Mexico.

Related species: Backcrosses with *S. arundinaceum* gave *S. drummondii* cultivated for forage; *S. halepense* (Mediterranean) is a widely naturalized fodder plant, often weedy.

Genetic base: Wide variation in landraces. Genetic erosion reported in Sudan. Modern varieties have not been widely popular for use as human food. Two spp. listed as threatened in 1997.

Breeding: Wild relatives may be an important source for disease resistance.

Germplasm collections: 168 500 accessions. 21 percent in the International Crops Research Institute for the Semi-Arid Tropics. 80 percent of landraces, 10 percent wild spp. represented.

CEREALS

Wheat
Triticum aestivum, T. turgidum (Gramineae)

Main uses: Most widely cultivated crop. Grain is gluten rich and highly valued for bread making. 90 percent of wheat grown is *T. aestivum*. Wheatgerm oil is highly unsaturated and high in vitamin E. Durum wheat, *T. turgidum*, has higher protein content. Used for making pasta. High nutritive value; easy processing, transport and storage. Main producers are China, India, United States, France and Russia.

Origins: Mediterranean and Near East. Origin is complex and not fully understood, probably involving *Aegilops* spp. Total of four spp. in the genus, distributed from the Mediterranean to Iran.

Genetic base: Large variation in the crop, around 25 000 different cultivars. However, large areas are planted with genetically uniform crops and the inflow of landrace material into breeding programs is low (8 percent). Genetic erosion is reported in China, Uruguay, Chile and Turkey.

Germplasm collections: Approximately 850 000 accessions. Largest collection (13 percent) in Centre for Maize and Wheat Improvement. 95 percent of landraces and 60 percent of wild spp. collected.

Maize
Zea mays (Gramineae)

Main uses: Mostly grown for human consumption in parts of Africa and Latin America; elsewhere mainly for animal fodder. Starch may be extracted and used in food processing. The germ oil is important. Also used in the brewing industry. Main producers are the United States, China, Brazil, Mexico, France.

Origins: Probably derived from teosinte, *Zea mays* ssp. *mexicana*. Total of four spp. in the genus, confined to Central America.

Genetic base: Most of the world's maize crop is derived from a few inbred lines. Landraces represent 40 percent of the crop grown in developing countries. Genetic erosion is reported in Mexico, Costa Rica, Chile, Malaysia, Philippines, Thailand. *Z. perennis* was presumed extinct in the wild until its rediscovery in 1977. *Z. diploperennis* was recently discovered and is now protected in the Sierra de Manantlan Biosphere Reserve, Mexico. Three spp. were listed as threatened in 1997.

Breeding: Teosinte crosses readily with maize to produce fertile offspring, *Tripsacum* crosses with less success. Neither has been widely used in breeding programs.

Germplasm collections: 277 000 accessions, the largest existing at Indian Agricultural Research Institute. 95 percent of landraces and 15 percent wild spp. are represented.

TUBERS

Yams

Dioscorea spp. *D. alata, D. bulbifera, D. cayenensis, D. dumetorum, D. esculenta, D. rotundata, D. trifida* (Dioscoreaceae)

Main uses: Edible stem tuber, 28 percent starch and limited vitamin C. Important staple in the humid and subhumid tropics. Also major ingredient in oral contraceptives. Religious and cultural role. Good storage properties. West Africa produces 90 percent of world production; Nigeria alone produces 70 percent.

Origins: Three main independent centers of diversity or domestication in Asia, Africa and America. Approximately 850 spp. exist in the genus, distributed in tropical and warm regions.

Genetic base: Predominantly a subsistence crop. Apparently little genetic erosion. Large genetic variability in wild edible forest yams. 68 spp. were listed as threatened in 1997.

Breeding: New World and Old World spp. show strong genetic isolating barriers and crosses between them are not successful.

Cassava

Manihot esculenta (Dioscoreaceae)

Main uses: Edible tuber containing 35 percent starch and vitamin C. Cultivated in almost all tropical and subtropical countries, mainly by smallholders. One of the most efficient crops for biomass production. A good famine reserve, able to withstand harsh conditions. Also animal feed. Main producers are Brazil, Nigeria, Dem. Rep. of Congo, Thailand, Indonesia.

Origins: Unknown in the wild state. Total of 98 spp. exist in the genus, occurring between southwest United States and Argentina. Most diversity occurs in northeast Brazil and Paraguay and in west and south Mexico.

Related species: *M. glaziovii* is the source of Ceara or Manicoba rubber and oilseeds.

Genetic base: Estimated 7 000 landraces. Local preferences in flavor, root texture and growth habit vary greatly; many farmers retain traditional cultivars despite improvements in new cultivars. Genetic erosion reportedly a risk in South and Central America, Thailand and China. 65 spp. were listed as threatened in 1997.

Breeding: Variability of cultivated forms has probably increased through crosses with wild forms. *M. glaziovii* and *M. melanobasis* have contributed to improvement of cultivated form. High diversity in germplasm provides good improvement potential. Interspecific crossing with wild relatives may be employed further to broaden tolerance of different conditions.

Germplasm collections: 28 000 accessions, mostly in international centers of research. 35 percent of landraces and 5 percent wild spp. collected were listed as threatened in 1997.

Potato
Solanum tuberosum (Dioscoreaceae)

Main uses: One of the most important world crops. Cultivated in 150 countries, mainly for local consumption. Little international trade. Tubers are cooked or processed into a range of products. Starch, alcohol, glucose and dextrin are also major products. Tubers also make animal feed. Potatoes are 80 percent water, 18 percent carbohydrates, with range of minerals, and a good source of vitamin C. Main producers are Russia, China, Poland, Germany, India.

Origins: Maximum diversity in cultivated and wild spp. on the high plateau of Bolivia and Peru. A number of ancestral spp. involved. The gene pool consists of *S. tuberosum* ssp. *andigenum* and *tuberosum*, *S. stenotomum*, *S. ajanhuiri*, *S. goniocalyx*, *S. x chauca*, *S. x juzepczukii*, *S. x curtilobum*, *S. phureja*. Total of 1 700 spp. in the genus, distributed worldwide.

Related species: *S. melongena* (India) (eggplant). *S. centrale* (arid Australia) and *S. muricatum* (pepino) (Andes) have edible fruit; *S. quitoense* (naranjillo) (Andes) is used for fruit juice; *S. melanocerasum* (?cultigen) (cultivated in tropical West Africa) fruit; *S. hyporhodium* (upper Amazon); *S. americanum* (yerba mora).

Genetic base: Between 3 000 and 5 000 varieties of potato are recognized by farmers in the Andes. Genetic erosion is reported in centers of origin, including Chile and Bolivia. In Peru, of the 90 wild potato spp. 35 are now extinct in the wild. Wild spp. and ancient cultivars largely replaced by modern varieties. Attempts to broaden the narrow genetic base have been slow. 125 spp. were listed as threatened in 1997.

Breeding: Much introgression from wild relatives has been attempted, improving disease resistance and other traits.

Germplasm collections: 31 000 accessions worldwide. 20 percent are held by the Centro Internacional de la Papa, Lima, Peru. 95 percent of landraces and 40 percent of wild spp. are collected.

TUBERS

Cabbage
Brassica oleracea (Cruciferae)

Main uses: Mainly temperate vegetable crop, but grown worldwide. Large number of edible and ornamental varieties, including cauliflower, calabrese and kohlrabi. Important component of human nutrition throughout the world: a good source of fiber, vitamins E, B and C, and also Vitamin A in the greener parts. Main production in Europe including Russia.

Origins: The wild cabbage is native to Europe; development of cultivars took place in the Mediterranean region. Total of 35 spp. exist in the genus, distributed in Eurasia.

Related species: Wide range of crops (variously leaves, buds, florets, stems and roots) eaten; also used for oil production. *B. campestris* and *B. napus* (rapeseed); *B.carinata* (Texsel greens) (northeast Africa); *B. hirta* (white and yellow mustard) (Mediterranean); *B. juncea* (Indian mustard) (Eurasia); *B. juncea* var. *crispifolia* (Chinese mustard).

Genetic base: Outbreeding nature. Large amounts of genetic variation in most crops, where not highly selected. Continuing emphasis on uniformity in recent decade and controls on release of new cultivars have led to significant reduction in genetic variation in commercial cultivars. Wild relatives in Mediterranean are threatened; 14 spp. listed as threatened in 1997.

Germplasm collections: Efforts made to ensure different crops are represented, including obsolete and locally popular varieties. Cultivars from southern Europe are less well collected.

LEAF VEGETABLES

BEANS

Beans
Lablab purpureus, Phaseolus lunatus, Phaseolus vulgaris, Vigna unguiculata (Leguminosae)

Main uses: *L. purpureus*: Young pods and young and mature seeds of lablab are eaten; pulse contains 25 percent protein, little fat and 60 percent carbohydrate. Main producers: India, Southeast Asia, Egypt, Sudan. *P. lunatus*: Dried or immature seeds of lima bean are used as pulses; seeds contain 20 percent protein, 1.3 percent fat, 60 percent carbohydrate; flour also obtained from seed. Main producer is the United States. *P. vulgaris*: Most widely cultivated of all beans; in temperate areas grown mainly for the pod, which contains 2 percent protein and 3 percent carbohydrate with vitamins A, B, C and E; seeds have 22 percent protein, 50 percent carbohydrate, 1.6 percent fat and vitamins B and E. Main producer: Brazil. *V. unguiculata*: Cowpea is a nutritionally important minor crop in subsistence agriculture in Africa; dry and green seeds, green pods and leaves are eaten; highly resistant to drought.

Origins: *Lablab*: African or Asian origin; only one species in the genus (previously *Dolichos lablab*). *Phaseolus*: It is thought that separate domestications occurred in Central and South America from conspecific races; total of 36 spp. in the genus, found in tropical and warm America. *Vigna*: Center of diversity of wild relatives in southern Africa; greatest diversity of cultivated form exists in West Africa; subspecies *dekindtiana* is probable progenitor; total of 150 spp. in the genus, mainly in the Old World tropics.

Related species: *Phaseolus*: Five cultigens exist in the genus: apart from *P. lunatus* and *P. vulgaris*, there are *P. acutifolius* (tepary bean, America); *P. coccineus* (scarlet runner, Central America) and *P. polyanthus* (year bean, Central America). Various other spp. are important pulse crops, previously listed as *Vigna* spp. *Vigna*: Other spp. are used for forage and green manure, etc. Other pulses include: *V. aconitifolia* (moth bean, South Asia; *V. angularis* (Aduki bean, Asia); *V. mungo* (urad, tropical Asia); *V. radiata* (mung bean, ?Indonesia); *V. subterranea* (Bambara groundnut, West Africa); *V. umbellata* (rice bean, southern Asia); *V. unguiculata* (cowpea, Old World); *V. vexillata* (tropical Old World) which has edible roots.

Genetic base: *Lablab*: Mainly grown in small plots and home gardens; larger areas under cultivation in Australia. No threat of genetic erosion. *Phaseolus*: Much dry bean cultivation in the United States depends on very small germplasm base; improved varieties also widely adopted by smallholder farmers. Relatively wide genetic base provided by landrace groups, if conserved; most wild relatives widespread but populations of several taxa being lost to overgrazing in southwest United States and northern Mexico. Two spp. listed as threatened in 1997. *Vigna*: Breeding relies on narrow genetic base and hybridization with other *Vigna* spp. is important; more variability in wild relatives in the primary gene pool than in cultivated cowpea. Four spp. were listed as threatened in 1997.

Breeding: *Phaseolus*: Several wild relatives are fully or partly compatible; populations of wild lima bean with larger seeds recently discovered in northwest Peru and Ecuador. *Vigna*: cowpea crosses successfully with wild subspecies of *V. unguiculata*.

Germplasm collections: *Lablab*: 11 500 accessions in Africa and Caribbean. *Phaseolus*: 268 500 accessions of *Phaseolus* spp. in total. 15 percent are held by Centro Internacional de Agricultura Tropical, Cali, Colombia. On average 50 percent diversity in the genus is represented.

OIL CROPS

Groundnut
Arachis hypogaea (Leguminosae)

Main uses: The edible nut contains 50-55 percent oil, 30 percent protein, and is good source of essential minerals and E and B vitamins. Cultivated for the nut or for oil in many tropical and subtropical countries. Seed residue useful as animal feed. Nutshells are used as fuel and in industry. Stems and leaves used as forage. Main producers are India, China, United States, Argentina, Brazil, Nigeria, Indonesia, Myanmar, Mexico, Australia.

Origins: Mato Grosso in Brazil is the primary center of origin and diversity for the genus. The cultivated groundnut is thought to have originated in southern Bolivia and northwest Argentina. Total of 22 spp. in the genus, all from South America.

Genetic base: Cultivated as a marginal crop with relatively little selection pressure. Many varieties exist worldwide with broad adaptability.

Breeding: *A. monticola* freely crosses with *A. hypogaea*. Wild *Arachis* material confers resistance on domestic form.

Germplasm collections: 13 000 accessions at the International Crop Research Institute for the Semi-Arid Tropics.

Coconut
Cocos nucifera (Palmae)

Main uses: The endosperm of the nut contains 65 percent saturated oil, used in manufacture of margarine, soap, cosmetics and confectionery. Also eaten fresh, desiccated or as a coconut milk. Residue is a high-protein animal feed. There are many more uses: source of naturally sterile water, fiber, wood, thatch. Mainly a smallholders' crop. Main producers are the Philippines, Indonesia, India, Sri Lanka, Malaysia, Mexico, Pacific Islands.

Origins: Possibly originated in Melanesian area of Pacific. Wild types predominate on the African and Indian coasts of the Indian Ocean, and scattered in Southeast Asia and the Pacific. Single species in the genus.

Genetic base: Tendency to plant uniform, improved hybrids is reducing genetic variation particularly in domesticated types.

Breeding: Wild and domestic coconuts are fully compatible. Hybridization with wild types has increased genetic diversity of cultivated crops.

Oil palm
Elaeis guineensis (Palmae)

Main uses: The mesocarp on the fruit yields oil for human consumption. Unrefined oil is high in vitamin A. Oil may also be extracted from the kernel. An export crop and important for local consumption. Very high yielding. Malaysia supplies 70 percent of world exports.

Origins: West Africa, originally a species of the transition zone between savannah and rainforest. Only 2 spp. exist in the genus.

Related species: *E. oleifera* (tropical America) is less important as an oil crop than *E. guineensis*.

Genetic base: Populations in Africa are semi-wild. They are being thinned to make way for other crops. Plantations in Malaysia were based on material from only four specimens. New material is being introduced to broaden the genetic base.

Breeding: Fertile offspring produced with *E. oleifera*.

OIL CROPS

Soybean
Glycine max (Leguminosae)

Main uses: The most important oil crop and grain legume in terms of production and international trade. An important basis of Asian cuisine, developed into various forms of food from soy sauce to tofu. Immature green beans and sprouts also eaten. Seeds contain 18-23 percent oil and 39-45 percent protein. Oil is used in various forms. Most of the meal is used as a high-protein animal feed. Main producers are the United States, Brazil, China, Argentina, India.

Origins: A cultigen, not known in the wild. Soybean is thought to have arisen as a domesticate in the eastern half of northern China about 3 000 years ago probably from *G. soja*. Total of 18 spp. exist in the genus, distributed from Asia to Australia.

Genetic base: The genetic base of varieties is narrow worldwide. Conservation of traditional landraces is urgently needed. Two spp. listed as threatened in 1997.

Breeding: Wild spp. are increasingly used for improvement and there is good potential for further valuable characteristics to be found in wild *Glycine* spp. *G. soja* easily crosses with soybean.

Germplasm collections: 174 500 accessions, 9 percent in Institute of Crop Germplasm Resources, Chinese Academy of Agricultural Sciences, Beijing, China. 60 percent of landraces and 30 percent wild spp. are represented.

Cotton seed
Gossypium barbadense, G. hirsutum (Malvaceae)

Main uses: Cotton is the second most valuable oil crop, as well as being the most important textile fiber. Crop development is concentrated on fiber production because value is three or four times greater. New World cottons took over from Old World forms after the European exploration of the Americas. Main producers of *G. barbadense* are Russia, Egypt, Sudan, India, United States, China.

Origins: Unique in that four spp. were domesticated independently for the same use as a fiber and oil crop, in Africa and India: *G. arboreum* and *G. herbaceum*; in Central and South America: *G. hirsutum* and *G. barbadense*. Total of 39 spp. in the genus, found in warm temperate to tropical zones.

Related species: *G. arboreum* is still important in India and Pakistan. *G. herbaceum* is grown only on a small scale in Africa and Asia.

Genetic base: Modern cultivars of *G. hirsutum* are responsible for over 90 percent of world production. New *Gossypium* spp. possibly occur in Arabia and Africa. Wild forms of *G. herbaceum*, *G. hirsutum* and *G. barbadense* are known. Past breeding involved much introduction of genetic material from different geographic regions, but a severe narrowing of the genetic base has occurred in the production of modern *G. hirsutum* varieties. Large amounts of fertilizers and pesticides required in modern cotton production. Eight spp. listed as threatened in 1997.

Breeding: At least six related spp. have contributed genes of importance to the cultivated crop. Material from wild gene pool used in genetic engineering. *G. herbaceum* and *G. arboreum* are able to interbreed, although later generations have a high probability of failing reproductively.

Sunflower seed
Helianthus annuus (Compositae)

Main uses: Seeds contain 27-40 percent polyunsaturated oil and 13-20 percent protein. Oils and margarines used for human consumption, and for industrial uses, and waste products useful in animal feed. Pollinating bees frequently used for honey production. Main producers are Russia, United States, Argentina.

Origins: Probably originated in southwest North America. Total of 50 spp. exist in the genus, distributed in North America.

Related species: Also ornamental; *H. tuberosus*

(Jerusalem artichoke) is also eaten. *H. petiolaris* used for hybridization.

Genetic base: Increased yields in hybrids led to increased interest and production in 1960s. Large gene pool exists in wild and weed sunflowers in North America, although habitat loss is resulting in population declines. 16 spp. listed as threatened in 1997.

Breeding: Resistance to several diseases was secured through hybridization with *H. tuberosus*.

Olive
Olea europaea (Oleaceae)

Main uses: Fruit with 40 percent oil content. Highly superior oil for cooking, margarines, dressing; also used in cosmetics and pharmaceutical industry. Fruit eaten pickled. Despite competition with more modern oil-producing crops, olive oil still commands premium price. Recent rise in popularity and recognition of nutritional value. Main producers are Spain, Italy, Greece, Turkey, Tunisia.

Origins: Olive is a cultigen, evolved in eastern Mediterranean. *O. europaea* ssp. *oleaster* recognized as progenitor. Total of 30 spp. in the genus, in tropical and warm temperate parts of Old World.

Related species: Related species provide good timber.

Genetic base: A long-lived tree. The turnover of clones should be slow. Hundreds of distinct cultivars, found in different geographic groups. Olive production still relies on traditional cultivars. Few new varieties have been released. Decline in area under cultivation. Marginal groves have been abandoned with serious consequences for Mediterranean wildlife. Wild populations outside area of cultivation under pressure from cutting and land clearance; two spp. were listed as threatened in 1998.

Breeding: Closely related to wild subspecies in the Mediterranean, Africa, Arabia, Iran and Afghanistan.

SUGAR CROPS

Sugarcane
Saccarhum officinarum (Gramineae)

Main uses: Major source of calories worldwide. Cultivated in about 70 countries, mainly in tropics. Requires good rainfall and rich soil for successful growth. Stems are easily transported. Main producers are Brazil, India, China, Thailand, Pakistan.

Origins: A cultigen with origin and center of diversity in New Guinea. Between 35 and 40 spp. in the genus, distributed in tropical and warm zones.

Related species: Other cultivated sugar canes include *S. barberi*, *S.edule* and *S. sinense*. *S.robustum* and *S. spontaneum* are wild sugar canes.

Genetic base: Risk of genetic erosion reported in Assam and suspected in Indonesia, Papua New Guinea and Thailand, where monocrop plantations have taken over from indigenous spp. Modern hybrids have a narrow genetic base. Plantations are prone to severe pest and disease epidemics. Attempts to incorporate more genetic diversity is slowly having effect. Only 10 percent wild germplasm used in breeding. *S. robustum* exhibits the most genetic diversity, but has had little application in breeding.

Breeding: Commercial varieties are derived from interspecific crosses with other wild and cultivated sugarcane spp.

Germplasm collections: 19 000 accessions in total, nearly a quarter of them in Centro Nacional de Pesquisa de Recursos Genéticos e Biotecnologia, Brasilia, Brazil. 70 percent of landraces, 5 percent of wild spp. represented.

FRUITS

Banana, plantain
Musa spp. (Musaceae)

Main uses: One of the most popular dessert fruits in industrial nations; a major source of calories and export earnings in developing countries. Bananas and plantains are high in carbohydrates and potassium; bananas are a good source of vitamins C and B6, and plantains contain high levels of vitamin A. Numerous other uses. Main producers are India, Brazil, Ecuador, Philippines and China for the banana; Uganda, Colombia, Rwanda, Dem. Rep. of Congo and Nigeria for the plantain.

Origins: Bananas evolved in Southeast Asia from *M. acuminata* or combinations of *M. acuminata* and *M. balbisiana*. Plantains probably originated in southern India. Primary areas of diversity exist in Southeast Asia. Secondary areas also occur in tropical Africa, Indian Ocean islands and the Pacific. Fe'i bananas (2n), thought to be derived from *M. maclayi* and possibly other related spp. Greatest diversity of fe'i bananas is on Tahiti. Total of 35 spp. in the genus, distributed throughout the tropics.

Related species: Fe'i bananas are a significant source of food in New Guinea and the Pacific. *M. textilis* recent domesticate in Philippines used for Manila hemp. Related *Ensete ventricosum* cultivated in Ethiopia for starchy pseudostem. *M. balbisiana* produces edible fruit and contributed to present-day cultivars.

Genetic base: About 500 genetically distinct cultivars. 90 percent of global banana production is from smallholdings. International trade in bananas relies on very few cultivars, based on the Cavendish type. Dangerously narrow genetic base and very susceptible to diseases. Increased disease resistance is extremely important given the economic importance of the export crop. The number of Fe'i banana cultivars has declined severely as a result of human demographic changes in the Pacific and the spread of pests. Banana is an aggressive weed. Wild populations of *Musa* benefit from forest clearance if succession is allowed to take place. Three spp. listed as threatened in 1997.

Breeding: An extensive contact zone between cultivated and weedy types exists in several areas, e.g. Sri Lanka. Much introgression is believed to have enriched the gene pool of cultivated types. *M. balbisiana* has valuable traits. Several other wild relatives have useful characteristics. Germplasm collections have been poorly used; better selections could be made to suit subsistence farmers.

Germplasm collections: Edible bananas, being seedless, are not storable. Seeds from wild spp. may be stored. Field gene banks hold collections. 10 500 accessions in total. The International Network for the Improvement of Bananas and Plantains holds 10 percent. Most of the diversity of wild and cultivated bananas is thought to be covered.

Cocoa
Theobroma cacao (Sterculiaceae)

Main uses: Seeds are fermented and roasted to produce cocoa powder and chocolate. Waste goes to produce animal feed, mulch or fertilizer. Cocoa is a nutritional beverage; the powder is 25 percent saturated fat, 16 percent protein and 12 percent carbohydrates. Main producers are West African countries, Brazil, Malaysia.

Origins: Upper Amazon basin. Center of cultivation in Central America. Total of 20 spp. in the genus, confined to tropical America.

Related species: All the following are cultivated: *T. grandiflorum* (cupuaçu, Amazonia); *T. speciosum* (cacaui, Central and South America); *T. subincanum* (South America); *T. obovatum* (Amazon); *T. angustifolium* (Central America); *T. bicolor* (Central and South America); *T. glaucum* (Amazonia).

Genetic base: Undoubted genetic erosion has occurred in recent years. Currently cacao plantations are established by seed with varying degrees of genetic heterogeneity. Production in West Africa is based on a particularly narrow gene pool. Originally three main cultivated types. Criollo yields the most superior chocolate but has been largely replaced because of low yields. Forastero dominates world production. Wild cacao is highly variable, especially in its core area. Dramatic increase in plantations of coca and pulp-producing spp. in various parts of the Amazon, agricultural expansion and movement of human populations have caused severe losses to the wild gene pool.

Breeding: Little use of or research into wild genetic reserves because they are relatively hard to cross.

Germplasm collections: Seeds do not remain viable for long. 4 000 to 5 000 accessions kept in field gene banks. International Cocoa Genebank in Trinidad has the most comprehensive collection. Close relatives are poorly represented. Vegetative germplasm is collected.

FOOD CROPS OF SECONDARY OR LOCAL IMPORTANCE

These species and groups of species are those that, according to national food supply data maintained by the FAO[1,2], provide a significant amount of the per capita supply of calories, protein or fat, but on the criteria followed here are not of equal importance to the crops in the previous table

(i.e. they provide below 5 percent of the total per capita supply and/or do so in fewer than ten countries).

Source: List of crops from Prescott-Allen and Prescott-Allen[1]; other information from Mabberley[3], Smartt and Simmonds[4], Smith *et al.*[5], Vaughan and Geissler[6]; conservation status from Walter and Gillett[7].

CEREALS AND PSEUDO-CEREALS

Oats
Avena sativa (Graminae)

Origin: West and north Europe from weed oat components of wheat and barley crops

One of the major temperate cereals, although currently declining in production and generally regarded as a secondary crop. Mostly used for animal feed. Oat kernel is higher in high-quality protein and fat than any other cereal. Oat bran is a good dietary fiber.

A. byzantina also cultivated. Genetic erosion from intensive breeding has resulted in efforts to conserve landraces and early varieties. Crosses between *A. sativa* and *A. byzantina* have led to numerous cultivars. Fertile hybrids obtained from crosses between cultivated oats and weed species. Some success in incorporating desirable genes from more distant relatives.

Quinoa
Chenopodium quinoa (Chenopodiaceae)

Origin: High Andes

An important and sacred pseudo-cereal in Inca times. Remains a staple in large parts of South America. Nutrient composition is superior to other cereals, being high in lysine and other essential amino acids, calcium, phosphorus, iron and vitamin E. Can grow in marginal conditions. Greatest diversity of genotypes in the

highlands of southern Peru and Bolivia. Cultivation declined with Spanish conquest until 1970s when grown as a sole crop only in parts of Peru and Bolivian-Peruvian Altiplano. Agricultural and nutritional benefits have now been recognized and acreage has increased significantly. Improvement so far has been based on inbred populations and pure lines. Considerable potential for improvement, both in the crop and its use.

Fonio
Digitaria exilis (Gramineae)

Origin: West Africa; thought to be a cultigen

Popular cereal in parts of West Africa. Adapted to marginal agricultural land. Several species are harvested as cereals during times of famine.

ROOTS AND TUBERS

Taro
Colocasia esculenta (Araceae)

Origin: India

Edible tuber; 25 percent starch, low protein, good vitamin C source. Probably cultivated before rice. Widely cultivated in China and staple in many Pacific islands. Also used by food and beverage industries, and in pasta products. Young leaves eaten as spinach. Tolerates high temperatures and poor soils. More than 1 000 cultivars have arisen through subsistence farming. Lack of interest and germplasm exchange at a more commercial level. Serious danger of genetic erosion.

Carrot
Daucus carota (Umbelliferae)

Origin: Afghanistan

Root crop, grown worldwide, and eaten raw, cooked or processed. The best plant source of provitamin A; low in other nutrients. Numerous wild and cultivated subspecies. Open pollinated crops almost entirely replaced by hybrids in United States, Japan and Europe. Environmental health concerns over level of pesticide has led to interest in genetic source of pest resistance. *D. capillifolius* has passed some pest resistance to cultivated crop.

Sweet potato
Ipomoea batatas (Convolvulaceae)

Origin: Not known in the wild. Greatest species diversity occurs between Yucatan and the mouth of the Orinoco. Major variation is found in Guatemala, Colombia, Ecuador and Peru.

The tuberous root is an important staple in the tropics. Able to grow in high temperatures with low water and fertilizer input. Good source of fiber, energy and vitamins A and C. Also industrial source of starch and ethanol. Although acreage has declined, increases are likely as a crop able to respond to population growth in marginal areas. China accounts for 80 percent of production. Until recently, material used in breeding programs represented a fraction of existing diversity. Genetic base has now broadened but requires further increase. Little work has been done on cultivar improvement in areas of highest production (i.e. where sweet potato is a staple). Countries where breeding programs exist have replaced native cultivars with improved varieties. Sweet potato is thought to have more potential for yield improvement than any other major crop in Asia.

Tannia
Xanthosoma sagittifolium (Araceae)

Origin: New World

Similar use and nutritional composition as taro, but starch is more difficult to digest. Used in preparation of fufu in West Africa.

BEANS AND OTHER LEGUMES

Pigeonpea
Cajanus cajan (Leguminosae)

Origin: Cultigen, India

One of the major pulse crops of the tropics. Mature seeds contain 20 percent protein, 60 percent carbohydrate and little fat. Important in small-scale farming in mainly semi-arid regions. A multipurpose species with good potential in agroforestry systems and on marginal lands. India contributes more than 90 percent of world production. Domestication has not altered the species as much as other crops. *C. cajanifolius* is closest relative;12 spp. may be crossed with pigeon pea.

Chickpea
Cicer arietinum (Leguminosae)

Origin: Southeast Turkey; *C. reticulatum* is probably the progenitor

One of the most important pulse crops. The seeds contain less protein (17 percent or more) but more fat (5 percent) than other pulses. Grown over large area from Southeast Asia to Mediterranean. Only 2-4 percent of world production is exported. Recently discovered *C. reticulatum* is confined to ten populations in Turkey. Two main cultivars have emerged. Traditional landraces have been selected to suit local ecological conditions. Commercial breeding is a recent phenomenon. *C. reticulatum* and *C. echinospermum* are compatible with the chickpea.

Lentil
Lens culinaris (Leguminosae)

Origin: Near East; wild progenitor *L. orientalis*

Seeds contain 25 percent protein, 56 percent carbohydrate and 1 percent fat. Young pods also eaten. Seeds are commercial source of starch for textiles and printing. Residues used as animal feed. Unique assemblages of landraces in different geographic regions. The crop has been altered little by modern breeding. Much variation in the crop unexploited.

Pea
Pisum sativum (Leguminosae)

Origin: Wild progenitor is unknown. Possible centers of origin are Ethiopia, the Mediterranean and Central Asia.

The second most important pulse. 90 percent production as dried peas. Seed coats are source of protein, used in bread or health foods. Russia and China produce 80 percent of the world production of dried peas; United States and United Kingdom are largest producers of green peas. Breeding relies on a fairly narrow genetic resource base and efforts to conserve genetic variability of the cultivated crop have been fairly limited.

Lupin
Lupinus mutabilis (Leguminosae)

Origin: Andes; other lupins originated in two main centers of genetic diversity in Mediterranean and in Americas

A relatively minor pulse crop, obtained from several *Lupinus* spp. Seed contains 44 percent protein, 17 percent oil. Seed is human food in subsistence agriculture. Also used as coffee substitute and high-protein animal feed. Seed flour used as soya. Species also important in soil improvement. Related spp. have ornamental value, used as fodder, coffee substitute, green manure or to stabilize sand dunes. May act as substitute for soybean, where climate is unsuitable for soybean growth. Other *Lupinus* spp. potentially suitable for cultivation.

Broad bean
Vicia faba (Leguminosae)

Origin: Cultigen; wild ancestor unknown, possibly from central Asia

A temperate pulse crop. Both immature seeds and dry mature seeds are eaten. The latter are also used as animal feed. Dried seed contains 25 percent protein, 1.5 percent fat, 49 percent carbohydrate; the immature bean has much less of these nutrients but more vitamin A and vitamin C. Also used as green manure.

Mustard seed
Brassica juncea (Cruciferae)

Origin: Central Asia: Himalaya; probably *B. nigra* x *B. campestris* and other *Brassica* spp.

The most important spice in the world in terms of quantity. Four species contributing to mustard exist. *B. juncea* took over from *B. nigra* in 1950s as it allowed completes mechanization of harvesting. Also valuable as oil crop and salad crop, vegetable and fodder. Long-lived seed allows easy maintenance of large collections. Wild material is widely distributed.

Rapeseed
Brassica napus (Cruciferae)

Origin: Probably a hybrid of *B. oleracea* x *B. campestris*

The seed is an important, relatively recent source of oil, containing 40 percent unsaturated fat, with industrial and culinary applications. The root crop provides animal and human food (swede). Uncertain whether *B. napus* exists in wild form. Domestication was a relatively recent event. The crop is tolerant of inbreeding, and landraces have been replaced by improved cultivars since the 19th century. Swedes, of which there are only a few varieties, are the result of hybridization with *B. campestris*. Various valuable contributions to oilseed rape also from *B. campestris* and *B. oleracea*.

BEANS AND OTHER LEGUMES

OIL CROPS

OIL CROPS

Safflower seed
Carthamus tinctorius (Compositae)

Origin: Turkestan, Turkey, Iran, Iraq, to Israel and Jordan

Oilseed crop, produces two types of oil for margarine and also cooking oils. Ingredient of animal feeds. Dried flowers serve as substitute for saffron. Applications in cosmetic industry and as medicine. Originally domesticated for use as dye plant. Much diversity developed as the species was cultivated over a wide area. Large-scale cultivation in few countries. No reported genetic erosion. Related species cross easily with cultivated crop and form natural hybrids.

Sesame seed
Sesamum orientale (Pedaliaceae)

Origin: Origin and ancestors unknown, possibly Ethiopia or peninsular India

An ancient oilseed crop. Seeds contain 50 percent unsaturated oil and 20-25 percent protein, and are used widely in bread and confectionery. Oil used in cooking, margarine, soaps and other industries. Residues are valuable animal feed. Interest in the crop is in decline as difficulty mechanizing harvesting and low seed yields compared with other oil crops. Good genetic diversity in related species. *S. malabaricum* produces fertile offspring with *S. indicum*.

Shea nut
Vitellaria paradoxa (Sapotaceae)

Origin: Dem. Rep. of Congo, Sudan, Uganda

The roasted kernels are used to make purified shea butter, rich in vitamin E, used in cooking and as an alternative to cocoa butter for chocolate manufacture. Also has commercial use in manufacture of soap, cosmetics, candles. Various local uses. Fruit is eaten and is a good source of carbohydrates, iron and B vitamins. A monospecific genus. Threatened by overexploitation as a timber and source of fuel, and also by land clearance. Stands may be conserved for their valuable seed, but no official protection exists. Mostly grown for local consumption.

LEAF AND FLOWER VEGETABLES

Artichoke
Cynara scolymus (Compositae)

Origin: Mediterranean, Canary Islands

Flowerheads and the receptacle are eaten. Small amount of vitamin C.

Lettuce
Lactuca sativa (Compositae)

Origin: Probably evolved in Asia Minor or Middle East from *L. serriola*

Lettuce leaves are a useful source of fiber, minerals (especially potassium), vitamins A, E and C. May be grown year round. Stem is boiled as a vegetable in China. A highly variable crop, resulting probably from long history of selection. Increasing diversity of lettuce types consumed. Wild species, including *L. serriola, L. saligna* and *L. virosa* have been used in breeding programs.

Spinach
Spinacia oleracea (Chenopodiaceae)

Origin: Southwest Asia

Edible leaves contain range of minerals, vitamins A, E and C and the B vitamin range.

Pineapple
Ananas comosus (Bromeliaceae)

Origin: Obscure origin in South America, probably on fringes of Amazon

Seedless fiber-rich fruit, source of vitamins C, A and B. Highly suited for canning and as a juice. Unique in that timing of harvest can be controlled by externally applied growth hormone. Over 65 countries grow pineapple for domestic consumption and export. No wild populations. Genetically variable species, but genetic base of commercial plantations very narrow. 70 percent of world production and 96 percent of cannery industry comes from one variety. Highest diversity of near relatives in Paraguay and Brazil. Poorly known, but *A. ananassoides* has contributed several characteristics to cultivated crop. *A. erectifolius* also considered for improvement programs.

Papaya
Carica papaya (Caricaceae)

Origin: Obscure, probably hybrid of several *Carica* spp., arising in lowland tropical forest in eastern Andes or Central America

Easily digested fruits produced all year round. Good source of vitamin C; red-fleshed fruits also rich in vitamin A. Papain extract is exported as a meat tenderizer; also used medicinally, to tan leather and in brewing beer. May be produced by biotechnology in future. Commercially produced in over 30 countries, mostly for domestic consumption. High diversity in eastern Andes. At least six other spp. domesticated and 12 spp. are harvested for their fruit. Several commercial cultivars come from highly inbred hermaphrodite lines. Most production from backyard papaya trees, where local variation is high. Many wild species have desirable characteristics, useful in breeding. Hybridization already carried out with five wild *Carica* spp. Highly susceptible to viral and fungal diseases; some resistance detected in wild relatives but conventional crossing impossible.

FRUITS

Lime, lemon, pomelo, tangerine, sweet orange
Citrus aurantiifolia, C. limon, C. maxima, C. x paradisi, C. reticulata, C. sinensis (Rutaceae)

Origin: Lime: cultivated hybrid with obscure origins, possibly a hybrid of *C. medica* with another sp. Lemon: probably a hybrid of lime with *C. medica*. Pomelo: probably a native of the Malay peninsula. Grapefruit: probably hybrid between orange and pomelo, arising in the Caribbean. Tangerine: possibly Indo-China. Orange: probably introgressed hybrids of *C. maxima* and *C. reticulata*, perhaps originating in China.

Fruits contain nearly 90 percent water, potassium, vitamins A, B, E and high vitamin C. They are eaten fresh, used as a flavoring and in marmalade. Orange accounts for 70 percent of *Citrus* production. Various other spp. are cultivated. Wild populations located in northern India. Wild species threatened by forest clearance. In Southeast Asia wild groves are being replaced by oil palm and cacao plantations. *C. taiwanica* is critically endangered in Taiwan, mainly because of extensive habitat loss but also because of use as a rootstock for citrus plantations. Wide variation within the genus. Can be crossed with several genera. Economic *Citrus* spp. are highly interfertile.

Pumpkin, squash, gourd
Cucurbita moschata, C. maxima, C. argyrosperma, C. pepo, C. ficifolia (Cucurbitae)

Origin: *C. moschata* is most like the wild species and was domesticated independently in Central and South America.

Fruits, containing 90 percent water, small amounts of starch, sugars, protein, fat and vitamins A, B and C, are used as vegetables and as animal feed. Leaves and flowers may be cooked. Seeds eaten and sometimes processed for oil. Grown worldwide in temperate and tropical zones, commonly in home gardens and as subsistence crops as well as commercially. Long shelf life. Broad gene pool because of wide use of traditional or unimproved varieties in subsistence farming and home gardens. Many *Cucurbita* spp. have restricted geographical ranges. Disease resistance is found in wild relatives, with some transfer to cultivated species through interspecific crosses. Crosses between crop species and wild or feral relatives have occurred and genetic exchange takes place where their ranges overlap.

Strawberry
Fragaria x ananassa (Rosaceae)

Origin: A hybrid between two American species, *F. chiloensis* and *F. virginiana*

Soft fruit, 90 percent water, high vitamin C, eaten fresh or in jams and confectionery. Grown in most temperate and subtropical countries. All *Fragaria* spp. produce palatable fruit. Hundreds of cultivars with wide ecological adaptability. Considerable genetic diversity lost in cultivated strawberries in the last 100 years. Attempts are being made to extend the genetic base of the crop. Much unused genetic variation in wild species.

FRUITS

Fig
Ficus carica (Moraceae)

Origin: Eastern Mediterranean

Fruits contain 10 percent sugar when fresh and 50 percent when dried. Also substantial amounts of potassium, especially in the dried fruit. Most world trade as dried figs. Figs are widely distributed in tropical, subtropical and warm temperate areas throughout the world. *Ficus* spp. are also source of rubber, fibers, paper, medicines and ornamental plants. Fig is largely grown for domestic consumption using traditional cultivars, hundreds of which exist, with local clones occurring in distinct geographical groups in the Mediterranean basin. Closely related wild forms are distributed throughout the Mediterranean basin. Fig culture is in decline. Many old groves have been abandoned or cleared. A number of wild relatives are considered threatened. 27 *Ficus* spp. were listed as threatened in 1998. Reproductive isolation is dependent solely on the specificity of the wasp pollinator. Artificial crosses can be made between species.

Apple
Malus domestica (Rosaceae)

Origin: An aggregate of over 1 000 cultivars, of ancient and complex hybrid origin

Apples, with pears, are the most important fruit crops of cooler temperate regions. Fruit is eaten fresh or cooked, as a juice or brewed as cider. Potassium is the main mineral with small amounts of vitamin C. Breeders in the 19th century used wild species in breeding. Genetic diversity accumulated in North America was greater than in Europe because propagation was by seed rather than by grafting. Current trend depending on few varieties has caused rapid loss in genetic diversity and potential breeding material. Widespread elimination of wild stands is also taking place. Three spp. were listed as threatened in 1998. Hybridization with many wild species within the genus occurs readily.

Mango
Mangifera indica (Anacardiaceae)

Origin: Northeast India

Fleshy edible fruit; a good source of vitamins A and C. Thrives on infertile marginal soils. Important tree in Hindu mythology and religion. Kernel oil may be used in chocolate manufacture. Demand for the fruit and its juice is increasing worldwide. India accounts for two thirds of production. Fruits of more than 12 wild spp. collected. Several are cultivated. The majority of fruit-bearing trees are more or less wild. Genetically highly heterogeneous. Over 1 000 cultivars exist, many in Borneo and the Malay peninsula. Feral populations are distributed throughout the tropics. Of the 40 to 60 spp. in the genus, many are poorly known, severely threatened or possibly extinct. 35 spp. were listed as threatened in 1998. Logging, forest clearance and replacement with commercial species in Southeast Asia are largely responsible for population extinctions. Various species are suitable for cultivation given further selection. Many display valuable traits, such as tolerance of waterlogged soils and more regular fruiting.

FRUITS

Avocado
Persea americana (Lauraceae)

Origin: Mexico to northwest Colombia

A highly nutritious fruit, containing 15-25 percent monounsaturated fat and vitamins C, B and E. The oil is used in cosmetics. Trees fruit year round. Importance has increased over recent decades and the crop is now grown in most tropical and subtropical countries. Most production is for domestic consumption. Other *Persea* spp. are used for timber and fruit. There are three geographically distinct varieties, which are able to interbreed. Commercially important cultivars have arisen mostly in private orchards by chance rather than as a result of germplasm manipulation. Increasing use of grafting and uniform varieties. Serious genetic erosion in traditional varieties. Diversity appears greater in traditional growing areas, where farmers still propagate by seed. Genetic exchange occurs between cultivated forms and wild populations. Wild populations of the avocado and its close relatives are small and becoming increasingly isolated. Deforestation poses a severe threat to their survival. 15 spp. were listed as threatened in 1998. A number of wild relatives show resistance or tolerance to disease, drought and frost.

Date
Phoenix dactylifera (Palmae)

Origin: Western India or Arabian Gulf

Edible fruit with sugar content of 30 to 80 percent, corresponding to soft and dry dates. Vitamins are relatively low in quantity. Eaten as an ingredient in a variety of foods or as a juice. A staple where produced. Good storage. One of the oldest cultivated tree crops. Current cultivars resulted from thousands of years of selection. Perhaps over 3 000 cultivars exist; only 60 grown widely. All commercial cultivars are female. Wild populations of some related spp. are highly restricted in geographical range. All *Phoenix* spp. intercross freely.

Apricot, cherry, plum, almond
Prunus armenica, P. avium, P. domestica, P. dulcis (Rosaceae)

Origin: Apricot: west China. Cherry: west Asia. Plum: an ancient 6n cultigen with complex origin, possibly in southwest Asia; North American plums may be native American spp. or hybrids with *P. salicina*. Almond: central to west Asia. Peach: west China, possibly a cultigen derived from *P. davidiana*.

Apricot, cherry, plum and peach are soft fruit with up to 10 percent sugar, good potassium and Vitamin A in the case of apricots, but low vitamin C. Consumed fresh, dried or as an ingredient in jams and confectionery. Almond is the most important tree nut crop. The kernel contains 40-60 percent unsaturated oil and 20 percent protein. Eaten as a dessert nut and in confectionery and marzipan. A major trading commodity. Many other *Prunus* spp. have edible fruit. Many cultivars and much genetic diversity. Plums are genetically central to the genus and harbor the most useful genetic material. Narrow breeding has led cherries to be more isolated from the rest of the genus. Increasing loss of genetic diversity. Developing countries are tending to replace indigenous types and wild stands with western varieties, e.g. the switch from seed to vegetatively propagated almonds in Turkey. A number of wild relatives are confined to narrow ranges. 23 spp. were listed as threatened in 1998.

FRUITS

Pear
Pyrus communis (Rosaceae)

Origin: Asia minor, the Caucasus, central Asia and China. Cultivars have come from *P. bretschneiderii, P. pyrifolia, P. sinkiangensis* and *P. ussuriensis. P. nivalis* for perry production.

With apples, pears are the most important fruit crops of cooler temperate regions. The fruit is eaten fresh or cooked, as a juice or brewed as perry. The fruit are a good source of dietary fiber, potassium and reasonable amounts of vitamin C. Currently about 20 spp. and

5 000 recorded cultivars. Major loss of genetic diversity through concentration on few varieties. Several wild species in Turkey are under threat. Five spp. were listed as threatened in 1998. Hybridization with high proportion of wild species within the genus is possible, providing useful rootstocks and possibly disease resistance. Much use of wild species in breeding in the past. Evolution of new varieties will be seriously limited unless stands of wild species conserved.

Blackcurrants, redcurrants
Ribes nigrum, R. rubrum (Grossulariaceae)

Origin: Europe and northern Asia, with the blackcurrant extending to the Himalayas

Fruits with high vitamin C content. A luxury crop, largely produced for processing into juice. Many spp. with edible fruits, cultivated and wild. Wide use has been made of wild or near-wild relatives.

Grape
Vitis vinifera (Vitaceae)

Origin: Eurasia

The fruit has high sugar content. 68 percent of grape production is for the manufacture of wine, 20 percent for dessert grapes, 11 percent for dried fruit – raisins, sultanas, currants – 1 percent for juice. Other

commercial products include grapeseed oil and vine leaves. Various other spp. produce edible grape. One estimate suggests there are 10 000 cultivars of grape. Wild species still occurs in Middle Asia. Wild relatives are suffering genetic erosion in the United States. All known *Vitis* spp. produce fertile offspring.

FRUIT VEGETABLES

Melon seed/watermelon
Citrullus lanatus (Cucurbitae)

Origin: Tropical and sub-tropical Africa; domestication took place in Mediterranean

The flesh of the fruit is 90 percent water; also contains vitamin C and A. The seeds contain 40 percent

unsaturated oil and 40 percent protein. Wild plants still harvested in Kalahari. *C. colocynthis* is fertile with the watermelon. An African watermelon with extraordinarily long storage life has been identified.

FRUIT VEGETABLES

Melon, cucumber
Cucumis melo, C. sativus (Cucurbitae)

Origin: Wild melon populations appear to be distributed south of the Sahara to Transvaal in South Africa. Cucumber's wild or feral relative and possible progenitor, var. *hardwickii*, is native to the southern Himalayan foothills.

Melon is grown worldwide in temperate and tropical countries. 90 percent water some sugar and vitamin C. Pink or orange-colored fruit have a high percentage of vitamin A. Also grown for their fragrance or ornamental value. Cucumber produces edible fruits, containing 96 percent water, some vitamin C and reasonable amounts of Vitamin A. Also used in production of fragrances, cosmetics and medicines. Young leaves and shoots may be cooked. Also cultivated *C. anguria* (West Indian gherkin) and *C. metuliferus* (African horned cucumber or jelly melon). Wild and feral populations of melon occur throughout Africa and southern Asia. Cucumber produces fertile hybrids with its wild counterpart *C. sativus* var. *hardwickii*. No interspecific hybridizations have been used to improve crops.

Tomato
Lycopersicon esculentum (Solanaceae)

Origin: Cultigen, from Mexico

There are few growing areas, from the tropics to the Arctic circle, where the tomato is absent. Fruit, containing potassium, vitamins A, B, C and E, is eaten fresh, dried or cooked as a vegetable, or processed in a wide range of food products. Disease is a common threat. The wild relatives of the tomato have limited ranges. Wild gene pools are prone to erosion by habitat destruction. Tomato can be hybridized with all spp. in the genus and wild relatives have been used as source of numerous useful traits, including disease resistance.

Eggplant
Solanum melongena (Solanaceae)

Origin: India; wild progenitor, *S. incanum*, occurs throughout Africa and Asia.

Fruit is eaten as a vegetable, contains over 90 percent water, large amount of potassium, some vitamins A, E, B, C. Highly productive and useful smallholders' crop. Various spp. cultivated and used as grafting stock.

BEVERAGE CROPS

Tea
Camellia sinensis (Theaceae)

Origin: Probably lower Tibetan mountains or central Asia

Tender shoots are used to make tea. Important plantation and smallholder crop throughout the tropics and subtropics. Planted commercially in at least 30 countries. Increasing consumption in developing countries. High diversity of forms or species in East and Southeast Asia. Many distinct forms, hybrids and species continue to be discovered. Recent trend to propagate the plant vegetatively, which has led to large areas being planted with one or few clones. No threat yet of genetic erosion in the crop. 11 *Camellia* spp. in China and Viet Nam were recorded as threatened in 1998.

Coffee
Coffea arabica, C. canephora (Rubiaceae)

Origin: Arabica coffee originated in montane forest in southwest Ethiopia and the neighboring Boma Plateau in Sudan; possibly also in Marsabit forest in Kenya. Robusta grows wild in West and Central Africa.

Important sources of foreign currency in many developing countries. Roasted seeds used for beverage containing 1-2.5 percent caffeine, and niacin and potassium. A number of other *Coffea* spp. also cultivated as coffee or as a source of edible berries. Commercial arabica cultivars have very narrow genetic base, especially in Latin America and the Caribbean, and are highly susceptible to disease. Robusta outcrosses and has wider variation. Much production by smallholders, but 40 percent of coffee from Americas and Caribbean from intensive monocrop plantations. Recent recognition of importance of conserving species-rich shade coffee systems. Significant percentage of Ethiopian coffee from uniform commercial cultivars; 400 000 ha remain of wild coffee, accounting for half of Ethiopia's coffee production. Several populations of wild species increasingly restricted in distribution and fragmented. Nine spp. from mainland Africa were listed as threatened in 1998.

Mate
Ilex paraguariensis (Aquifoliaceae)

Origin: South America

Tea is made from leaves; contains 2 percent caffeine. Little use in export market.

Onion, garlic
Allium cepa, Allium sativum (Alliaceae)

Origin: Onion exists only in cultivation; may have come from Afghanistan, Iran, and former USSR area. Possible progenitors of garlic are *A. longicuspis* or wild *A. ampeloprasum*. The greatest number of *Allium* spp. are in North Africa and Eurasia.

Underground bulbs are more important for their flavor and antimicrobial properties than nutritional value. Garlic contains large amounts of potassium and significant vitamin C. Important component in the diet of a wide range of cultures. Numerous medicinal functions. The allins contained in *Allium* spp., especially garlic, may protect against cancer and cardiovascular disease. Seven economically important cultivated *Allium* spp.; many other species consumed on a lesser scale. Open pollinated populations still represent most of the production in tropical and subtropical countries. The habitat of some wild *Allium* spp. is severely threatened. Poor results from interspecific hybridizations. The genetic variability available in wild and cultivated relatives has not been extensively used in crop improvement.

SPICES AND FLAVORS

Chili pepper, sweet pepper
Capsicum annuum (Solanaceae)

Origin: Tropical America

Fruits of varying pungency are used either fresh as a vegetable or dried or powdered as a spice. Fresh fruits contain large quantities of vitamin A, plus vitamin C.

Breeding has generally depended on pure lines. Wild peppers are still collected and sold locally. Some interfertility with other *Capsicum* spp. Wild spp. offer valuable new traits.

Cardamom
Elettaria cardamomum (Zingiberaceae)

Origin: India

Seeds used as a spice. Essential oil used in perfume and as flavor for liqueurs. Wild populations exist in

monsoon forests in south India and Sri Lanka. There are no essential differences between wild and cultivated forms. Collection from the wild contributes to the commercial trade. Wild populations are disappearing.

Pimento
Pimenta dioica (Myrtaceae)

Origin: West Indies and Central America

Plants grow in semi-wild state. Populations of wild relatives are confined to small areas of remaining dry

forest and coastal habitat in the Caribbean, especially Cuba. They are poorly studied and are severely threatened by habitat loss.

Pepper
Piper nigrum (Piperaceae)

Origin: Western Ghats, India

The dried fruits, high in alkaloid content, represent one of the oldest spice crops. Several other *Piper* spp. important for local pepper production. Grown as a

smallholder crop in tropical countries. Large-scale planting is based on one clone and is dangerously vulnerable to disease. Wild pepper still grows in the Western Ghats.

Sugar beet
Beta vulgari (Chenopodiaceae)

Origin: Evolved from sea-beet (*B. vulgaris* var. *maritima*) in Europe and west Asia.

Swollen taproot provides nearly half the world production of sucrose. Forms of the same species include leaf beets and chards used as garden vegetables, and other beets with swollen taproots, e.g. beetroot and mangold, for human consumption and animal feed. All forms within the species may be crossed. Wild relatives have already provided some disease resistance. The only source of resistance against the beet cyst nematode is detected in relatives from a different section in the genus.

Hazel and filbert
Corylus avellana, C. maxima (Betulaceae)

Origin: Europe and western Asia

Edible tree nut and ornamental. Kernels are used in confectionery; 18 percent protein and 68 percent oil. All *Corylus* spp. have edible nuts. *C. colurna* is also cultivated for nuts. Populations of *C. chinensis* in China have declined, largely because of overexploitation.

Walnut
Juglans regia (Juglandaceae)

Origin: Balkans to China

Edible nut, containing vitamins E, C and B. Use as dessert and in confectionery; oil also extracted. The kernel contains 15 percent protein and 70 percent unsaturated oil. Leaves make good fodder. The timber is highly valued. All *Juglans* spp. produce edible seeds, timber, ornamentals. No apparent threat of genetic erosion in the crop. Major cultivation in the United States but enormous unexplored potential elsewhere. Wild walnut forest has declined and become fragmented throughout its native range. Of the 21 species in the genus, seven were listed as threatened in 1997.

Pistachio
Pistacia vera (Anacardiaceae)

Origin: Near East and western Asia

Tree nut: low in sugar, more than 20 percent protein, 50 percent oil. Important food for nomads during migration in Iran and Afghanistan. Highly drought resistant. Trees used for ornamental and shade purposes, also as a source of resin, dye, turpentine, mastic and medicine. Cultivated in the Mediterranean and western Asia for 3 000-4 000 years. None of the related species has value as a nut crop, although seven spp. are used as rootstocks and also for pollination. Largely harvested from wild in Afghanistan and parts of Pakistan. Iran has had commercial plantations for hundreds of years. Wild species may have a role in future improvements. Many wild populations have been destroyed by forest clearance, over-cutting for charcoal and grazing. Three spp. were listed as threatened in 1998.

SUGAR CROPS

TREE NUTS

TREE NUTS

Brazil nut
Bertholletia excelsa (Lecythidaceae)

Origin: Tropical South America

Kernel contains 17 percent protein, 65-70 percent monounsaturated oil. Largely an export crop. Also a staple for indigenous people and important ecological component of rainforest in the Amazon basin. Oil used for cooking or as fuel or animal feed. Valuable timber. Attempts to establish plantations have generally failed. Well-managed plantations have the potential of producing yields far exceeding natural groves. Almost all nut production is from wild trees. Distribution and density of groves may have been largely influenced by indigenous groups in the past. Little information exists on genetic variation. Populations appear to tolerate different soil types. Sustainable system of harvesting in extractive reserves, but considerable habitat loss and illegal tree felling continues elsewhere. Development in the Tocantins valley where there is high concentration of brazil nut trees continues to cause population decline. Developments elsewhere are also resulting in serious genetic losses. The species was listed as vulnerable in 1998. Protected populations are found in biological reserves, Indian and extractive reserves and corporate property.

REFERENCES

1 Prescott-Allen, R. and Prescott-Allen, C. 1990. How many plants feed the world? *Conservation Biology* 4(4): 365-374.

2 FAO. FAOSTAT database. Available at Food and Agriculture Organization of the United Nations website, http://apps.fao.org/ (accessed February 2002).

3 Mabberley, D.J. 1997. *The plant-book. A portable dictionary of the vascular plants.* 2nd edition. Cambridge University Press, Cambridge.

4 Smartt, J. and Simmonds, N.W. 1995. *Evolution of crop plants.* 2nd edition. Longman Scientific and Technical, Harlow.

5 Smith, N.J.H. *et al.* 1992. *Tropical forests and their crops.* Cornell University Press, Ithaca and London.

6 Vaughan, J.G and Geissler, C.A. 1997. *The new Oxford book of food plants.* Oxford University Press, Oxford.

7 Walter, K.S. and Gillett, H.J. (eds) 1998. *1997 IUCN Red List of threatened plants.* Compiled by the World Conservation Monitoring Centre. IUCN-the World Conservation Union, Gland and Cambridge.

APPENDIX 3:
DOMESTIC LIVESTOCK

This table presents information on the major domestic mammals used in agriculture and related activities, such as hunting or transport, and on the number and status of closely related wild species.

At the local level, a great many wild animal species are used primarily to meet subsistence needs, the kind depending largely on availability, convenience and tradition. Far fewer are used in livestock production: breeds of goat, sheep, cattle and pigs are cosmopolitan in distribution and, along with domestic fowl, form the basis for most of the world's agricultural animal food production on land. These four principal mammalian livestock species have diversified into more than 4 000 recognized breeds. While intensification of production has typically gone hand in hand with narrowing of the genetic base, such that semen from individually documented and tested lines commands a premium, there is increasing recognition of the genetic potential resident in less commercially developed breeds and blood lines, and of the often neglected value of locally adapted stock in comparison with commercial stock from advanced industrial countries. The pool of genetic resources represented by domestic animal diversity is an essential basis for efficient and sustainable food production, and is likely to be of increasing importance in the more demanding production environments.

There is no universally accepted system for naming domestic stock. Some authorities apply the earliest valid name to both the wild species and to domestic stock derived from it; others prefer to retain separate names for domestic stock where such a name has been in common use, and apply the next available valid name to the wild species. In the first case, for example, *Capra hircus* Linn. 1758 would be applied to the wild goat and all domestic derivatives; in the latter case, that name would be restricted to domestic stock and *Capra aegagrus* Erxleben 1777 applied to the wild goat of Eurasia. The second approach is adopted below.

In 'Number and status of breeds' the first figure is the number of breeds given in the Food and Agriculture Organization of the United Nations database. Subsequent figures are, in order: the number of threatened breeds in the 'critical' category, the number 'endangered' and the number 'extinct'. See Scherf[3] for definitions. For both threatened categories, the counts include breeds maintained by active conservation programs and by institutions.

Source: For general information, see Clutton-Brock[1] and Mason[2]; for number and status of breeds see Scherf[3]; for number of congeneric wild species see Wilson and Reeder[4]; and for status of wild species see Hilton-Taylor[5].

Cattle

Bos taurus (humpless, mainly European cattle, taurine type)
Bos indicus (humped, mainly Asian cattle, zebu type)
(Bovidae)

Meat, milk, transport, draught, dung, etc.

Domestic longhorn cattle from around 8 000 years ago at several Middle East sites, later in Nile region, and circum-Mediterranean by 3 000 years before the present. European cattle probably of Middle East origin. Humps, assumed result of artificial selection, at base of neck or over shoulder in zebu type. Zebu generally heat and parasite resistant, dominant in Asia and Africa (some longhorns persist, e.g. trypanosome-resistant N'Dama in West Africa). Cattle were first draught farm animals; in Europe only specialized for meat or milk when replaced as power source by horse. Very high breed diversity, many now rare. British breeds to North America and Australia in the 19th century; Iberian breeds earlier to South America. Cattle certain to continue as major farm animals for meat and milk. Much potential in tropics for development of local stock, e.g. zebu dairy breeds. Several feral herds.

Number and status of breeds: 1 479; C 106, E 193, Ex 255

Derived from: *Bos primigenius* wild ox or aurochs (extinct)

Congeneric wild species: 4; all 4 listed as threatened in 2000.

Mithan or gayal

Bos frontalis (Bovidae)

Ceremonial sacrifice, barter

No firm evidence but probably of early origin. Restricted to Bhutan, hills in northeast India bordering China and Myanmar, and Chittagong hills of Bangladesh. Typically higher elevation than cattle and lower than yak. Kept mainly by hill tribes, usually by men of high status, for use in ceremonial sacrifice, exchange and trophy display. Not much used for draught or milk. Mithan generally forage freely in forest during day or for months, restrained at intervals, lack human control over breeding. May breed with cattle and gaur.

Number and status of breeds: None formally recognized

Derived from: *Bos gaurus* gaur, South and Southeast Asia

Congeneric wild species: 4; all 4 listed as threatened in 2000 (as for domestic cattle).

Yak

Bos grunniens (Bovidae)

Milk, transport, meat, dam of 'dzo' (cattle x yak hybrid draught animal)

Possibly domesticated at same time as cattle, probably on Tibetan plateau or the Himalayas. Most yak in west China, many in Mongolia, fewer in Tajikistan, Kyrgyzstan, Nepal, Bhutan, Afghanistan, India. Usually at 3 000-5 000 meters altitude. Variable size and pelage, usually smaller than wild yak. Yak tail in trade for centuries; white tips favored for ease of dyeing. Yak can graze where other livestock cannot. Much medical or religious use in Tibet, where milk and butter most important; used as meat source in Mongolia. Hair used for rope, felt; skin for leather; dung for fuel; important pack animal.

Number and status of breeds: 13; C 0, E 1, Ex 0

Derived from: *Bos mutus* yak, China: north of Tibet plateau (Altun Shan, Qilian Shan)

Congeneric wild species: 4; all 4 listed as threatened in 2000 (as for domestic cattle).

Bali cattle

Bos javanicus (Bovidae)

Draught, meat

Domestic cattle present in Southeast Asia ca 5 500 years before present. Banteng possibly domesticated in prehistory in Southeast Asia or Java. Now in many parts of Indonesia; small herds in Malaysia, Philippines, Australia. Uniform in type. Organized selection in 20th century: no entire males exported, no crossing with other cattle. Small size, highly fertile, little fat, uses poor pasture in hot humid conditions. Good draught animal for small fields and terraced slopes; much potential as meat or crossing stock. Feral herd in Cobourg Peninsula (Australia).

Number and status of breeds: No data

Derived from: *Bos javanicus* banteng, Southeast Asia

Congeneric wild species: 4; all 4 listed as threatened in 2000 (as for domestic cattle).

Water buffalo

Bubalus bubalis (Bovidae)

Draught, milk

Probably domesticated earlier than 4 500 years ago in Middle East. Wild ancestor occurred from Mesopotamia east to Southeast Asia; by the 19th century restricted to India and adjacent areas, where local. Domestic buffalo reached southeast Europe by 12th century where from 14th century much used in Muslim communities; later taken to the Americas and Australia, and Africa in 20th century. Breed development centered in India and Pakistan. Broadly divided into swamp buffalo in Southeast Asia, mainly for draught, and river buffalo in South Asia, mainly for milk. Do better than cattle on swamp and floodplain grazing. Much potential for development as meat producer. Milk rich in fat. Large feral herds in Australia.

Number and status of breeds: 86; C 3, E 8, Ex 0

Derived from: *Bubalus arnee* wild water buffalo, Bhutan, India, Nepal, Thailand

Congeneric wild species: 4; all 4 listed as threatened in 2000.

Goat

Capra hircus (Bovidae)

Meat, milk, hair

Goats and sheep next to be domesticated after dog. Domesticated around 10 000 ago in Zagros Mountains of western Iran; to Europe by mid-Neolithic. Worldwide distribution. Great variety in form of horns and ears, hair color, etc. Highest numbers in South Asia. Milk breeds developed in Switzerland have influenced many milk breeds worldwide. The Boer (South Africa) is major meat breed. Two fleece breeds: angora (Turkey) and cashmere (central Asia). Many feral populations, where often adverse impact on native biota. Much potential for further breed development, e.g. for specialized tropical dairy animals. Ruminant physiology allows efficient use of coarse vegetation in semi-arid and arid regions.

Number and status of breeds: 587; C 31, E 70, Ex 17

Derived from: *Capra aegagrus* wild goat, southwest Asia: Turkey east to Pakistan

Congeneric wild species: 9; 7 listed as threatened in 2000.

Sheep

Ovis aries (Bovidae)

Meat, milk, wool

Sheep and goats next to be domesticated after dog. Sheep in use in Mesolithic; evidence for domestication around 11 000 years ago in Middle East; to North Africa (where no wild sheep) by 6 000 years ago; to Americas in 16th century. Worldwide distribution; important in Europe, Middle East, Central Asia. Coat of wild sheep has outer hairs over woolly inner coat; hairs lost during domestication to produce fine fleece breeds. Wool and milk often more important than meat. Wool trade basis of great wealth in medieval and early modern Europe. Many breeds: some multipurpose, others specialized for milk, fleece or meat. Sheep numbers in decline in some developed countries e.g. United States and Australia, but elsewhere provide vital support to human life in marginal and rangeland environments. Ruminant physiology allows efficient use of coarse vegetation in semi-arid and arid regions.

Number and status of breeds: 1 495; C 68, E 199, Ex 181

Derived from: *Ovis orientalis* mouflon, southwest Asia: Turkey east to Iran; Mediterranean populations (Corsica, Sardinia, Cyprus) possibly feral primitive domestic stock

Congeneric wild species: 6; 4 listed as threatened in 2000.

Pig

Sus domesticus (Suidae)

Meat

First evidence of domestic pigs by 9 000 years ago in Anatolia; widespread in Eurasia, incl. Egypt, by 5 000 years ago. Worldwide; nearly half the world's pigs occur in non-Muslim Asia, mostly in China. Management varied: may free-range in woodland or be sty-fed. Pigs introduced to the Americas from Europe; few in Africa or Australia, New Zealand. Several feral herds. Large number of breeds. Commercial production now dominated by few lines. Production increasingly specialized, but still an important role for local varieties in utilizing household waste and wild foods. Pigs have a major cultural significance in parts of Southeast Asia and Melanesia.

Number and status of breeds: 649; C 58, E 106, Ex 151

Derived from: *Sus scrofa*, Eurasian wild pig, North Africa, Europe, Asia

Congeneric wild species: 10; 6 taxa listed threatened in 2000

Bactrian camel

Camelus bactrianus (Camelidae)

Draught, transport, meat, milk, wool, dung

Fossil camels known from North America (where no extant camels) and Eurasia west to North Africa. Rock drawing in Mongolia of two-humped camel may be 10 000 years old. First evidence of domestication in Iran and Turkmenistan about 5 000 years before the present. Widespread in central Asia by ca 3 000 years ago. Main transport on Silk Route between Mesopotamia and China but replaced by dromedary in west and south from ca 2 000 years ago. Restricted to Central Asia, incl. Mongolia and China.

Numbers probably in decline.

Number and status of breeds: 11; C 1, E 1, Ex 0

Derived from: A domesticated form of wild Bactrian camel, southwest Mongolia, northwest China

Congeneric wild species: Non-domesticated populations in central Asia, listed threatened (endangered) in 2000.

Dromedary

Camelus dromedarius (Camelidae)

Transport (draught, meat, milk, wool, dung)

Remains of dromedary or similar species at Paleolithic sites in North Africa about 80 000 years of age. Wild camels apparently extinct in Africa by 5 000 years ago but persisted in Saudi Arabia (where perhaps first domesticated) until ca 2 000 years ago. Domestic camel to Horn of Africa around 4 000 years ago. Reached present importance with rise and spread of Arab power from 7th century onwards. Most camels in northeast Africa and Afghanistan/Pakistan/India, where numbers rising; fewer and decreasing in Middle East. Primarily for transport; specialized pack and riding breeds exist. Introduced to Canaries and Australia (where feral herds). Ability to withstand long periods without drinking and use thorny browse key to human use of hot deserts.

Number and status of breeds: 52; C 1, E 2, Ex 0

Derived from: Ancestral form unknown, presumed extinct *Camelus* species; Bactrian camel closest living relative

Congeneric wild species: See Bactrian camel.

Llama

Lama glama (Camelidae)

Transport, wool (coarse), meat, dung

Domesticated by 6 000 years before the present in high-altitude Andean pastures, possibly centered around Lake Titicaca basin of south Peru and west Bolivia. Alpaca textiles known from 2 500 years ago. Domestic camelids spread to lower altitudes and along Andean chain by 4 000 years ago and reached greatest extent during Inca period; in decline since Spanish conquest in early 16th century and introduction of European stock. Remain important to Andean culture and for superior adaptation to poor high-altitude grazing. Pad feet may cause less pasture damage than hoofs of sheep. Llamas and most alpacas held by small-scale pastoralists on communal grazing; some alpaca kept in large herds by cooperatives in Peru. Not milked. Alpaca wool has high commercial value. Llama flocks in the United States and Europe.

Number and status of breeds: 8; C 0, E 0, Ex 0

Derived from: Probably *Lama guanicoe* guanaco, south Peru, west Bolivia, northwest Argentina

Congeneric wild species: 1; not listed threatened.

Alpaca

Lama pacos (Camelidae)

Wool (fine)

See llama for background.

Number and status of breeds: 6; C 0, E 1, Ex 0

Derived from: Unknown in wild, presumed *Lama* sp. or *Lama* x *Vicugna* hybrid

Congeneric wild species: See llama.

Reindeer

Rangifer tarandus (Cervidae)

Meat, milk, transport

Fossil evidence for use of reindeer from 80 000 years ago, domesticated by 2 500 years before the present. Management varies: riding or milk animals may be separated from herd and fed, or herds may roam widely and be gathered annually for marking or slaughter. Reindeer industry important in north Scandinavia, northwest Russia and Siberian Russia, less so in North America. Reindeer exploitation key to settling the far north. Wild reindeer include four major types, all used in husbandry systems. Some potential for better use; numbers have been increasing but with local indications of overgrazing. Lichens, the main winter feed, are vulnerable to atmospheric pollution.

Number and status of breeds: No data

Derived from: Semi-domesticated forms of Eurasian *Rangifer tarandus*, reindeer (Eurasia), caribou (North America)

Congeneric wild species: See above; 1 North American subspecies listed threatened in 2000.

Ass or donkey

Equus asinus (Equidae)

Transport, draught, sire of mule (ass x horse hybrid)

Probably domesticated in northeast Africa; records from about 6 000 years ago in Egypt. The only domestic animal certainly of African origin. Widespread in Middle East by ca 2 000 years ago. To Americas in 16th century. Much more important than horse in Africa where present in north and west. Common in south and central Asia; also present in south Europe. Mostly for transport; specialized riding and pack breeds exist. Formerly milked, meat sometimes used. Feral asses widespread incl. Socotra, Galapagos, United States, Australia, Sahara, etc. Numbers worldwide likely to decline, but because of hardiness and low cost will retain importance in less developed areas.

Number and status of breeds: 103; C 5, E 16, Ex 6

Derived from: *Equus africanus*, African wild ass, North Africa to Somalia

Congeneric wild species: 9; 8 taxa listed threatened in 2000.

Horse

Equus caballus (Equidae)

Transport, draught, sport, dam of mule (ass x horse hybrid)

Some evidence of domestic horses 6 000 years ago in central Eurasia (Ukraine). Spread through Eurasia during Bronze and Iron Ages. Important early military use, to draw chariots and for riding, especially after invention of stirrups before the start of the 6th century. Wild horses present with Amerindians in North America but extinct there by 12 000 years ago; domestic horses introduced by European colonists. Most horses occur in South America where numbers also highest in relation to humans; numbers high in North America and Asia. Specialized for draught or riding, but both uses in decline. Feral horses on all continents (except Antarctica).

Number and status of breeds: 820; C 127, E 178, Ex 94

Derived from: *Equus ferus* (*E. przewalskii*) wild horse, formerly Americas, Europe, Asia

Congeneric wild species: See ass.

Dog

Canis familiaris (Canidae)

Companion, hunting, security, food

Domestication may have begun long before that of agricultural stock, possibly 100 000 years ago; first direct evidence 14 000 years ago in Middle East; distinct kinds of dog evident by 7 000 years ago. The dingo is a feral domestic dog taken to Australia around 12 000 years ago.

Number and status of breeds: A few hundred, no data on status

Derived from: *Canis lupus* wolf, North America, Europe, Asia

Congeneric wild species: 7; 3 taxa listed threatened in 2000.

Guinea pig

Cavia porcellus (Caviidae)

Meat, laboratory, companion

One of the few domestic animals of South American origin. Probably domesticated 3 000-6 000 years ago, but in use long before. Taken to Caribbean and Europe by mid-16th century. Some planned selective breeding during past 30 years. Potential for more development as meat source, especially in original Andean range, but broiler fowl increasingly used instead.

Number and status of breeds: No data

Derived from: *Cavia aperea*, widespread in South America, or *C. tschudii* Peru, south Bolivia, northwest Argentina, north Chile

Congeneric wild species: 4; 1 listed threatened in 2000

European rabbit

Oryctolagus cuniculus (Leporidae)

Meat, fur, laboratory, companion

Kept enclosed (in *leporaria* by Romans) for more than 2 000 years. Kept by medieval monks; newborn or unborn young were permissible food during Lent. Distributed worldwide by mariners; many feral populations. Some development of meat breeds since the second world war; much potential as low-cost converter of surplus vegetation into meat.

Number and status of breeds: 10; C 0, E 4, Ex 0

Derived from: *Oryctolagus cuniculus* European rabbit, west and south Europe to northwest Africa

Congeneric wild species: None; a monospecific genus (several related genera listed threatened).

REFERENCES

1   Clutton-Brock, J. 1999. *A natural history of domesticated mammals*, 2nd edition. Cambridge University Press, Cambridge, and the Natural History Museum, London.

2   Mason, I.L. (ed.) 1984. *Evolution of domesticated animals*. Longman, London and New York.

3   Scherf, B.D. (ed.) 2000. *World Watch List for domestic animal diversity*, 3rd edition. Food and Agriculture Organization of the United Nations, Rome.

4   Wilson, D.E. and Reeder, D.M. (eds) 1993. *Mammal species of the world: A taxonomic and geographic reference*, 2nd edition. Smithsonian Institution Press, Washington DC and London.

5   Hilton-Taylor, C. (compiler) 2000. *2000 IUCN Red List of threatened species*. IUCN-the World Conservation Union, Gland and Cambridge. Available online at http://www.redlist.org/ (accessed April 2002).

APPENDIX 4:
RECENT VERTEBRATE EXTINCTIONS

This list provides information on vertebrate species that have been listed as globally extinct. It is based in part on lists produced using the IUCN threat category system, but mammals, birds and fishes have been revised following more rigorous criteria developed by the Committee on Recently Extinct Organisms (CREO). The CREO system relates to extinctions since AD1500. Where information meets all CREO criteria, the extinction event is considered fully resolved; these species appear on a darker background. To date, extinct and possibly extinct mammals and fishes have been evaluated in this way, and data are available at the CREO website. The bird list is based on *Threatened Birds of the World*[1] which itself informally follows CREO guidelines. The reptiles and amphibians listed have been regarded as extinct, but in several cases data needed to meet CREO criteria (e.g. on taxonomy or survey effort) appear to be unavailable. A number of species have been removed, as new evidence suggests that extant populations of that taxon persist or because extinction took place before AD1500. The Lake Victoria cichlid fishes in the genus *Haplochromis* (*sensu lato*) at the end have been widely regarded as extinct, but data available leave some uncertainty whether these species are extinct in the wild, or persist in small numbers outside sampled areas. These fishes are listed as questionably extinct and are distinguished by a paler background. The Period column indicates, very approximately, the probable period during which extinction occurred. 'Early' refers to the first four decades of a century (C), 'mid' the next three decades, 'late' the final three decades. Where a more precise date is available, this is given in parentheses.

Source: Data on mammals and fishes, CREO[2]; on birds, BirdLife International[1]; on reptiles and amphibians, Hilton-Taylor[3].

Species	Common name	Place	Period
Class MAMMALIA			
Order DASYUROMORPHIA			
Family Thylacinidae			
Thylacinus cynocephalus	Thylacine	Tasmania	Early 20th C (1936)
Order PERAMELEMORPHIA			
Family Peramelidae			
Chaeropus ecaudatus	Pig-footed bandicoot	Australia	Mid 20th C
Macrotis leucura	Lesser bilby	Australia	Mid 20th C (1960s)
Perameles eremiana	Desert bandicoot	Australia	Mid 20th C (1960s)
Order DIPROTODONTIA			
Family Macropodidae			
Caloprymnus campestris	Desert rat-kangaroo	Australia	Early 20th C (1932)
Lagorchestes asomatus	Central hare-wallaby	Australia	Mid 20th C (1960s)
Lagorchestes leporides	Eastern hare-wallaby	Australia	Late 19th C (1890)
Macropus greyi	Toolache wallaby	Australia	Late 20th C (1972)
Onychogalea lunata	Crescent nailtail wallaby	Australia	Mid 20th C (1956)
Potorous platyops	Broad-faced potoroo	Australia	Late 19th C (1875)
Order INSECTIVORA			
Family Nesophontidae			
Nesophontes hypomicrus	Atalaye island-shrew	Hispaniola	Early 16th C?
Nesophontes longirostris	Long-nosed island-shrew	Cuba	Post-1500

Species	Common name	Place	Period
Nesophontes major		Cuba	Post-1500
Nesophontes micrus	Western Cuban island-shrew	Cuba	Post-1500
Nesophontes paramicrus	St Michel island-shrew	Hispaniola	Early 16th C?
Nesophontes submicrus		Cuba	Post-1500
Nesophontes superstes		Cuba	Early 16th C?
Nesophontes zamicrus	Haitian island-shrew	Hispaniola	Early 16th C?
Nesophontes sp. 1	Grand Cayman island-shrew	Grand Cayman	Post-1500
Nesophontes sp. 2	Cayman Brac island-shrew	Cayman Brac	Post-1500
Family Solenodontidae			
Solenodon marcanoi	Marcano's solenodon	Hispaniola	Early 16th C?
Order CHIROPTERA			
Family Molossidae			
Mystacina robusta	New Zealand lesser short-tailed bat	New Zealand	Mid 20th C (1965)
Family Pteropodidae			
Dobsonia chapmani	Dobson's fruit bat	Negros I. (Philippines)	Late 20th C (1970s)
Nyctimene sanctacrucis	Nendo tube-nosed fruit bat	Santa Cruz (Solomons)	Late 19th C (1892)
Pteropus brunneus	Dusky flying fox	Percy I. (Queensland, Australia)	Late 19th C (1874)
Pteropus pilosus	Large Palau flying fox	Palau	Late 19th C (1874)
Pteropus subniger	Réunion flying fox	Mauritius, Réunion	Mid 19th C (1860s)
Pteropus tokudae	Guam flying fox	Guam	Mid 20th C (1968)
Family Vespertilionidae			
Kerivoula africana	Dobson's painted bat	Tanzania	Pre-1878
Pharotis imogene	Large-eared nyctophilus	Papua New Guinea	Late 19th C (1890)
Order PRIMATES			
Family Megaladapidae			
Megaladapis cf. *M. edwardsi*	Tretretretre	Madagascar	Early 16th C?
Family Paleopropithecidae			
Palaeopropithecus ingens	Large sloth lemur	Madagascar	Early 16th C?
Family Xenotrichidae			
Xenothrix mcgregori	Jamaican monkey	Jamaica	Early 16th C?
Order CARNIVORA			
Family Canidae			
Dusicyon australis	Falkland island dog	Falkland Is	Late 19th C (1876)

Species	Common name	Place	Period
Family Mustelidae			
Mustela macrodon	Sea mink	USA	Late 19th C (1860)
Family Phocidae			
Monachus tropicalis	Caribbean monk seal	Caribbean Sea	Mid 20th C (1950s)
Order SIRENIA			
Family Dugongidae			
Hydrodamalis gigas	Steller's sea cow	Russia, USA	Mid 18th C (1768)
Order ARTIODACTYLA			
Family Bovidae			
Gazella rufina	Red gazelle	Algeria	Late 19th C (1894)
Hippotragus leucophaeus	Bluebuck	South Africa	Early 19th C (1800)
Family Hippopotamidae			
Hippopotamus lemerlei	Malagasy dwarf hippo	Madagascar	Early 16th C
Hippopotamus madagascariensis	Common Malagasy hippo	Madagascar	Post-1500
Order RODENTIA			
Family Capromyidae			
Capromys sp. 1	Cayman hutia	Cayman Is	Post-1500
Geocapromys colombianus	Cuban coney	Cuba Little	Early 16th C?
Geocapromys thoractus	Swan Island hutia	Swan I. (Honduras)	Mid 20th C (1950s)
Geocapromys sp. 1	Great Cayman coney	Grand Cayman	Post-1500
Geocapromys sp. 2	Cayman Brac coney	Cayman Brac	Post-1500
Hexolobodon phenax	Imposter hutia	Hispaniola	Early 16th C?
Isolobodon montanus	Montane hutia	Hispaniola	Early 16th C?
Isolobodon portoricensis	Allen's hutia	Hispaniola	Early 16th C?
Plagiodontia ipnaeum	Johnson's hutia	Hispaniola	Early 16th C?
Rhizoplagiodontia lemkei	Lemke's hutia	Hispaniola	Early 16th C?
Family Echimyidae			
Boromys offella	Cuban esculent spiny rat	Cuba	Post-1500
Boromys torrei	De la Torre's esculent spiny rat	Cuba	Post-1500
Brotomys voratus	Muhoy	Hispaniola	Early 16th C?
Family Heptaxodontidae			
Quemisia gravis	Quemi	Hispaniola	Post-1500
Family Muridae			
Conilurus albipes	White-footed rabbit rat	Australia	Late 19th C (1875)
Crateromys paulus	Ilin bushy-tailed cloud-rat	Philippines	1953
Leimacomys buettneri	Groove-toothed forest mouse	Ghana	1890
Leporillus apicalis	Lesser stick-nest rat	Australia	Late 20th C (1970)

Species	Common name	Place	Period
Malpaisomys insularis	Volcano mouse	Canary Is	Post-1500
Megalomys audreyae	Barbuda giant rice rat	Barbuda and Antigua	Pre-1890
Megalomys desmarestii	Martinique rice rat	Martinique	Early 20th C (1902)
Megalomys luciae	St Lucia giant rice rat	Saint Lucia	Late 19th C (1881)
Megaoryzomys curioi	Curio's giant rice rat	Santa Cruz, Galapagos	Mid 18th C (1740)
Megaoryzomys sp. 1	Isabela giant rice rat	Isabela, Galapagos	Post-1500
Nesoryzomys darwini	Darwin's rice rat	Santa Cruz, Galapagos	Mid 20th C (1940s)
Nesoryzomys sp. 2	Isabela Island rice rat 'B'	Isabela, Galapagos	Post-1500
Noronhomys vespuccii	Vespucci's rice rat	Fernando da Noronha I. (Brazil)	Early 16th C?
Notomys amplus	Short-tailed hopping-mouse	Australia	Late 19th C (1896)
Notomys longicaudatus	Long-tailed hopping-mouse	Australia	Early 20th C (1901)
Notomys macrotis	Big-eared hopping-mouse	Australia	1843
Notomys mordax	Darling Downs hopping-mouse	Australia	1846
Notomys sp. 1	Great hopping-mouse	Australia	Pre-1900
Oligoryzomys victus	St Vincent pygmy rice rat	Saint Vincent	Late 19th C (1892)
Oryzomys antillarum	Jamaican rice rat	Jamaica	Late 19th C (1877)
Oryzomys hypenemus	Barbuda rice rat	Barbuda and Antigua	Post-1500
Oryzomys nelsoni	Nelson's rice rat	Maria Madre I. (Mexico)	Late 19th C (1897)
Oryzomys sp. 1	Barbados rice rat	Barbados	Pre-1890
Peromyscus pembertoni	Pemberton's deer mouse	San Pedro Nolasco I. (Mexico)	Early 20th C (1931)
Pseudomys gouldii	Gould's mouse	Australia	Early 20th C (1930)
Rattus macleari	Maclear's rat	Christmas I. (Australia)	Early 20th C (1903)
Rattus nativitatis	Bulldog rat	Christmas I. (Australia)	Early 20th C (1903)
Uromys imperator	Giant naked-tailed rat	Guadalcanal (Solomons)	Mid 20th C (1960s)
Uromys porculus	Little pig rat	Guadalcanal (Solomons)	Late 19th C (1887)

Order LAGOMORPHA
Family Leporidae

Sylvilagus insonus	Omilteme cottontail	Mexico	Late 20th C (1991)

Family Ochotonidae

Prolagus sardus	Sardinian pika	Corsica, Sardinia	Late 18th C (1777)

Species	Common name	Place	Period
Class AVES			
Order CASUARIIFORMES			
Family Dromaiidae			
Dromaius ater	King Island emu	King I. (Australia)	Early 19th C
Dromaius baudinianus	Kangaroo Island emu	Kangaroo I. (Australia)	Early 19th C?
Order PODICIPEDIFORMES			
Family Podicipedidae			
Podiceps andinus	Colombian grebe	Colombia	Late 20th C (1977)
Podilymbus gigas	Atitán grebe	Guatemala	Late 20th C (1986)
Order PROCELLARIIFORMES			
Family Procellariidae			
Bulweria bifax	St Helena bulwer's petrel	St Helena	16th C
Oceanites maorianus	New Zealand storm petrel	South I. (New Zealand)	19th C
Pterodroma rupinarum	St Helena gadfly petrel	St Helena	16th C
Order PELECANIFORMES			
Family Ardeidae			
Ixobrychus novaezelandia	New Zealand little bittern	New Zealand	Late 19th C?
Nycticorax duboisi	Réunion night-heron	Réunion	Late 17th C?
Nycticorax mauritianus	Mauritius night-heron	Mauritius	Early 18th C?
Nycticorax megacephalus	Rodrigues night-heron	Rodrigues	Mid 18th C
Family Phalacrocoracidae			
Phalacrocorax perspicillatus	Pallas's cormorant	Bering Straits (Russia)	Mid 19th C (1850s)
Family Threskiornithidae			
Threskiornis solitarius	Réunion flightless ibis	Réunion	Early 18th C
Order ANSERIFORMES			
Family Anatidae			
Alopochen mauritiania	Mauritian shelduck	Mauritius	Late 17th C (1698)
Anas marecula	Amsterdam Island duck	Amsterdam I. (France)	Early 19th C? (1793)
Anas theodori	Mauritian duck	Mauritius, Réunion?	Early 18th C? (1710)
Camptorhynchus labradorius	Labrador duck	Canada, USA	Late 19th C (1878)
Mergus australis	Auckland island merganser	New Zealand	Early 20th C (1902)

Species	Common name	Place	Period
Order FALCONIFORMES			
Family Falconidae			
Cracara lutosus	Guadalupe caracara	Guadalupe (Mexico)	Early 20th C? (1900)
Order GALLIFORMES			
Family Phasianidae			
Argusianus bipunctatus	Double-banded argus	Southeast Asia?	Late 19th C?
Coturnix novaezelandiae	New Zealand quail	New Zealand	Late 19th C (1875)
Order GRUIFORMES			
Family Rallidae			
Aphanapteryx bonasia	Red rail	Mauritius	Early 18th C? (1700)
Aphanapteryx leguati	Rodrigues rail	Rodrigues	Mid 18th C (1761)
Aramides gutturalis	Red-throated wood rail	Peru	19th C?
Atlantisia elpenor	Ascension flightless crake	Ascension I.	Early 19th C?
Atlantisia podarces	St Helena crake	St Helena	Early 16th C?
Cabalus modestus	Chatham rail	New Zealand	Late 19th C (1900?)
Fulica newtoni	Mascarene coot	Mauritius, Réunion	Early 18th C? (1693)
Gallinula nesiotis	Tristan moorhen	Tristan da Cunha	Late 19th C
Gallirallus dieffenbachii	Dieffenbach's rail	Chatham Is (New Zealand)	Late 19th C (1872)
Gallirallus pacificus	Tahiti rail	Society Is (French Polynesia)	Early 20th C (1930s)
Gallirallus sharpei	Sharpe's rail	Indonesia?	Late 19th or early 20th C?
Gallirallus wakensis	Wake Island rail	Wake I. (USA)	Mid 20th C (1945)
Nesoclopeus poecilopterus	Bar-winged rail	Fiji	Late 20th C (1973)
Porphyrio albus	Lord Howe swamphen	Lord Howe I. (Australia)	Early 19th C?
Porphyrio coerulescens	Réunion gallinule	Réunion	Early 18th C (1730)
Porphyrio hochstetteri	North Island takahe	North I., New Zealand	Late 19th C?
Porphyrio kukwiedei	New Caledonia gallinule	New Caledonia	17th C?
Porzana astrictocarpus	St Helena rail	St Helena	Early 16th C
Porzana monasa	Kosrae crake	Fed. States Micronesia	Mid 19th C
Porzana nigra	Miller's rail	Society Is (French Polynesia)	Late 18th C
Porzana palmeri	Laysan crake	Hawaii	Mid 20th C (1944)
Porzana sandwichensis	Hawaiian crake	Hawaii	Late 19th C (1884)
Order CHARADRIIFORMES			
Family Charadriidae			
Haematopus meadewaldoi	Canary Islands oystercatcher	Canary Is	Mid or late 20th C

Species	Common name	Place	Period
Family Laridae			
Pinguinus impennis	Great auk	North Atlantic coasts	Mid 19th C (1852)
Family Scolopacidae			
Prosobonia ellisi	White-winged sandpiper	Society Is (French Polynesia)	Late 18th C
Prosobonia leucoptera	Tahitian sandpiper	Society Is (French Polynesia)	Late 18th C?
Order COLUMBIFORMES			
Family Columbidae			
Alectroenas nitidissima	Mauritius blue pigeon	Mauritius	Early 19th C (1830s)
Alectroenas rodericana	Rodrigues pigeon	Rodrigues (Mauritius)	Early 18th C?
Columba duboisi	Réunion pigeon	Réunion	Early 18th C?
Columba jouyi	Ryukyu pigeon	Nansei-shoto (Japan)	Early 20th C (1936)
Columba versicolor	Bonin wood pigeon	Ogasawara-shoto (Japan)	Late 19th C (1889)
Dysmoropelia dekarchiskos	St Helena dove	St Helena	Early 16th C?
Ectopistes migratorius	Passenger pigeon	USA	Early 20th C (1900)
Gallicolumba ferruginea	Tanna ground dove	Vanuatu	Late 18th C
Gallicolumba norfolciensis	Norfolk Island ground dove	Norfolk I. (Australia)	Late 18th C?
Microgoura meeki	Choiseul pigeon	Choiseul (Solomon Is)	Early 20th C (1904)
Ptilinopus mercierii	Red-moustached fruit-dove	Marquesas Is (French Polynesia)	Early 20th C (1922)
Family Raphidae			
Pezophaps solitaria	Rodrigues solitaire	Rodrigues	Late 18th C (1760s)
Raphus cucullatus	Dodo	Mauritius	Late 17th C (1665)
Order PSITTACIFORMES			
Family Psittacidae			
Amazona martinicana	Martinique parrot	Martinique	Late 18th C
Amazona violacea	Guadeloupe parrot	Guadeloupe	Late 18th C (1779)
Ara atwoodi	Dominican green-and-yellow macaw	Dominica	Early 19th C?
Ara erythrocephala	Jamaican green-and-yellow macaw	Jamaica	Early 19th C?
Ara gossei	Jamaican red macaw	Jamaica	Late 18th C?
Ara guadeloupensis	Lesser Antillean macaw	Guadeloupe, Martinique	Late 18th C?

Species	Common name	Place	Period
Ara tricolor	Cuban red macaw	Cuba, Hispaniola	Late 19th C (1885)
Aratinga labati	Guadeloupe parrot	Guadeloupe	Late 18th C
Conuropsis carolinensis	Carolina parakeet	USA	Early 20th C (1904)
Cyanoramphus ulietanus	Raiatea parakeet	Society Is (French Polynesia)	Late 18th C? (1773)
Cyanoramphus zealandicus	Black-fronted parakeet	Society Is (French Polynesia)	Mid 19th C (1844)
Lophopsittacus bensoni	Mauritius gray parrot	Mauritius	Late 18th C (1764)
Lophopsittacus mauritianus	Broad-billed parrot	Mauritius	Late 17th C (1675)
Mascarinus mascarinus	Mascarene parrot	Réunion	Late 18th C (1770s)
Necropsittacus rodericanus	Rodrigues parrot	Rodrigues (Mauritius)	Late 18th C (1761)
Nestor productus	Norfolk Island kaka	Phillip I. (Australia)	Early 19th C
Psephotus pulcherrimus	Paradise parrot	Australia	Mid 20th C (1927)
Psittacula exsul	Newton's parakeet	Rodrigues (Mauritius)	Late 19th C (1875)
Psittacula wardi	Seychelles parrot	Seychelles	Late 19th C (1880s)
Order CUCULIFORMES			
Family Cuculidae			
Coua delalandei	Snail-eating coua	Madagascar	Mid 19th C (1834)
Nannococcyx psix	St Helena cuckoo	St Helena	18th C?
Order STRIGIFORMES			
Family Strigidae			
Mascarenotus grucheti	Réunion owl	Réunion	Early 17th C?
Mascarenotus murivorus	Rodrigues owl	Rodrigues	Early 18th C (1726)
Mascarenotus sauzieri	Mauritius owl	Mauritius	Mid 19th C (1837)
Sceloglaux albifacies	Laughing owl	New Zealand	Early 20th C (1914)
Order APODIFORMES			
Family Trochilidae			
Chlorostilbon bracei	Brace's emerald	Bahamas	Late 19th C (1877)
Chlorostilbon elegans	Gould's emerald	Bahamas? Jamaica?	Late 19th C?
Order UPUPIFORMES			
Family Upupidae			
Upupa antaois	St Helena hoopoe	St Helena	Early 16th C?
Order PASSERIFORMES			
Family Acanthisittidae			
Traversia lyalli	Stephens Island wren	Stephens I. (New Zealand)	Late 19th C (1894)
Xenicus longipes	Bush wren	New Zealand	Late 20th C (1972)

Species	Common name	Place	Period
Family Callaeidae			
Heteralocha acutirostris	Huia	New Zealand	Early 20th C (1907)
Family Drepanididae			
Akialoa ellisiana	Oahu 'akialoa	Oahu, Hawaii	Early 20th C ? (1837)
Akialoa lanaiensis	Maui Nui 'akialoa	Lanai, Hawaii	Early 20th C ? (1892)
Akialoa obscura	'Akialoa	Hawaii	Mid 20th C ? (1940)
Akialoa stejnegeri	Kaua'i 'akialoa	Kaua'i, Hawaii	Late 20th C ? (1969)
Ciridops anna	Ula-ai-hawane	Hawaii	Late 19th C (1892)
Drepanis funerea	Black mamo	Hawaii	Early 20th C (1907)
Drepanis pacifica	Hawaii mamo	Hawaii	Late 19th C (1899)
Dysmorodrepanis munroi	Lana'i hookbill	Hawaii	Early 20th C (1920)
Paroreomyza flammea	Kakawihie	Hawaii	Mid 20th C (1963)
Psittirostra kona	Kona grosbeak	Hawaii	Late 19th C (1894)
Rhodacanthis flaviceps	Lesser koa-finch	Hawaii	Late 19th C (1891)
Rhodacanthis palmeri	Greater koa-finch	Hawaii	Late 19th C (1896)
Viridonia sagittirostris	Greater 'amakihi	Hawaii	Early 20th C (1901)
Family Fringillidae			
Chaunoproctus ferreorostris	Bonin grosbeak	Ogasawara-shoto (Japan)	Late 19th C (1890?)
Family Icteridae			
Quiscalus palustris	Slender-billed grackle	Mexico	Early 20th C (1910)
Family Meliphagidae			
Chaetoptila angustipluma	Kioea	Hawaii	Late 19th C (1859)
Moho apicalis	Oahu oo	Hawaii	Mid 19th C (1837)
Moho bishopi	Bishop's oo	Hawaii	Early 20th C (1904)
Moho braccatus	Kaua'i oo	Kaua'i, Hawaii	Late 20th C (1987)
Moho nobilis	Hawaii oo	Hawaii	Early 20th C? (1934)
Family Muscicapidae			
Anthornis melanocephala	Chatham Island bellbird	New Zealand	Early 20th C (1906)
Bowdleria rufescens	Chatham Island fernbird	Chatham Is (New Zealand)	Early 20th C ? (1900)
Gerygone insularis	Lord Howe gerygone	Lord Howe I., Australia	Early 20th C
Myadestes oahensis	'Amaui	Hawaii	Mid 19th C? (1825)
Myiagra freycineti	Guam flycatcher	Guam	Late 20th C (1983)
Nesillas aladabrana	Aldabra bush-warbler	Aldabra (Seychelles)	Late 20th C (1983)
Pomarea pomarea	Maupiti monarch	Maupiti, French Polynesia	Early 19th C? (1823)
Turdus ravidus	Grand Cayman thrush	Cayman Is	Early 20th C (1938)
Turnagra capensis	South Island piopio	New Zealand	Mid 20th C (1963)
Turnagra turnagra	North Island piopio	New Zealand	Mid 20th C (1955)

Species	Common name	Place	Period
Zoothera terrestris	Bonin thrush	Ogasawara-shoto (Japan)	Mid 19th C? (1828)
Family Sturnidae			
Aplonis corvina	Kosrae mountain starling	Kosrae (Fed. States Micronesia)	Mid 19th C (1828)
Aplonis fusca	Norfolk Island starling	Norfolk I. (Australia)	Early 20th C (1923)
Aplonis mavornata	Mysterious starling	Cook Is	Mid 19th C (1825)
Fregilupus varius	Réunion starling	Réunion	Mid 19th C (1850s)
Necrospar rodericanus	Rodrigues starling	Rodrigues	Early 18th C (1726)
Family Zosteropidae			
Zosterops strenuus	Robust white-eye	Lord Howe I. (Australia)	Early 20th C (1928)
Class REPTILIA			
Order SAURIA			
Family Anguidae			
Celestus occiduus	Jamaican giant galliwasp	Jamaica	Mid 19th C (1840)
Family Gekkonidae			
Hoplodactylus delcourti		New Zealand (?)	Mid 19th C?
Phelsuma gigas	Giant day gecko	Rodrigues	Late 19th C
Family Iguanidae			
Leiocephalus eremitus		Navassa I. (USA)	Early 20th C (1900)
Leiocephalus herminieri		Martinique	Early 19th C (1830s)
Family Scincidae			
Leiolopisma mauritiana		Mauritius	Early 17th C (1600)
Macroscincus coctei	Cape Verde giant skink	Cape Verde	Early 20th C
Tachygia microlepis	Tongan ground skink	Tonga	17th C ?
Tetradactylus eastwoodae	Eastwood's longtailed seps	South Africa	Early 20th C ?
Family Teiidae			
Ameiva cineracea		Guadeloupe	Early 20th C
Ameiva major	Martinique giant ameiva	Martinique	17th C ?
Order SERPENTES			
Family Boidae			
Bolyeria multocarinata		Round I. (Mauritius)	Late 20th C (1975)
Family Colubridae			
Alsophis sancticrucis	St Croix racer	Virgin Is (USA)	Mid 20th C

Species	Common name	Place	Period
Family Typhlopidae			
Typhlops cariei		Mauritius	17th C ?
Order TESTUDINES			
Family Testudinidae			
Cylindraspis borbonica		Réunion	Early 19th C (1800)
Cylindraspis indica		Réunion	Early 19th C (1800)
Cylindraspis inepta		Mauritius	Early 18th C
Cylindraspis peltastes		Rodrigues	Early 19th C (1800)
Cylindraspis triserrata		Mauritius	Early 18th C
Cylindraspis vosmaeri		Rodrigues	Early 19th C (1800)
Class AMPHIBIA			
Order ANURA			
Family Discoglossidae			
Discoglossus nigriventer	Israel painted frog	Israel	Mid 20th C (1940)
Family Myobatrachidae			
Uperoleia marmorata	Marbled toadlet	Australia	Late 19th C
Family Ranidae			
Arthroleptides dutoiti		Kenya	Early 20th C
Rana fisheri	Relict leopard frog	USA	Mid 20th C (1960)
Rana tlaloci		Mexico	Late 20th C (1990s)
Class ACTINOPTERYGII			
Order CYPRINIFORMES			
Family Catostomidae			
Chasmistes muriei	Snake River sucker	USA	Early 20th C (1927)
Moxostoma lacerum	Harelip sucker	USA	Late 19th C (1893)
Family Cyprinidae			
Acanthobrama hulensis		Lake Huleh (Israel)	Late 20th C ?
Anabarilius alburnops		Lake Dianchi (China)	20th C
Anabarilius polylepis		Lake Dianchi (China)	20th C
Barbus microbarbis		Lake Luhondo (Rwanda)	Mid 20th C
Cephalakompsus pachycheilus		Lake Lanao (Philippines)	Mid 20th C ? (post 1921)
Chondrostoma scodrensis		Lake Skadar (Albania, Yugoslavia)	20th C ? (post 1881)
Cyprinus yilongensis		Lake Yilong (China)	Late 20th C (1977)

Species	Common name	Place	Period
Evarra bustamantei	Mexican dace	Mexico	Mid 20th C (1957)
Evarra eigenmanni	Plateau dace	Mexico	Mid 20th C (1954)
Evarra tlahuacensis	Endorheic dace	Mexico	Mid 20th C (1957)
Gila crassicauda	Thicktail chub	USA	Mid 20th C (1957)
Hybopsis amecae	Ameca shiner	Mexico	Late 20th C (1969)
Hybopsis aulidion	Durango shiner	Mexico	Late 20th C
Lepidomeda altivelis	Pahranagat spinedace	USA	Mid 20th C (1938)
Mandibularca resinus	Bagangan	Lake Lanao (Philippines)	Early 20th C (1922)
Notropis amecae	Ameca shiner	Mexico	Late 20th C (1970s?)
Notropis aulidion	Durango shiner	Mexico	Mid 20th C (1961)
Notropis orca	Phantom shiner	Mexico, USA	Late 20th C ?
Ospatulus palaemorphagus		Lake Lanao (Philippines)	Early 20th C (1924)
Ospatulus trunculatus	Bitungu	Lake Lanao (Philippines)	Early 20th C (1921)
Phoxinellus egridiri	Yag baligi	Lake Egridir (Turkey)	Mid 20th C (1955)
Phoxinellus handlirschi	Çiçek	Lake Egridir (Turkey)	Mid 20th C (1955)
Pogonichthys ciscoides	Clear Lake splittail	USA	Late 20th C (1970)
Puntius amarus	Pait	Lake Lanao (Philippines)	Early 20th C (1910)
Puntius baoulan	Baolan	Lake Lanao (Philippines)	Early 20th C (1926)
Puntius clemensi	Bagangan	Lake Lanao (Philippines)	Early 20th C (1921)
Puntius disa	Disa	Lake Lanao (Philippines)	Early 20th C (1932)
Puntius flavifuscus	Katapa-tapa	Lake Lanao (Philippines)	Early 20th C (1921)
Puntius herrei		Lake Lanao (Philippines)	Early 20th C (1908)
Puntius lanaoensis	Kandar	Lake Lanao (Philippines)	Early 20th C (1922)
Puntius manalak	Manalak	Lake Lanao (Philippines)	Early 20th C (1924)
Puntius sirang	Sirang	Lake Lanao (Philippines)	Early 20th C (1932)
Puntius tras	Tras	Lake Lanao (Philippines)	Early 20th C (1925)
Rhinichthys deaconi	Las Vegas dace	USA	Mid 20th C (1940)
Sprattlicypris palata	Palata	Lake Lanao (Philippines)	Early 20th C (1922)
Stypodon signifer	Stumptooth minnow	Mexico	Early 20th C (1903)

Species	Common name	Place	Period
Order CHARACIFORMES			
Family Characidae			
Brycyon acuminatus		Brazil	20th C
Order OSMERIFORMES			
Family Retropinnidae			
Prototroctes oxyrhynchus	New Zealand grayling	New Zealand	Early 20th C (1923)
Order SILURIFORMES			
Family Schilbeidae			
Platytropius siamensis		Thailand	Late 20th C (1966)
Family Trichomycteridae			
Rhizosomichthys totae		Lake Tota (Colombia)	Mid 20th C (1957)
Order SALMONIFORMES			
Family Salmonidae			
Coregonus alpenae	Longjaw cisco	Great Lakes (Canada, USA)	Late 20th C (1975)
Coregonus confusus	Pärrig	Lake Morat (Switzerland)	Mid 20th C
Coregonus fera	Féra	Lake Geneva (Switzerland)	Mid 20th C
Coregonus gutturosus	Kilch	Lake Constance (Switzerland)	Mid 20th C
Coregonus hiemalis	Gravenche	Lake Geneva (Switzerland)	Mid 20th C
Coregonus johannae	Deepwater cisco	Great Lakes (Canada, USA)	Mid 20th C (1952)
Coregonus restrictus	Férit	Lake Morat (Switzerland)	Mid 20th C
Salvelinus agassizi	Silver trout	Dublin Pond (USA)	Early 20th C (1930)
Salvelinus inframundis	Orkney char	Hoy I. (UK)	Early 20th C (1909)
Salvelinus scharffi	Scharff's char	Lough Owel (Ireland)	Early 20th C (1908)
Order ATHERINIFORMES			
Family Atherinidae			
Chirostoma compressum	Cuitzeo silverside	Lake Cuitzeo (Mexico)	Mid 20th C ? (1940)
Rheocles sikorae	Zona	Madagascar	Mid 20th C (1966)
Order CYPRINODONTIFORMES			
Family Cyprinodontidae			
Cyprinodon ceciliae	Cachorrito de la Presa	Mexico	Late 20th C (1988)
Cyprinodon inmemoriam	Cachorrito de la Trinidad	Mexico	Late 20th C (1984)

Species	Common name	Place	Period
Cyprinodon latifasciatus	Perrito de Parras	Mexico	Early 20th C (1903)
Cyprinodon sp. ?	Monkey Spring pupfish	USA	Late 20th C
Leptolebias marmoratus	Ginger pearlfish	Brazil	Mid 20th C (1944)
Orestias cuvieri	Lake Titicaca orestias	Lake Titicaca (Bolivia, Peru)	Mid 20th C (1937)
Family Fundulidae			
Fundulus albolineatus	Whiteline topminnow	USA	Early 20th C ? (1899)
Family Goodeidae			
Characodon garmani	Parras characodon	Mexico	Late 19th C (1889)
Empetrichthys merriami	Ash Meadows killifish	USA	Mid 20th C (1948)
Family Poeciliidae			
Gambusia amistadensis	Amistad gambusia	USA	Mid 20th C (1968)
Order BELONIFORMES			
Family Adrianichthyidae			
Adrianichthys kruyti	Duck-billed buntingi	Lake Poso (Indonesia)	Late 20th C (1983)
Order GASTEROSTEIFORMES			
Family Gasterosteidae			
Pungitius kaibarae	Kyoto ninespine stickleback	Japan	Mid 20th C (1959)
Order SCORPAENIFORMES			
Family Cottidae			
Cottus echinatus	Utah Lake sculpin	Utah Lake (USA)	Early 20th C (1928)
Order PERCIFORMES			
Family Cichlidae			
Tristramella magdalenae		Israel	Late 20th C (1997)
Family Gobiidae			
Weberogobius amadi	Poso bungu	Lake Poso, Sulawesi (Indonesia)	Late 20th C (1985)
Haplochromis altigenis		Lake Victoria	Mid/late 20th C ?
Haplochromis apogonoides		Lake Victoria	Mid/late 20th C ?
Haplochromis arcanus		Lake Victoria	Mid/late 20th C ?
Haplochromis argenteus		Lake Victoria	Mid/late 20th C ?
Haplochromis artaxerxes		Lake Victoria	Mid/late 20th C ?
Haplochromis barbarae		Lake Victoria	Mid/late 20th C ?
Haplochromis bareli		Lake Victoria	Mid/late 20th C ?
Haplochromis bartoni		Lake Victoria	Mid/late 20th C ?
Haplochromis bayoni		Lake Victoria	Mid/late 20th C ?
Haplochromis boops		Lake Victoria	Mid/late 20th C ?

Species	Common name	Place	Period
Haplochromis cassius		Lake Victoria	Mid/late 20th C ?
Haplochromis cinctus		Lake Victoria	Mid/late 20th C ?
Haplochromis cnester		Lake Victoria	Mid/late 20th C ?
Haplochromis decticostoma		Lake Victoria	Mid/late 20th C ?
Haplochromis dentex group		Lake Victoria	Mid/late 20th C ?
Haplochromis diplotaenia		Lake Victoria	Mid/late 20th C ?
Haplochromis estor		Lake Victoria	Mid/late 20th C ?
Haplochromis flavipinnis		Lake Victoria	Mid/late 20th C ?
Haplochromis gilberti		Lake Victoria	Mid/late 20th C ?
Haplochromis gowersi		Lake Victoria	Mid/late 20th C ?
Haplochromis guiarti		Lake Victoria	Mid/late 20th C ?
Haplochromis heusinkveldi		Lake Victoria	Mid/late 20th C ?
Haplochromis hiatus		Lake Victoria	Mid/late 20th C ?
Haplochromis iris		Lake Victoria	Mid/late 20th C ?
Haplochromis longirostris		Lake Victoria	Mid/late 20th C ?
Haplochromis macrognathus		Lake Victoria	Mid/late 20th C ?
Haplochromis maculipinna		Lake Victoria	Mid/late 20th C ?
Haplochromis mandibularis		Lake Victoria	Mid/late 20th C ?
Haplochromis martini		Lake Victoria	Mid/late 20th C ?
Haplochromis megalops		Lake Victoria	Mid/late 20th C ?
Haplochromis michaeli		Lake Victoria	Mid/late 20th C ?
Haplochromis microdon		Lake Victoria	Mid/late 20th C ?
Haplochromis mylergates		Lake Victoria	Mid/late 20th C ?
Haplochromis nanoserranus		Lake Victoria	Mid/late 20th C ?
Haplochromis nigrescens		Lake Victoria	Mid/late 20th C ?
Haplochromis nyanzae		Lake Victoria	Mid/late 20th C ?
Haplochromis obtusidens		Lake Victoria	Mid/late 20th C ?
Haplochromis pachycephalus		Lake Victoria	Mid/late 20th C ?
Haplochromis paraguiarti		Lake Victoria	Mid/late 20th C ?
Haplochromis paraplagiostoma		Lake Victoria	Mid/late 20th C ?
Haplochromis parorthostoma		Lake Victoria	Mid/late 20th C ?
Haplochromis percoides		Lake Victoria	Mid/late 20th C ?
Haplochromis pharyngomylus		Lake Victoria	Mid/late 20th C ?
Haplochromis prognathus		Lake Victoria	Mid/late 20th C ?
Haplochromis pseudopellegrini		Lake Victoria	Mid/late 20th C ?
Haplochromis pyrrhopteryx		Lake Victoria	Mid/late 20th C ?
Haplochromis spekii		Lake Victoria	Mid/late 20th C ?
Haplochromis teegelaari		Lake Victoria	Mid/late 20th C ?
Haplochromis thuragnathus		Lake Victoria	Mid/late 20th C ?
Haplochromis tridens		Lake Victoria	Mid/late 20th C ?
Haplochromis victorianus		Lake Victoria	Mid/late 20th C ?
Haplochromis xenostoma		Lake Victoria	Mid/late 20th C ?
Haplochromis 'bartoni-like'		Lake Victoria	Mid/late 20th C ?
Haplochromis 'bicolor'		Lake Victoria	Mid/late 20th C ?
Haplochromis 'big teeth'		Lake Victoria	Mid/late 20th C ?
Haplochromis 'black cryptodon'		Lake Victoria	Mid/late 20th C ?
Haplochromis 'back pectoral'		Lake Victoria	Mid/late 20th C ?

Species	Common name	Place	Period
Haplochromis 'chlorocephalus'		Lake Victoria	Mid/late 20th C ?
Haplochromis 'citrus'		Lake Victoria	Mid/late 20th C ?
Haplochromis 'coop'		Lake Victoria	Mid/late 20th C ?
Haplochromis 'elongate rockpicker'		Lake Victoria	Mid/late 20th C ?
Haplochromis 'filamentus'		Lake Victoria	Mid/late 20th C ?
Haplochromis 'fleshy lips'		Lake Victoria	Mid/late 20th C ?
Haplochromis 'gray pseudo-nigricans'		Lake Victoria	Mid/late 20th C ?
Haplochromis 'large eye guiarti'		Lake Victoria	Mid/late 20th C ?
Haplochromis 'lividus-frels'		Lake Victoria	Mid/late 20th C ?
Haplochromis 'longurius'		Lake Victoria	Mid/late 20th C ?
Haplochromis 'macrops like'		Lake Victoria	Mid/late 20th C ?
Haplochromis 'micro-obesus'		Lake Victoria	Mid/late 20th C ?
Haplochromis 'morsei'		Lake Victoria	Mid/late 20th C ?
Haplochromis 'orange cinereus'		Lake Victoria	Mid/late 20th C ?
Haplochromis 'orange macula'		Lake Victoria	Mid/late 20th C ?
Haplochromis 'orange yellow big teeth'		Lake Victoria	Mid/late 20th C ?
Haplochromis 'orange yellow small teeth'		Lake Victoria	Mid/late 20th C ?
Haplochromis 'paropius-like'		Lake Victoria	Mid/late 20th C ?
Haplochromis 'pink paedophage'		Lake Victoria	Mid/late 20th C ?
Haplochromis 'pseudo-morsei'		Lake Victoria	Mid/late 20th C ?
Haplochromis 'purple head'		Lake Victoria	Mid/late 20th C ?
Haplochromis 'purple miller'		Lake Victoria	Mid/late 20th C ?
Haplochromis (?) 'purple rocker'		Lake Victoria	Mid/late 20th C ?
Haplochromis 'red empodisma'		Lake Victoria	Mid/late 20th C ?
Haplochromis 'red eye scraper'		Lake Victoria	Mid/late 20th C ?
Haplochromis 'reginus'		Lake Victoria	Mid/late 20th C ?
Haplochromis 'regius'		Lake Victoria	Mid/late 20th C ?
Haplochromis 'short supramacrops'		Lake Victoria	Mid/late 20th C ?
Haplochromis 'small blue zebra'		Lake Victoria	Mid/late 20th C ?
Haplochromis 'small empodisma'		Lake Victoria	Mid/late 20th C ?
Haplochromis 'smoke'		Lake Victoria	Mid/late 20th C ?
Haplochromis 'soft gray'		Lake Victoria	Mid/late 20th C ?
Haplochromis 'stripmac'		Lake Victoria	Mid/late 20th C ?
Haplochromis 'supramacrops'		Lake Victoria	Mid/late 20th C ?
Haplochromis 'theliodon-like'		Lake Victoria	Mid/late 20th C ?
Haplochromis 'tigrus'		Lake Victoria	Mid/late 20th C ?
Haplochromis 'too small'		Lake Victoria	Mid/late 20th C ?
Haplochromis 'twenty'		Lake Victoria	Mid/late 20th C ?
Haplochromis 'two stripe white lip'		Lake Victoria	Mid/late 20th C ?
Haplochromis 'wyber'		Lake Victoria	Mid/late 20th C ?
Haplochromis 'xenognathus-like'		Lake Victoria	Mid/late 20th C ?
Haplochromis 'yellow'		Lake Victoria	Mid/late 20th C ?
Haplochromis 'yellow-blue'		Lake Victoria	Mid/late 20th C ?
Hoplotilapia retrodens		Lake Victoria	Mid/late 20th C ?
Psammochromis cryptogramma group		Lake Victoria	Mid/late 20th C ?

REFERENCES

1   BirdLife International 2000. *Threatened birds of the world*. Lynx Edicions and BirdLife International, Barcelona and Cambridge.

2   CREO. Committee on Recently Extinct Organisms (CREO) website: http://creo.amnh.org/ (accessed January 2002).

3   Hilton-Taylor, C. (compiler) 2000. *2000 IUCN Red List of threatened species*. IUCN, Gland and Cambridge. Online at http://www.redlist.org/ (accessed April 2002).

APPENDIX 5:
BIODIVERSITY AT COUNTRY LEVEL

This table includes estimates of the number of
mammals, breeding birds and vascular plants in
each country of the world, together with estimates
of the number of these endemic to each country.
The threat status of mammals and birds has been
comprehensively assessed, and for these classes
the number and percentage of globally threatened
species present in each country are given. In the
mammals, most estimates of the total number
present relate to native non-marine species only,
but the threatened species counts for many
countries include marine species (a few island
countries have a small number of non-marine
mammals but a higher number of threatened
species – the percentage figure is omitted in such
cases). The richness and endemism data should

be taken as provisional; they are subject to
change with new taxonomic treatments and
new surveys. The columns headed DI and AI,
respectively, contain the unweighted national
diversity indices, and the diversity indices adjusted
for area (see methodology notes below for
details). The index is based on data for the groups
shown here, plus reptiles and amphibians, not
included here. The AI column is the basis for
Map 5.4.

Source: WCMC database; data derived from a large number of
published and unpublished sources, including country reports and
regional checklists. Numbers of threatened species retrieved
from the online version of the *2000 IUCN Red List of threatened
species*, at http://www.redlist.org/ (accessed March 2002).

Methodology notes

Numerical indices representing national biodiversity
have been derived from available estimates of species
total and species endemism. Taxonomic groups covered
are: mammals, birds, reptiles, amphibians and vascular
plants (ferns to flowering plants). For birds alone, the
database includes estimates of both the total number
recorded (i.e. including non-breeding migrants and
accidental visitors) and the number of breeding bird
species; the latter has been used in the analysis. All
countries below 5 000 km^2 in area have been excluded
from the analysis, leaving 169 countries.

Four arbitrary assumptions are made: the four
vertebrate classes included are of equal importance;
plants are of equal importance to the vertebrates
combined; richness and endemism are reasonably
well correlated at country level across the taxonomic
groups covered; vertebrates plus plants provide
a valid surrogate for biodiversity in general.

Because interest lies in relative biodiversity rather
than absolute values, data in each column are first
normalized. Each estimate Ni for each country i is
divided by $Nmax$ where $Nmax$ is the highest value for

that parameter. This transforms the data to within the
range 0-1, with the most important country having the
value $Nmax/Nmax$ = 1, and the least important having
the value closest to zero. Estimates for mean vertebrate
richness (VR) and mean vertebrate endemism (VE) are
derived by averaging the figures for all classes, and
estimates for combined richness (R) and combined
endemism (E) are derived for each country by
averaging figures for vertebrates and plants. Inspection
of the data shows that PR and VR correlate quite
closely, while PE tends to be approximately half VE;
where estimates of PR and PE are missing the
appropriate vertebrate-based value has been inserted
before calculating R and E.

An overall diversity index (DI) is calculated for each
country as the mean of R and E. This treats richness
and endemism equally and so makes fewest
assumptions about their relative significance in terms of
overall biodiversity, but the DI could be weighted to give
greater importance to either. Because species richness
tends to increase with area, and with proximity to the
humid tropics, DI is strongly affected by country area

and geographical position, but it also takes account of levels of endemism, which are shaped by several factors, including topography, geographic isolation and tectonic history.

Because area is an important determinant of species number, there is much interest in evaluating relative levels of biodiversity per unit area, i.e. how much more or less rich in species is any given area (or country). This may be addressed by the Arrhenius equation describing the species-area relationship ($\log S = g + z \log A$), where S = number of species, A = area, z is the slope of the line, and g another

Country	Area km²	DI	AI	Mammals total	Mammals endemic	Mammals no. theatened
Afghanistan	652 225	0.063	− 0.296	119	2	13
Albania	28 750	0.035	− 0.019	68	0	3
Algeria	2 381 745	0.045	− 1.003	92	2	13
American Samoa	197			3	0	3
Andorra	465			44	0	1
Angola	1 246 700	0.176	0.544	276	7	18
Anguilla	91			3	0	0
Antigua and Barbuda	442			7	0	0
Argentina	2 777 815	0.196	0.423	320	49	32
Armenia	29 800	0.042	0.153	84	3	7
Aruba	193			-	0	1
Australia	7 682 300	0.608	1.268	252	206	63
Austria	83 855	0.036	− 0.293	83	0	9
Azerbaijan	86 600	0.05	0.027	99	0	13
Bahamas	13 865	0.017	− 0.503	12	3	5
Bahrain	661			17	0	1
Bangladesh	144 000	0.059	0.058	125	0	21
Barbados	430			6	0	0
Belarus	207 600	0.029	− 0.771	74	0	5
Belgium	30 520	0.023	− 0.441	58	0	11
Belize	22 965	0.056	0.526	125	0	4
Benin	112 620	0.08	0.437	188	0	7
Bermuda	54			3	0	2
Bhutan	46 620	0.058	0.366	160	0	20
Bolivia	1 098 575	0.239	0.882	316	16	23
Bosnia and Herzegovina	51 129	0.034	− 0.200	72	0	10
Botswana	575 000	0.062	− 0.287	164	0	5
Brazil	8 511 965	0.74	1.436	394	119	79
British Ind. Oc. Terr.				-	0	0
Brunei	5 765	0.071	1.145	157	0	9
Bulgaria	110 910	0.044	− 0.167	81	0	15
Burkina Faso	274 122	0.068	0.011	147	0	7
Burundi	27 835	0.072	0.723	107	0	5
Cambodia	181 000	0.059	0.001	123	0	21
Cameroon	475 500	0.167	0.762	409	14	37
Canada	9 922 385	0.067	− 1.014	193	7	14
Cape Verde	4 035			5	0	3
Cayman Islands	259			8	0	0
Central African Republic	624 975	0.08	− 0.058	209	2	12

constant. If it is assumed, given its area dependence, that DI is scaled in the same way as S, a regression analysis of log DI and log A allows the constants g and z to be calculated. The regression line establishes the expected biodiversity value of each country for its area, and the distance of each country point from the line

gives a measure (AI) of how much more (+ve) or less (-ve) diverse is a given country than expected. AI attempts to assess diversity per unit area, rather than overall biodiversity value per country, and it is noticeable that several smaller mainland countries and island states move up in rank order, as would be expected.

0 = zero; - = no data.

Mammals % threatened	Birds breeding total	Birds endemic	Birds no. threatened	Birds % threatened	Plants total	Plants endemic
11	235	0	11	5	4 000	800
4	230	0	3	1	3 031	24
14	192	1	6	3	3 164	250
100	34	0	2	6	471	15
2	113	0	0	0	1 350	-
7	765	12	15	2	5 185	1 260
0	-	0	0		321	1
0	49	0	1	2	1 158	22
10	897	19	38	4	9 372	1 100
8	242	0	4	2	3 553	-
	48	0	0	0	460	25
25	649	350	32	5	15 638	14 074
11	213	0	3	1	3 100	35
13	248	0	8	3	4 300	240
42	88	3	4	5	1 111	118
6	28	0	6	21	195	-
17	295	0	23	8	5 000	-
0	24	0	1	4	572	3
7	221	0	3	1	2 100	-
19	180	0	2	1	1 550	1
3	356	0	2	1	2 894	150
4	307	0	2	1	2 500	-
67	8	1	2	25	167	15
13	448	0	12	3	5 468	75
7	-	18	27		17 367	4 000
14	218	0	3	1	-	-
3	386	1	7	2	2 151	17
20	1 492	185	113	8	56 215	-
	14	0	0	0	101	-
6	359	0	15	4	6 000	7
19	240	0	10	4	3 572	320
5	335	0	2	1	1 100	-
5	451	0	7	2	2 500	-
17	307	0	19	6	-	-
9	690	8	15	2	8 260	156
7	426	5	8	2	3 270	147
60	38	4	2	5	774	86
0	45	0	1	2	539	19
6	537	1	3	1	3 602	100

Country	Area km²	DI	AI	Mammals total	Mammals endemic	Mammals no. theatened
Chad	1 284 000	0.049	− 0.739	134	1	17
Chile	751 625	0.112	0.229	91	16	21
China	9 597 000	0.392	0.767	394	83	76
Colombia	1 138 915	0.538	1.685	359	34	36
Comoros	1 860			12	2	2
Congo, Dem. Rep.	2 345 410	0.218	0.579	450	28	40
Congo, Republic	342 000	0.128	0.589	200	2	12
Cook Islands	233			1	0	1
Costa Rica	50 900	0.162	1.358	205	7	14
Côte d'Ivoire	322 465	0.116	0.507	230	0	17
Croatia	56 538	0.036	− 0.169	76	0	9
Cuba	114 525	0.12	0.829	31	12	11
Cyprus	9 250	0.017	− 0.429	21	1	3
Czech Republic	78 864	0.033	− 0.356	81	0	8
Denmark	43 075	0.021	− 0.643	43	0	5
Djibouti	23 000	0.02	− 0.528	61	0	4
Dominica	751			12	0	1
Dominican Republic	48 440	0.076	0.625	20	0	5
Ecuador	461 475	0.353	1.519	302	25	31
Egypt	1 000 250	0.038	− 0.936	98	7	12
El Salvador	21 395	0.048	0.393	135	0	2
Equatorial Guinea	28 050	0.084	0.869	184	1	15
Eritrea	117 600	0.057	0.088	112	0	12
Estonia	45 100	0.025	− 0.483	65	0	5
Ethiopia	1 104 300	0.145	0.383	277	31	34
Falkland Islands	15 931	0.004	− 2.040	0	0	4
Faroe Islands				-	0	3
Fiji	18 330	0.028	− 0.100	4	1	5
Finland	337 030	0.023	− 1.145	60	0	6
France	543 965	0.051	− 0.473	93	0	18
French Guiana	91 000	0.079	0.483	150	3	9
French Polynesia	3 940			0	0	3
French S. and Antarctic Terr.	7 241	0.001	− 3.261	-	0	0
Gabon	267 665	0.116	0.56	190	3	15
Gambia	10 690	0.036	0.308	117	0	3
Georgia	69 700	0.051	0.111	107	2	14
Germany	356 840	0.033	− 0.770	76	0	12
Ghana	238 305	0.114	0.571	222	1	13
Gibraltar	7			7	0	0
Greece	131 985	0.062	0.129	95	3	14
Greenland	2 175 600	0.007	− 2.821	9	0	7
Grenada	345			15	0	0
Guadeloupe	1 780			11	4	5
Guam	450			2	0	2
Guatemala	108 890	0.142	1.014	250	3	6
Guinea	245 855	0.094	0.373	190	1	11
Guinea-Bissau	36 125	0.05	0.289	108	0	2

Mammals % threatened	Birds breeding total	Birds endemic	Birds no. threatened	Birds % threatened	Plants total	Plants endemic
13	370	0	5	1	1 600	-
23	296	16	15	5	5 284	2 698
19	1 100	70	73	7	32 200	18 000
10	1 695	67	77	5	51 220	15 000
17	50	14	9	18	721	136
9	929	24	28	3	11 007	1 100
6	449	0	3	1	6 000	1 200
100	27	6	7	26	284	3
7	600	6	13	2	12 119	950
7	535	2	12	2	3 660	62
12	224	0	4	2	4 288	-
35	137	21	18	13	6 522	3 229
14	79	2	3	4	1 682	-
10	199	0	2	1	1 900	-
12	196	0	1	1	1 450	1
7	126	1	5	4	826	6
8	52	2	3	6	1 228	11
25	136	0	15	11	5 657	1 800
10	1 388	37	60	4	19 362	4 000
12	153	0	7	5	2 076	70
1	251	0	0	0	2 911	17
8	273	3	5	2	3 250	66
11	319	0	7	2	-	-
8	213	0	3	1	1 630	-
12	626	28	16	3	6 603	1 000
	64	4	3	5	165	14
	71	0	0	0	236	1
	74	24	12	16	1 518	760
10	248	0	3	1	1 102	-
19	269	1	5	2	4 630	133
6	-	1	0		5 625	144
	60	25	22	37	959	560
	48	3	3	6	-	-
8	466	1	5	1	6 651	-
3	280	0	2	1	974	-
13	-	0	3		4 350	380
16	239	0	5	2	2 682	6
6	529	0	8	2	3 725	43
0	34	0	1	3	600	-
15	251	0	7	3	4 992	742
78	62	0	0	0	529	15
0	50	1	1	2	1 068	4
45	52	2	1	2	1 400	26
100	18	2	2	11	330	69
2	458	1	6	1	8 681	1 171
6	409	0	10	2	3 000	88
2	243	0	0	0	1 000	12

0 = zero; - = no data.

Country	Area km²	DI	AI	Mammals total	Mammals endemic	Mammals no. theatened
Guyana	214 970	0.133	0.758	193	1	9
Haiti	27 750	0.071	0.71	20	0	4
Honduras	112 085	0.094	0.597	173	2	9
Hungary	93 030	0.031	– 0.457	83	0	9
Iceland	102 820	0.006	– 2.080	11	0	6
India	3 166 830	0.326	0.896	390	44	86
Indonesia	1 919 445	0.731	1.844	515	222	140
Iran	1 648 000	0.091	– 0.194	140	6	23
Iraq	438 445	0.041	– 0.629	81	2	10
Ireland	68 895	0.013	– 1.248	25	0	5
Israel	20 770	0.043	0.285	116	4	14
Italy	301 245	0.065	– 0.056	90	3	14
Jamaica	11 425	0.051	0.619	24	2	5
Japan	369 700	0.124	0.536	188	42	37
Jordan	96 000	0.036	– 0.310	71	0	8
Kazakhstan	2 717 300	0.071	– 0.581	178	4	18
Kenya	582 645	0.145	0.56	359	23	51
Kiribati	684			-	0	0
Korea, DPR	122 310	0.025	– 0.775	-	0	13
Korea, Republic	98 445	0.03	– 0.518	49	0	13
Kuwait	24 280	0.007	– 1.564	21	0	1
Kyrgyzstan	198 500	0.036	– 0.537	83	1	7
Lao PDR	236 725	0.081	0.229	172	0	27
Latvia	63 700	0.025	– 0.553	83	0	5
Lebanon	10 400	0.031	0.145	57	0	6
Lesotho	30 345	0.025	– 0.354	33	0	3
Liberia	111 370	0.059	0.132	193	0	16
Libya	1 759 540	0.029	– 1.343	76	5	9
Liechtenstein	160			64	0	3
Lithuania	65 200	0.026	– 0.544	68	0	5
Luxembourg	2 585			55	0	6
Macedonia, FYR	25 713	0.037	0.077	78	0	11
Madagascar	594 180	0.298	1.277	141	93	50
Malawi	94 080	0.079	0.473	195	0	8
Malaysia	332 965	0.254	1.28	300	36	47
Maldives	298			3	0	0
Mali	1 240 140	0.053	– 0.658	137	0	13
Malta	316			22	0	3
Marshall Islands	181			0	0	1
Martinique	1 079			9	0	0
Mauritania	1 030 700	0.041	– 0.856	61	1	10
Mauritius	1 865			4	1	3
Mayotte	376			-	0	0
Mexico	1 972 545	0.589	1.621	491	140	69
Micronesia, Fed. States	702			6	3	6
Moldova	33 700	0.025	– 0.396	68	0	3
Monaco	2			-	0	0

0 = zero; - = no data.

Mammals % threatened	Birds breeding total	Birds endemic	Birds no. threatened	Birds % threatened	Plants total	Plants endemic
5	678	0	2	0	6 409	-
	75	1	14	19	5 242	1 623
5	422	1	5	1	5 680	148
11	205	0	8	4	2 214	38
55	88	0	0	0	377	1
22	923	58	70	8	18 664	5 000
27	1 519	408	113	7	29 375	17 500
16	323	1	13	4	8 000	-
12	172	1	11	6	-	-
20	142	0	1	1	950	-
12	180	0	12	7	2 317	-
16	234	0	5	2	5 599	712
21	113	26	12	11	3 308	923
20	250	21	32	13	5 565	2 000
11	141	0	8	6	2 100	-
10	396	0	15	4	6 000	-
14	844	9	24	3	6 506	265
	26	1	3	12	60	2
	115	1	19	17	2 898	107
27	112	0	25	22	2 898	224
5	20	0	7	35	234	-
8	-	0	4		4 500	-
16	487	1	19	4	8 286	-
6	217	0	3	1	1 153	-
11	154	0	7	5	3 000	-
9	58	0	7	12	1 591	2
8	372	1	11	3	2 200	103
12	91	0	1	1	1 825	134
5	124	0	1	1	1 410	-
7	202	0	4	2	1 796	-
11	126	0	1	1	1 246	-
14	210	0	3	1	3 500	-
35	202	105	27	13	9 505	6 500
4	521	0	11	2	3 765	49
16	501	18	37	7	15 500	3 600
0	23	0	1	4	583	-
9	397	0	4	1	1 741	11
14	26	0	1	4	914	5
	17	0	1	6	100	5
0	52	1	2	4	1 287	30
16	273	0	2	1	1 100	-
75	27	8	9	33	750	325
	27	2	3	11	500	-
14	769	92	38	5	26 071	12 500
100	40	18	5	13	1 194	293
4	177	0	5	3	1 752	-
	-	0	0		-	-

Country	Area km²	DI	AI	Mammals total	Mammals endemic	Mammals no. theatened
Mongolia	1 565 000	0.051	– 0.767	133	0	12
Montserrat	104			7	0	1
Morocco	458 730	0.057	– 0.304	105	4	16
Mozambique	784 755	0.09	0.005	179	2	15
Myanmar	678 030	0.141	0.493	300	6	36
Namibia	824 295	0.102	0.116	250	3	14
Nauru				-	0	0
Nepal	141 415	0.096	0.549	181	2	27
Netherlands	41 160	0.022	– 0.599	55	0	11
Netherlands Antilles	800			-	0	3
New Caledonia	19 105	0.078	0.904	11	3	6
New Zealand	265 150	0.065	– 0.017	2	2	8
Nicaragua	148 000	0.098	0.555	200	2	6
Niger	1 186 410	0.061	– 0.512	131	0	11
Nigeria	923 850	0.107	0.131	274	4	25
Niue	259			1	0	0
Northern Marianas	477			-	0	2
Norway	386 325	0.024	– 1.107	54	0	10
Oman	271 950	0.03	– 0.812	56	2	9
Pakistan	803 940	0.08	– 0.121	188	4	18
Palau	492			2	0	3
Panama	78 515	0.162	1.236	218	16	20
Papua New Guinea	462 840	0.271	1.254	214	65	58
Paraguay	406 750	0.115	0.429	305	2	9
Peru	1 285 215	0.396	1.344	460	49	47
Philippines	300 000	0.225	1.188	153	102	50
Pitcairn Islands				0	0	0
Poland	312 685	0.032	– 0.761	84	0	15
Portugal	92 390	0.045	– 0.088	63	1	17
Puerto Rico	8 960	0.033	0.259	16	0	2
Qatar	11 435	0.005	– 1.770	11	0	0
Réunion	2 510			2	0	3
Romania	237 500	0.039	– 0.490	84	0	17
Russia	17 075 400	0.179	– 0.179	269	22	42
Rwanda	26 328	0.087	0.925	151	0	8
San Marino				13	0	1
São Tomé and Príncipe	964			8	4	3
Saudi Arabia	2 400 900	0.04	– 1.129	77	0	7
Senegal	196 720	0.057	– 0.065	192	0	11
Seychelles	404			6	2	4
Sierra Leone	72 325	0.083	0.588	147	0	11
Singapore	616			85	1	3
Slovakia	14 035	0.037	0.252	85	0	9
Slovenia	20 251	0.036	0.106	75	0	9
Solomon Islands	29 790	0.049	0.316	53	21	21
Somalia	630 000	0.087	0.025	171	12	19
South Africa	1 184 825	0.252	0.915	247	35	41

Mammals % threatened	Birds breeding total	Birds endemic	Birds no. threatened	Birds % threatened	Plants total	Plants endemic
9	426	0	16	4	2 823	229
14	37	1	2	5	671	2
15	210	0	9	4	3 675	625
8	498	0	16	3	5 692	219
12	867	4	35	4	7 000	1 071
6	469	3	9	2	3 174	687
	9	1	2	22	50	1
15	611	2	26	4	6 973	315
20	191	0	4	2	1 221	–
	77	0	1	1	–	–
55	107	22	9	8	3 250	3 200
	150	74	49	33	2 382	1 942
3	482	0	5	1	7 590	40
8	299	0	3	1	1 460	–
9	681	2	9	1	4 715	205
0	15	0	1	7	178	1
	28	2	8	29	315	81
19	243	0	2	1	1 715	1
16	107	0	10	9	1 204	73
10	375	0	17	5	4 950	372
	45	10	2	4	–	–
9	732	9	16	2	9 915	1 222
27	644	94	32	5	11 544	–
3	556	0	26	5	7 851	–
10	1 538	112	71	5	17 144	5 356
33	196	186	67	34	8 931	3 500
	19	5	8	42	76	14
18	227	0	4	2	2 450	3
27	207	2	7	3	5 050	150
13	105	12	8	8	2 493	235
0	23	0	6	26	355	–
	18	4	5	28	546	165
20	247	0	8	3	3 400	41
16	628	13	38	6	11 400	–
5	513	0	9	2	2 288	26
8	–	0	0		–	–
38	63	25	9	14	895	134
9	155	0	15	10	2 028	–
6	384	0	4	1	2 086	26
67	38	11	10	26	250	182
7	466	1	10	2	2 090	74
4	118	0	7	6	2 282	2
11	209	0	4	2	3 124	92
12	207	0	1	0	3 200	22
40	163	43	23	14	3 172	30
11	422	11	10	2	3 028	500
17	596	8	20	3	23 420	–

0 = zero; – = no data.

Country	Area km²	DI	AI	Mammals total	Mammals endemic	Mammals no. theatened
Spain	504 880	0.067	− 0.172	82	4	24
Sri Lanka	65 610	0.082	0.606	88	15	20
St Helena and dep.	411			2	0	1
St Kitts and Nevis	261			7	0	0
St Lucia	619			9	0	1
St Vincent	389			8	1	2
Sudan	2 505 815	0.137	0.093	267	11	24
Suriname	163 820	0.092	0.471	180	2	11
Swaziland	17 365	0.044	0.353	47	0	4
Sweden	440 940	0.026	− 1.067	60	0	8
Switzerland	41 285	0.033	− 0.173	75	0	6
Syria	185 680	0.046	− 0.265	63	2	4
Taiwan	36 960	0.058	0.418	63	11	13
Tajikistan	143 100	0.033	− 0.536	84	1	9
Tanzania	939 760	0.189	0.693	316	15	43
Thailand	514 000	0.162	0.709	265	7	34
Togo	56 785	0.094	0.781	196	0	9
Tokelau				0	0	0
Tonga	699			2	0	1
Trinidad and Tobago	5 130	0.045	0.729	100	1	1
Tunisia	164 150	0.033	− 0.572	78	1	11
Turkey	779 450	0.114	0.237	116	2	17
Turkmenistan	488 100	0.044	− 0.572	103	0	13
Turks and Caicos Islands	430			-	0	0
Tuvalu	25			-	0	0
Uganda	236 580	0.12	0.624	345	6	19
Ukraine	603 700	0.05	− 0.509	108	1	17
United Arab Emirates	75 150	0.019	− 0.883	25	0	3
United Kingdom	244 880	0.024	− 1.003	50	0	12
United States of America	9 372 614	0.342	0.638	428	105	37
Uruguay	186 925	0.05	− 0.186	81	1	6
US Pacific Islands	658			-	0	0
Uzbekistan	447 400	0.051	− 0.413	97	0	9
Vanuatu	14 765	0.014	− 0.728	11	2	4
Venezuela	912 045	0.379	1.398	323	19	25
Viet Nam	329 565	0.147	0.737	213	9	37
Virgin Islands (British)	153			3	0	0
Virgin Islands (US)	352			-	0	1
Wallis and Futuna	255			1	0	0
Western Sahara				32	1	3
Western Samoa	2 840			3	0	2
Yemen	477 530	0.041	− 0.654	66	1	4
Yugoslavia	102 173	0.046	− 0.086	96	0	11
Zambia	752 615	0.096	0.074	233	3	12
Zimbabwe	390 310	0.099	0.298	270	0	12

0 = zero; - = no data.

Mammals % threatened	Birds breeding total	Birds endemic	Birds no. threatened	Birds % threatened	Plants total	Plants endemic
29	278	5	7	3	5 050	941
23	250	24	14	6	3 314	890
50	53	9	9	17	165	50
0	32	0	1	3	659	1
11	50	4	5	10	1 028	11
25	108	2	2	2	1 166	-
9	680	1	6	1	3 137	50
6	603	0	1	0	5 018	-
9	364	0	5	1	2 715	4
13	249	0	2	1	1 750	1
8	193	0	2	1	3 030	1
6	204	0	8	4	3 000	-
21	160	14	20	13	3 568	-
11	-	0	7		5 000	-
14	822	24	33	4	10 008	1 122
13	616	2	37	6	11 625	-
5	391	0	0	0	3 085	-
	5	0	1	20	26	-
50	37	2	3	8	463	25
1	260	1	1	0	2 259	236
14	173	0	5	3	2 196	-
15	302	0	11	4	8 650	2 675
13	-	0	6		-	-
	42	0	3	7	448	9
	9	0	1	11	-	-
6	830	3	13	2	4 900	-
16	263	0	8	3	5 100	-
12	67	0	8	12	-	-
24	230	1	2	1	1 623	16
9	650	67	54	8	19 473	4 036
7	237	0	11	5	2 278	40
	-	0	1		-	-
9	-	0	9		4 800	400
36	76	9	7	9	870	150
8	1 340	40	24	2	21 073	8 000
17	535	10	35	7	10 500	1 260
0	70	0	2	3	-	-
	70	0	2	3	-	-
0	25	0	1	4	475	7
9	60	0	0	0	330	-
67	40	8	6	15	737	-
6	143	8	12	8	1 650	135
11	224	0	5	2	4 082	-
5	605	2	11	2	4 747	211
4	532	0	10	2	4 440	95

APPENDIX 6:
IMPORTANT AREAS FOR
FRESHWATER BIODIVERSITY

This table presents information on areas identified as of special importance for diversity (species richness, and/or endemism) in the inland water groups treated (fishes, mollusks ['Moll.' in table], crabs, crayfish, fairy shrimps).

This is a preliminary synthesis, designed to represent expert opinion on relative levels of diversity for each taxon at continent level. In the absence of global criteria for relative importance, areas on different continents do not represent strictly equivalent levels of diversity.

The rows of data in this list are sorted first by continent, and secondly by the taxon concerned.

Source: This information was prepared for WCMC[74] on the basis of data kindly provided by a number of expert ichthyologists and members of the IUCN/SSC Specialist Groups for Inland Water Crustaceans and Molluscs. The particular source of information is indicated by letters in square brackets in the Remarks text. Please note that numerical estimates and other information may have been superceded by later survey and taxonomic work.

For North American and African fishes, where source is UNEP-WCMC, information has been extracted from available literature, with additional data and advice for Africa from Christian Lévêque and Guy Teugels.

In most instances (MK, SK, MSG, NC/RvS, DB) contributors indicated the approximate location of the important areas concerned on a series of A3-sized base maps provided. These areas, and those identified from literature by UNEP-WCMC, were digitized for purposes of presentation (Maps 7.2, 7.3, 7.4) and analysis.

GA	Gerald Allen, *in litt.* March 1998.
MK	Maurice Kottelat, report compiled for WCMC[74].
SK	Sven Kullander, report compiled for WCMC[74].
MSG	Adapted from a report by IUCN/SSC Mollusc Specialist Group, primarily by Philippe Bouchet and Olivier Gargominy, and also by Arthur Bogan and Winston Ponder.
KC	Keith Crandall, information on distribution of crayfish genera and species.
DB	Denton Belk, summary of fairy shrimp distribution patterns.
NC/RvS	Neil Cumberlidge and R. von Sternberg, report compiled for WCMC[74].

AFRICA

	Area name	Group	Remarks
1	L. Tanganyika	Crabs	L. Tanganyika is the only East African great lake where endemic species of freshwater crabs occur: of the 9 species and 2 genera present, 1 genus and 7 species are endemic. [NC/RvS]
2	Lower Congo	Crabs	Diversity is marked in the Congo R. basin, but appears highest in 2 areas, the lower parts of the basin (including Congo, Cabinda and DR Congo (former Zaire)) and the upper reaches (including Rwanda/Burundi and parts of DR Congo). [NC/RvS]
3	Madagascar	Crabs	4 genera and 10 species of freshwater crabs, all endemic, occur in Madagascar. [NC/RvS]
4	Niger-Gabon	Crabs	Southeast Nigeria, southern Cameroon and Gabon: 3 endemic genera and more than 10 endemic species of freshwater crabs[1]. [NC/RvS]
5	Upper Congo	Crabs	Diversity is marked in the Congo R. basin, but appears highest in 2 areas, the lower parts of the basin (including Congo, Cabinda and DR Congo (former Zaire)) and the upper reaches (including Rwanda/Burundi and parts of DR Congo). [NC/RvS]
6	Upper Guinea	Crabs	Upper Guinean rainforest, centered on Guinea, Sierra Leone, Liberia, and western Côte d'Ivoire (including Mount Nimba): 2 endemic genera and 5 endemic species of gecarcinucids[2,3,1,4]. [NC/RvS]
7	Southern Africa	Fairy shrimp	2 endemic genera, 45 species, 38 endemic. South Africa proper: 34 species, 22 endemic. [DB]
8	Cape rivers	Fishes	With 4 families and 33 species the fish fauna of southern Africa is rather poor in comparison with most other parts of the continent; most species are cyprinids. However, there is marked local endemism; most rivers in the southern Cape region have 3 or 4 native endemics (several species are threatened). [UNEP-WCMC]

AFRICA

	Area name	Group	Remarks
9	Congo (Zaire) basin	Fishes	General region of very high richness; second only to the Amazon basin in species richness. 25 families and 686 species have been reliably reported from the Congo/Zaire basin, excluding L. Tanganyika and L. Moero[5]. Around 548 of the species present (ca 70%) are endemic to this basin. The basin can be divided into 4 sections: Upper Lualaba, Cuvette Centrale, Luapula-Mweru and the rapids. [UNEP-WCMC]
10	Congo 'Cuvette Centrale'	Fishes	High richness plus marked endemism. Around 690 species occur in the Congo system; the Cuvette Centrale section possibly has the highest species richness owing to the great diversity of freshwater habitats available. [UNEP-WCMC]
11	Congo rapids	Fishes	High richness plus marked endemism. The rapids between Kinshasa and the sea have a high concentration of fish species (150 species), 34 of which are endemic to this section. The caves near Thysville are fed by the Congo system and support one of Africa's few true hypogean fishes *Caecobarbus geertsi*. *Caecomastacembelus brichardi* and *Gymnanallabes tihoni*, not strictly cave fishes, have been collected in the Stanley Pool in riffles under flagstones or in crevices. [UNEP-WCMC]
12	Cross R.	Fishes	Nigeria-Cameroon. 42 families, 166 species[6]. Very high species diversity compared to the relatively modest catchment area, and marked endemism. Transitional ichthyofauna between the Nile-Sudan province and the Lower Guinea province. [UNEP-WCMC]
13	L. Barombi-Mbo	Fishes	This small (ca 4.5 km²) crater lake in Cameroon has 15 species (plus another 2 present in the inflow stream, not the lake proper). At least 12 of the species are endemic, notably the 11 cichlids that form 1 of the 2 recorded 'species flocks' in West Africa. 4 of the 5 cichlid genera are endemic: *Konia*, *Myaka*, *Pungu* and *Stomatepia*. This very important site is at risk from overfishing, the effects of introduced crustaceans and fishes, siltation from local deforestation and water pollution. [UNEP-WCMC]
14	L. Bermin	Fishes	A small (ca 0.5 km²) crater lake in southwest Cameroon with 2 non-endemic fishes and a remarkable species flock of 9 tilapiine cichlids. The cichlids are very small and not exploited; they are at some risk because of the small distribution and deforestation in the surrounding area. [UNEP-WCMC]
15	L. Malawi	Fishes	30 800 km². 12 families, more than 845 species, most of them endemic to the lake. Rich species flocks among Cichlidae, and a small species flock of Clariidae. [UNEP-WCMC]

AFRICA

	Area name	Group	Remarks
16	L. Tana	Fishes	The fish fauna of this large (3 150 km²) lake includes 21 species in 4 families and is dominated by lake endemic cyprinids. The large *Barbus* cyprinids form 1 of 2 recorded cyprinid species flocks (the other being that of *L. Lanao* in the Philippines, many species of which are severely threatened). [UNEP-WCMC]
17	L. Tanganyika	Fishes	32 000 km². In the lake itself, 16 families, around 250 cichlid species and 72 non-cichlid species. Several species flocks are present not only in Cichlidae, but also among Clariidae, Bagridae, Mochokidae, Centropomidae, Mastacembelidae.
18	L. Turkana	Fishes	6 750 km². 51 species, 35 genera, 17 families. High family and generic diversity, many of the species are lake endemic; cyprinids form the most diverse family. [UNEP-WCMC]
19	L. Victoria	Fishes	68 800 km². 12 families, around 545 species (many undescribed). High species diversity dominated by cichlids. The majority of species are lake endemic. [UNEP-WCMC]
20	Madagascar	Fishes	Around 140 fish species have been recorded from the brackish and freshwaters of Madagascar[7]; although species richness is not remarkable, endemism is high. Two endemic families (Bedotiidae and Anchariidae) have been recognized in Madagascar, as well as 13 endemic genera and 43 endemic species. Most endemic species are restricted to freshwater habitats, mainly in eastern forested regions. About one quarter of endemic species are known only from the type of locality. Blind cave fishes have been described from Madagascar: the gobiid *Glossogobius ankaranensis* and the elotrids *Typheleotris madagascarensis* and *T. pauliani*. [UNEP-WCMC]
21	Niger basin	Fishes	General region of high richness. 36 families, ca 243 species, with 225 primary freshwater species[8]. Endemism moderate: 20 species endemic to Niger. The basin includes 11 of the 13 primary freshwater families that are endemic to Africa. 164 primary freshwater fishes reported[8] from the Niger delta in Nigeria, based on reference specimens for each species; the high diversity (73% of the freshwater species in the entire basin) in this area is seriously threatened by oil pollution. [UNEP-WCMC]
22	Ntem R.	Fishes	Cameroon. High richness for area, plus marked endemism. 16 families, 94 species, 8 endemic. [UNEP-WCMC]

AFRICA

	Area name	Group	Remarks
23	Ogooue (Ogowe) R.	Fishes	Gabon. High richness for area, plus marked endemism. 23 families, 185 species, 48 species endemic to Ogowe. A relatively small drainage basin with a very high concentration of species. Many of the families represented are endemic to Africa. Available data certainly underestimate actual diversity (several new species are now being described, resulting from a project of Tervuren Museum, the American Museum of Natural History and Cornell University). [UNEP-WCMC]
24	Sanaga R.	Fishes	Cameroon. 21 families; high concentration of species in a small river basin; probably at least 135 (and this figure is believed to be a significant underestimate)[5]. Between 10 and 18 species endemic to the Sanaga. [UNEP-WCMC]
25	Upper Guinea rivers	Fishes	High richness for area, plus marked endemism. The Upper Guinea province includes coastal rivers from south of the Kogon R. in Guinea to Liberia, and has faunal affinities with the lower Guinea province and the Congo/Zaire. The fauna includes many taxa endemic to the area[9,10]. Many small river basins, many of them still poorly investigated. Konkoure R. (Guinea): 19 families, 85 species, at least 10 endemic. Kolente or Great Scarcies R. (Guinea-Sierra Leone): 19 families, 68 species. Jong R. (Sierra Leone): 20 families, 94 species. Saint-Paul R. (Liberia): 19 families, 76 species. Cess-Nipoué R. (Liberia-Côte d'Ivoire): 20 families, 61 species. [UNEP-WCMC]
26	Volta basin	Fishes	General region of high richness. 27 families, about 139 species, 8 endemic to Volta basin. High species richness, with 9 of the 13 African endemic primary freshwater fish families represented[11]. [UNEP-WCMC]
27	L. Malawi	Moll.	Gastropods: 28 species, 16 endemic. Bivalves: 9 species, 1 endemic. [MSG]
28	L. Tanganyika	Moll.	Gastropods: 68 species, 45 endemic. Bivalves: 15 species, 8 endemic. [MSG]
29	L. Victoria	Moll.	Gastropods: 28 species, 13 endemic. Bivalves: 18 species, 9 endemic. [MSG]

	Area name	Group	Remarks
30	Lower Congo basin	Moll.	The region downstream of Kinshasa in Congo and DR Congo (former Zaire). Gastropods: 96 species, 24 endemic. Endemic gastropods are almost all prosobranchs; 5 endemic 'rheophilous' (specialized for life in the rapids) genera, belonging to the Bithyniidae (*Congodoma, Liminitesta*) and Assimineidae (*Pseudogibbula, Septariellina, Valvatorbis*). Bivalves: no data. [MSG]
31	Madagascar	Moll.	Gastropods: 30 species, 12 endemic. Genus *Melanatria* endemic. Bivalves: no data. [MSG]
32	Western lowland forest and Volta basin	Moll.	Upper Guinea region in Ghana, Côte d'Ivoire, Sierra Leone, Liberia, Guinea. Around 28 gastropod species of which 19 endemic (and 9 near-endemic). Bivalves: no data. [MSG]
33	The Mollucas, New Guinea and northerm Australia	Crabs	More than 30 species of freshwater crabs belonging to 5 genera, all in Parathelphusidae. [NC/RvS]
34	Southeast Australia	Crayfish	Large area of high richness and endemism, centered on Victoria, 35 species, and Tasmania, 19 species. [KC]
35	Southwest Australia	Fairy shrimp	19 species, 12 endemic. [DB]
36	Fly R., Papua New Guinea	Fishes	High species richness, 103 species in Fly proper, and high local endemism, 12 endemics in system. [GA]
37	Kikori R., L. Kutubu, Papua New Guinea	Fishes	Headwaters of Kikori and Purari systems, with L. Kutubu. High richness, 103 species and high endemism, 16 species in Kikori; plus 14 species in L. Kutubu. [GA]
38	Kimberley District, Western Australia	Fishes	14 endemic species (a density second only in Australia to Tasmania and equal to southwest Western Australia), including 5 species within Prince Regent Reserve and 4 in the Drysdale R. area; and 47 species in total. [GA]
39	Aikwa (Iwaka) R., Irian Jaya	Fishes	Near Timiki, Irian Jaya. High species richness: ca 78 species. [GA]
40	Southeast Australia	Fishes	11 endemic species occur in coastal southeast Australia, a lower count per area than the other 3 areas cited here, and 42 species in total. [GA]
41	Southwest Western Australia	Fishes	9 endemic species (i.e. density similar to the Kimberleys), and 14 species in total. [GA]

AFRICA

AUSTRALASIA

AUSTRALASIA

	Area name	Group	Remarks
42	Tasmania	Fishes	12 endemic species, a greater number per area than anywhere else in Australia, including 6 concentrated in the Central Plateau area; and 24 species in total. [GA]
43	Vogelkop, Irian Jaya	Fishes	Moderate richness with high local endemism, ca 14 endemic species, including Triton and Etna Bay lakes. [GA]
44	Great Artesian basin, Australia	Moll.	Springs and underground aquifers. Important area of gastropod diversity. Bivalves: no data. [MSG]
45	New Caledonia	Moll.	Springs and underground aquifers. Gastropods: 81 species, 65 endemic. Bivalves: no data. [MSG]
46	Western Tasmania, Australia	Moll.	Springs and underground aquifers. Important area of gastropod diversity. Bivalves: no data. [MSG]
47	Indonesia	Crabs	The area containing Sumatra, Java, Borneo, Sulawesi and the southern Philippines has the greatest freshwater crab diversity in Indo-Australia, with representatives of the Parathelphusidae (10 genera and 71 species) and the Gecarcinucidae (5 genera and 21 species). [NC/RvS]
48	Myanmar-Malaysia	Crabs	Northeast India (Assam), Myanmar, Thailand, the Mekong basin in southern Indochina, to the Malaysian peninsula and Singapore. In this region there are an estimated 30 genera and more than 100 species of freshwater crabs in 3 families, the Potamidae, the Parathelphusidae and the Gecarcinucidae[12-15]. [NC/RvS]
49	South China	Crabs	Only the Potamidae occur in China, but more than 160 species and subspecies in 22 genera are present, most of which are endemic. The southern provinces of China represent the hotspot of biodiversity for this country[13,16-20]. [NC/RvS]
50	South India	Crabs	The freshwater crabs of the Indian peninsula south of the Ganges basin are all endemic to the subcontinent and belong to 2 families, the Gecarcinucidae and the Parathelphusidae[12,13]. The west coast of the peninsula and the south show most diversity: an estimated 7 endemic genera and about 20 endemic species in 2 families (the Parathelphusidae and Gecarcinucidae). A third freshwater crab family, the Potamidae, is found only in northern India but is not represented in the Indian peninsula. [NC/RvS]
51	Sri Lanka	Crabs	Sri Lanka has some 16 endemic species of freshwater crabs belonging to 3 genera, 1 of which (*Spiralothelphusa*) is endemic to the island[21,22]. [NC/RvS]

EURASIA

	Area name	Group	Remarks
52	Italy	Fairy shrimp	16 species, 7 endemic. [DB]
53	Borneo highlands	Fishes	The fish fauna of the highlands of Borneo seems to be poor in absolute number of species, but many of them have developed specialization for hill-stream habitats and are endemic to single basins. The area is still largely unsurveyed. About 50 known endemic species, but actual figure might be over 200[23]. [MK]
54	Caspian Sea	Fishes	Moderate species richness. Although many species are shared with the Black Sea region, and/or the Aral basin, there is marked endemism, including the monotypic lamprey *Caspiomyzon*, ca 12 gobies, including monotypic genera *Asra* and *Anatirostrum*, also 3 *Alosa*. [UNEP-WCMC]
55	Central Anatolia	Fishes	An arid plateau with several endorheic lakes. About 20 endemic species, apparently underestimated by inadequate taxonomy. Adjacent areas also have a number of endemics. In urgent need of critical reassessment; probably one of the most poorly known fish faunas in Eurasia. [MK]
56	Coastal peat swamps and swamp forests of Malaysia, Sumatra and Borneo	Fishes	Includes Bangka island. Extent along eastern coast of Borneo not known. Probably formerly present on Java but apparently cleared. About 100 endemic species in peat swamp forests, a habitat type often restricted to a narrow fringe along the coasts, still largely unsurveyed. Although peat swamps are traditionally considered as a habitat with poor diversity, good data for limited areas in Malay peninsula and Borneo indicate that up to 50 species may be found within a small area (less than 1 km²), about half of them endemic and stenotypic. Most species have small distribution ranges (some possibly only a few km²)[23-26]. [MK]
57	Coastal rainforest of Southeast Asia	Fishes	Thailand, Cambodia and southern Viet Nam. Southern extent not known accurately. This habitat is largely destroyed in Thailand, and virtually unsurveyed in Cambodia and Viet Nam. Endemic species expected in peat swamp forests[27,28]. [MK]
58	High Asia	Fishes	Boundaries not known with accuracy; includes the Tibetan plateau and probably parts of Chinese Turkestan. Distribution and ecological data are sparse outside the Chinese literature. About 150 known fish species, about half of them endemic to this area[29]. Survey probably still superficial as a result of difficulties of access. [MK]

EURASIA

	Area name	Group	Remarks
59	Karstic basins of Yunnan, Guizhou and Guangxi	Fishes	Boundary not known with accuracy. About 14 known species of cave fishes. Survey is still superficial and numerous additional species are expected[30]. [MK]
60	L. Baikal, Siberia	Fishes	A species flock of 36 species of the family Cottidae (sculpins) (including the endemic family Comephoridae), 4 'ecologically differentiated stocks' (many probably endemic species using Western concepts) of *Coregonus*, 2 of *Thymallus*, 2 of *Lota*[31]. Endemic mollusks, gammarids, sponges and Baikal seal. [MK]
61	L. Biwa, Japan	Fishes	Reportedly 4 endemic species[32]. [MK]
62	L. El'gygytgyn, Siberia	Fishes	An old lake formed on the site of a meteorite crater. 113 km². Total fish diversity: 5 species, including an endemic genus and species (*Salvethymus svetovidovi*), an endemic species (*Salvelinus elgyticus*), and 1 species endemic to eastern Siberia (*Salvelinus boganidae*). Endemic diatom species and apparently endemic invertebrate(s)[33,34]. [MK]
63	L. Inle, Myanmar	Fishes	About 25 native fish species, ca 10 of them endemic, including 3 endemic genera[35,36]. [MK]
64	L. Lindu, Sulawesi	Fishes	Very limited information. One native and endemic species; others might be expected[37]. [MK]
65	L. Poso, Sulawesi	Fishes	10 native and endemic species, 2 endemic genera (both extinct?) and with L. Lindu comprises the entire known distribution of the subfamily Adrianichthyinae[37-40]. Additional species might still be expected. [MK]
66	L. Thingvalla,	Fishes	5 native fish species, including 3 endemic *Salvelinus* (recent summary in Kottelat[41]). [MK]
67	Lakes of Isles	Fishes	A number of lakes host 1 or 2 species of British *Salvelinus*, although information on individual lakes is usually inadequate. At the beginning of the century up to 14 species were recognized; although generally not accepted under later systematic concepts, recent work suggests that this figure may be underestimated. Also at least 5 endemic *Coregonus*, 1 endemic Clupeidae and potential for endemic *Salmo* (recent summary in Kottelat[41]). [MK]

Area name	Group	Remarks
68 Lakes of Central Yunnan, China	Fishes	Lakes Dianchi, Fuxian, Er Hai, Yangling, Yangzong, Xingyun, etc., have a distinctive fauna; despite the lakes being now in different river basins (Mekong, Yangtze, Nanpangjiang), they have similar fauna, characterized by numerous endemic species in the genera *Cyprinus, Schizothorax, Anabarilius* and *Yunnanilus.* Exact up-to-date figures of the number of species are difficult to extract from the Chinese literature, but we have the following data: Dianchi: 25 native species, 11 endemic of which apparently all but 2 are extinct. The lake basin has 2 other endemics[42]; Fuxian: 25 native species, 12 endemic plus 2 endemic shared only with Xingyun[43]; Er Hai: 17 native species, 9 endemic, several apparently extinct[44]; Yangzong has (had) at least 2 endemics; Yangling and Xingyun at least 1 each. [MK]
69 Lough Melvin, Ireland	Fishes	Three endemic species of *Salmo* (recent summary in Kottelat[41]). [MK]
70 Lower Danube	Fishes	The lower Danube basin has a relatively richer fauna (especially more diverse communities) than any other European river. Endemics: about 6, possibly underestimated (counted in Kottelat 1997[41]). [MK]
71 Mainland Southeast Asian hills	Fishes	Northern boundary not clear as published data on fish distribution (and ground surveys) in southern China are too scanty. Could be subdivided into a) upper Song Hong (includes hills of Hainan and southern Nanpang Jiang); b) Annamite cordillera; c) upper Mekong, ChaoPhraya and Mae Khlong basins; d) Salween, upper Irrawaddy and southeastern Assam (including Tenasserim). Recorded fish fauna estimated to be over 1 000 species (with an estimated 200-500 species still awaiting discovery), 500 endemic to this area. Includes ca 400 known species endemic to headwaters of individual sub-basins. The fauna of the lower reaches of the main rivers (excluded from this polygon) is richer (in terms of the number of species that can be observed at a given locality) but most have wide distributions crossing several river basins[27,45,46]. [MK]

EURASIA

	Area name	Group	Remarks
72	Malili Lakes, Sulawesi	Fishes	Most important single site for aquatic biodiversity in Asia. A complex of 5 lakes (Towuti, Matano, Mahalona, Wawontoa, Masapi) with endemic radiations of fishes of the families Telmatherinidae (3 genera, 15 species, all but 1 endemic), Hemiramphidae (3 endemic species), Oryziidae (3 endemic species), Gobiidae (at least 8, all but 1 endemic), prawns (ca 12 species?), crabs (4 species?), mollusks (ca 60 endemic species), etc. The distribution of the fishes is not uniform within the lakes, all but 1 of the species of L. Matano are endemic, while the others (and 2 genera) are endemic to Towuti, Mahalona, Wawontoa. Masapi has not yet been surveyed. Only 2 species of the Telmatherinidae are known outside this area. [MK]
73	Maros karst, Sulawesi	Fishes	One endemic genus (possibly an artifact of limited collection; more surveys might show it to be present outside this area) and about 6 endemic species, including a cave species[23]. [MK]
74	Mindanao, Philippines	Fishes	About 30 endemic species of cyprinid fishes, including about 18 endemic species of *Puntius* in L. Lanao (all but 2 or 3 reportedly extinct)[47,48]. Cyprinids are fishes which live only in freshwater and cannot disperse in marine environments; several other families also occur in the island's freshwaters, but all are able to disperse through the seas. [MK]
75	Northwest Mediterranean drainage	Fishes	Includes Spain, Portugal, southern France and northern Italy. The total diversity in the whole area is quite low, the communities are quite poor, but this area holds 55 endemics, many with small distribution ranges. 3 of the Rhône endemics extend almost to the northern extremity of the basin. Endemics:1 Petromyzonidae, 1 Acipenseridae, 1 Clupeidae, 34 Cyprinidae, 5 Cobitidae, 6 Salmonidae, 1 Valenciidae, 1 Cyprinodontidae, 2 Cottidae, 1 Percidae and 4 Gobiidae (counted in Kottelat[41]). [MK]
76	Palawan, Philippines	Fishes	About 10 recorded species of cyprinid fishes, actual figure probably higher. [MK]

EURASIA

	Area name	Group	Remarks
77	Southwest Balkans	Fishes	The total diversity in the whole area is quite low, the communities are quite poor, but the area holds 84 endemics, most of them with restricted or very restricted distribution ranges: 1 Petromyzonidae, 2 Clupeidae, 48 Cyprinidae, 8 Cobitidae, 1 Balitoridae, 1 Siluridae, 13 Salmonidae, 1 Valenciidae, 1 Gasterosteidae, 1 Percidae and 7 Gobiidae. The systematics of many groups is still very poorly known and more species will be recognized or even discovered in the future (possibly 10-20). Noteworthy are L. Ohrid with apparently 4 endemic *Salmo*, L. Prespa with apparently 7 endemic species and the Vardar basin with at least 8 endemic species[41]. [MK]
78	Southwest Sri Lanka	Fishes	28 of the 91 native fish species of Sri Lanka are endemic to this area. Several of the species traditionally given the same name as Indian species are being revised and turn out to be specifically distinct, so that the figure will rise[49,50]. [MK]
79	Subalpine lakes	Fishes	Stretches from L. Bourget in the west to Traunsee in the east. Numerous endemic *Coregonus* (possibly 27, several already extinct), at least 2 endemic *Salvelinus* and possibly some endemic *Salmo*. Some lakes have more complex communities, e.g. L. Konstanz with 4 *Coregonus*, 2 *Salvelinus*, 1 *Salmo* and several other species (recent summary in Kottelat[48]. [MK]
80	Sundaic foothills and floodplains	Fishes	About 400 known species. Most of the floodplain species are widely distributed over the whole area, while those of foothill streams have more localized distributions and are of greater interest in terms of endemicity. Northern limit: Tapi basin in peninsular Thailand[23,28,51]. [MK]
81	Western Ghats, India	Fishes	About 100 endemic fish species (estimated from Talwar and Jhingran[52]; Pethiyagoda and Kottelat[53]). Difficult to give accurate figures. Many wide-ranging 'species' of fishes in South Asia are in fact complexes of species, so that the actual number of species is likely to increase significantly after adequate systematic revision. [MK]
82	Balkans region (Former Yugoslavia, Austria, Bulgaria, Greece)	Moll.	Springs and underground aquifers. Gastropods: ca 190 species, some 180 endemic. Bivalves: no data. [MSG]
83	Chilka L.	Moll.	Brackish water. Gastropods: 28 species, ca 11 endemic. Bivalves: 43 species, 25 endemic. [MSG]
84	L. Baikal	Moll.	Gastropods: 147 species, 114 endemic. Bivalves: 13 species, 3 endemic. [MSG]

EURASIA

	Area name	Group	Remarks
85	L. Biwa	Moll.	Gastropods: 38 species, 19 endemic. Bivalves: 16 species, 9 endemic. [MSG]
86	L. Inle	Moll.	Gastropods: 25 species, 9 endemic. Bivalves: 4 species, 2 endemic. [MSG]
87	L. Ohrid and Ohrid basin	Moll.	Gastropods: 72 species, 55 endemic. Bivalves: no data. [MSG]
88	L. Poso and Malili Lakes system	Moll.	Sulawesi. Gastropods: ca 50 species, ca 40 endemic. Bivalves: 5 species, 2 endemic. [MSG]
89	Lower Mekong R., Thailand, Laos, Cambodia	Moll.	River habitat. Only ca 500 km of the lower Mekong main course (with the tributary Mun R.) has been well studied. Gastropods: 121 species, 111 endemic. Two rissoacean groups dominate this entirely prosobranch assemblage of 120 plus species, the pomatiopsid Triculinae (92 endemic species, 11 endemic genera) and the Stenothyridae (19 endemic species). Bivalves: 39 species, 5 endemic. [MSG]
90	North Western Ghats	Moll.	River habitat. Gastropods: ca 60 species, 10 endemic. 2 endemic genera *Turbinicola*, *Cremnoconchus*. The succineid genus *Lithotis* is known from 2 species: *L. tumida* not collected since its description in 1870, and *L. rupicola* only known from a single locality. The highly localized genus *Cremnoconchus* is the only littorinid living in a freshwater/terrestrial environment. Bivalves: 11 species, 5 endemic. [MSG]
91	Zrmanja R., Croatia	Moll.	Gastropods: all are hydrobioid snails, 11 species, 5 endemic. Bivalves: no data. [MSG]

NORTH AMERICA

92	Southeast USA	Crayfish	Large area of high richness and endemism at generic and species level; including the eastern and southern Mississippi drainage (Ohio R., Tennessee R., to Ozark and Ouachita mountains); 72 species in Alabama, 71 in Tennessee. [KC]
93	Western USA	Fairy shrimp	26 species, 13 endemic. [DB]
94	Bear L.	Fishes	This lake is part of the Bonneville R. basin and contains 1 local endemic (*Prosopium gemmiferum*) and 2 species that are now restricted to this site (*Prosopium spilonotus* and *P. abyssicola*)[54]. [UNEP-WCMC]

NORTH AMERICA

	Area name	Group	Remarks
95	Colorado basin	Fishes	The largest basin of the western USA, this has high species richness and endemism, including 5 endemic genera of which only *Plagopterus* is monotypic[55]. About one third of the ichthyofauna of the Colorado is threatened, endangered or extinct due to dams and introduced species[56]. [UNEP-WCMC]
96	Cumberland Plateau (Cumberland and Tennessee rivers)	Fishes	This area has the highest species richness and local endemism in North America. It is part of the highly diverse Mississippi basin, with ca 240 species in total, 160 present in both the Tennessee and Cumberland drainages, 14 endemic to the 2 basins. Of these 14, 10 are darters, 3 are minnows and 1 is a topminnow. The Tennessee has the greatest species diversity with 224 species including 25 endemics (as well as 64 not found in the Cumberland). The Cumberland has 176 native species, including 9 endemics and 16 species not shared with the Tennessee[57]. [UNEP-WCMC]
97	Death Valley region	Fishes	There is a high level of local endemism associated with the dispersed pattern of springs and marshes. 4 families are present (Cyprinidae, Catostomidae, Cyprinodontidae and Goodeidae) with 9 species including an endemic species of Catostomidae[54]. Several are globally threatened, including 2 of the 5 *Cyprinodon* species (*Cyprinodon radiosus* and *C. diabolis*). [UNEP-WCMC]
98	Eastern USA	Fishes	This is a general area of high species richness and endemism which, with the possible exception of the incompletely known East Asian fish species, represents the most diverse of all the freshwater faunas of the temperate zone[55]. This includes a) the Ozark Plateau, b) the Ouachita Mountains, c) the South Atlantic Central Plain and d) the Tennessee-Cumberland Plateau. [UNEP-WCMC]
99	Klamath-upper Sacramento	Fishes	The Klamath R. basin contains 28 species in total with relatively high endemism. The 6 endemic species include 2 *Catostomus*, 1 *Chasmistes* and 1 *Gila*[54]. The ichthyofauna of the Sacramento differs from that of the Klamath and contains 4 genera that are confined to this river and a few neighboring drainages[55]. [UNEP-WCMC]
100	Ouachita Mountains	Fishes	This area includes parts of the lower Red and Ouachita rivers, each containing 133 species. The Ouachita and the Red river systems both contain 18 endemic species[58]. [UNEP-WCMC]

NORTH AMERICA

Area name	Group	Remarks
101 Ozark Plateau	Fishes	The Ozark Plateau is an area of high species diversity and particularly high local endemism in the southeast USA; it represents a concentration of the species-rich southwestern Mississippi drainage (more than 30 endemic fish species)[55]. [UNEP-WCMC]
102 Rio Grande-Pecos confluence	Fishes	The Rio Grande basin overall has more than 60 endemic species[58] and the Pecos, a tributary, has 5 [59]. Many of the endemics occur at the confluence of the 2 rivers, and many are globally threatened. [UNEP-WCMC]
103 Southern Oregon-California rivers	Fishes	These rivers share few family similarities with the eastern USA and have about 25% of the number of species, but the region is high in local endemism. [UNEP-WCMC]
104 Southern Atlantic coastal plain	Fishes	This includes the Alabama-Tombigbee basin with a species-rich fauna, including about 40 endemic taxa[60]. This region also contains the Pearl R., with 106 species[60] and the species-rich lower Mississippi. [UNEP-WCMC]
105 Arid/semi-arid western USA	Moll.	Springs and underground aquifers. Gastropods: all are hydrobioid snails, ca 100 species, at least 58 endemic. Great radiation in genus *Pyrgulopsis*. 3 extinct species, and all others are candidates for listing by US Fish and Wildlife Service. Bivalves: no data. [MSG]
106 Cuatro Cienegas basin, Mexico	Moll.	Springs and underground aquifers. Gastropods: all are hydrobiids; 12 species, more than 9 endemic. 5 genera (*Nymphophilus, Coahuilix, Paludiscala, Mexithauma, Mexipyrgus*) are endemic to this small area of 30 x 40 km. Bivalves: no data. [MSG]
107 Florida, USA	Moll.	Springs and underground aquifers. Gastropods: mostly hydrobiid snails. 84 species, ca 43 endemic. No bivalves. [MSG]
108 Mobile Bay basin	Moll.	Tombigbee-Alabama rivers. River habitat. Gastropods: 118 species, 110 endemic; 6 endemic genera; greatest species richness (76 species) in Pleurocercidae. 38 of the gastropod species believed extinct, 70 candidates for listing by US Fish and Wildlife Service. Bivalves: 74 species, 40 endemic, 25 extinct. [MSG]
109 Ohio-Tennessee rivers	Moll.	Eastern Mississippi drainage. River habitat. High species richness and endemism. [MSG]

CENTRAL AND SOUTH AMERICA

Area name	Group	Remarks
110 Central America	Crabs	The freshwater crabs of Central America belong to exclusively neotropical families, Pseudothelphusidae and Trichodactylidae. Central America from Mexico to Panama, including some of the Caribbean islands, holds at least 22 genera and over 80 species of pseudothelphusid crabs, and 4 genera and ca 10 species of trichodactylids. The Isthmus of Tehuantepec in central Mexico is a hotspot of biodiversity for freshwater crabs in Central America[61,62], and richness declines toward to the south and north. The 7 species of freshwater crab belonging to 1 genus found in Cuba are all endemic to that island[63]. [NC/RvS]
111 South America	Crabs	2 freshwater crab families (Pseudothelphusidae, Trichodactylidae) endemic to the neotropics occur here. Freshwater crabs do not extend to southern Chile or southern Argentina. There are an estimated 17 genera and over 90 species of pseudothelphusids, found mainly in the highland regions of Peru, Ecuador, Colombia, Venezuela and the Guianas, and on the islands of the southern Caribbean, and 12 genera and over 40 species of trichodactylids in the Amazon basin. The Cordilleras of Colombia[64,65], coastal Venezuela and the Guianas[62,65], and the highland areas of Ecuador and Peru[66] are all diversity hotspots for freshwater crabs. The Amazon basin is rich in species[67-71], but most are widespread in the basin, and it is not possible yet to delimit special areas. [NC/RvS]
112 Southern South America	Fairy shrimp	18 species, 14 endemic. [DB]
113 Altiplano of the Andes	Fishes	Species flock of *Orestias* with 43 or more species, representing an endemic subfamily, Orestiinae, of the Cyprinodontidae. [SK]
114 Amazon R. basin	Fishes	The Amazon (with adjacent Tocantins) basin probably has about 3 000 species, and is one gigantic hotspot. The Amazon fauna equals or exceeds other continental faunas in species richness. Endemism in tributaries and subtributaries makes up most of the overall diversity, rather than the main Amazon itself. Only a few of the constituent rivers have been studied in any detail. [SK]
115 Aripuanã R., a tributary of the Madeira	Fishes	Known to have a highly endemic but still little-studied fauna upstream of the lowermost falls, with at least 10 endemic species, some restricted to rapids. [SK]

**CENTRAL AND
SOUTH AMERICA**

	Area name	Group	Remarks
116	Central America between the Isthmus of Tehuantepec and the Isthmus of Panama	Fishes	280 freshwater fish species, all endemic. [SK]
117	Iguaçu R.	Fishes	On the border between Argentina and Brazil, tributary to the Paraná R. Its fish fauna is separated from the Paraná by the Iguaçu falls, which do not permit any migration; highly endemic, with ca 50 endemic species out of a total of 65 species (ca 80%). There are considerable difficulties with the nomenclature and systematic status of the Iguaçu fish species, most belonging to groups that have never been revised. Nonetheless, endemism will probably remain above 50%. The endemic fauna, mainly a running water one, is highly endangered by hydroelectric power projects, pollution and introduced species. The fauna is not protected. [SK]
118	La Plata basin: Uruguay, Paraguay and Paraná rivers	Fishes	Marked by numerous waterfalls providing isolation. Mainly endemic species, including numerous local endemics. Number of species unknown, but estimated fewer than 1 000, possibly ca 600. The tributaries of the Paraná down to about Encarnación have a very high number of local endemics, often restricted to a single river, mostly separated from the Paraná by one or more waterfalls near the mouth. Many of these have not been described or examined by a specialist, but are known only from occasional collections made before the Itaipu, Acaray and Yacyretá dams were constructed. Unfortunately, environmental impact assessment for those dams did not result in any significant collections to show what species were in the area before the dams were built. A lesser collection of pre-dam fishes is available in Museo Nacional de Historia Natural del Paraguay and the Muséum d'histoire naturelle de Genève. [SK]
119	L. Titicaca and smaller lakes of the Altiplano extending from Chile to Peru	Fishes	These lakes hold a large number of species of the genus *Orestias* (Cyprinodontidae), 23 are endemic to Titicaca. The genus, with a total of 43 species, has a narrow range from northern Chile to southern Peru. The lake species flocks may not be monophyletic[72], but the group certainly attained its present species richness in the area. The sister group is the North American Cyprinodontidae. Other highland Andean fish families include the Astroblepidae, ranging from Bolivia across Peru and Ecuador into Colombia, and many trichomycterid fishes (Trichomycteridae) occur. L. Titicaca with its *Orestias* fauna is the only identifiable hotspot. [SK]

Area name	Group	Remarks
120 Marowijne/Maroni R. drainage, Guyana and Suriname	Fishes	Known to have many endemic species above the falls, with the same genera as in the rest of the Guianas area. [SK]
121 Mata Atlántica	Fishes	Numerous endemic species in small mountain streams or in the few major river systems, most incompletely known. The Ribeira R. has 77 species, and similar numbers appear to be in the other rivers. The Jequitinhonha is notable for several endemic species, including 1 of *Rhamdia*, which is otherwise represented by only a few widespread species in South America. The Mata Atlántica fauna extends to eastern Uruguay and southeastern Paraguay as numerous fragmented habitat patches, and although not high in species richness (perhaps ca 150), has a large number of locally restricted species, with related species replacing each other from one river to another. [SK]
122 Mazaruni and Potaro rivers, Guyana highlands	Fishes	Separated from the rest of the Essequibo system by falls, with several endemic species, but little explored. [SK]
123 Mesa Central, Mexico	Fishes	Endemic subfamily Goodeinae of family Goodeidae with about 36 species. [SK]
124 Mexican Plateau	Fishes	
125 Negro R. and upper Orinoco R., Brazil, Colombia and Venezuela	Fishes	At least 700 species, probably nearer to 1 000, many of which are endemic to the clear and black waters distinguishing the basin. [SK]
126 Nicaraguan lakes	Fishes	The Nicaraguan great lakes (Nicaragua and Managua) in the San Juan basin do not have great numbers of species (about 16 cichlids), but endemism is high (9 endemic species, 2 endemic genera). [SK]
127 Orinoco R. basin	Fishes	More than 1 000 species, most of which may be endemic. There is much local endemism as habitats vary considerably, including lowland inundation savannahs, fast-flowing mountain rivers, etc. Includes thus different biogeographic regions. [SK]
128 Oyapock R., Brazil and French Guiana	Fishes	Known to have many endemic species, especially rheophilic, from the lowermost falls upstream. Still little studied. [SK]
129 Pacific coast of Colombia and Ecuador	Fishes	Although a high-rainfall region there are few large rivers. The fauna is poor, but species are highly endemic to the region and to particular rivers or portions of rivers. In particular, the Baudó R. and San Juan R. in Colombia seem to have numerous endemics. Possibly the Atrato R. should be included. [SK]

CENTRAL AND SOUTH AMERICA

	Area name	Group	Remarks
130	Patagonia (Argentina and Chile, from around the R. Negro southward – except the most arid areas)	Fishes	Low diversity but endemic relict fauna of more general southern hemisphere type, with families such as Geotriidae and Galaxiidae, and also the endemic catfish family Diplomystidae with 6 species, the monotypic catfish family Nematogenyidae, and 4 species of the percoid family Percicihtyidae. This is not a hotspot of species richness, but a region of considerable local endemism, and a fauna completely different from that of the rest of South America. The Nematogenyidae are related to the Loricariidae and Trichomycteridae of northern South America (the Brazilian fauna), but the Diplomystidae are the most primitive living catfish family. The scaleless characid *Gymnocharacinus bergii* represents the Brazilian fauna, but lives isolated in one Patagonian locality on the Sumuncurá Mountain which maintains about 22.5°C water temperature year-round. [SK]
131	Panuco R. basin	Fishes	A small drainage with ca 75 species, ca 30% endemic, including several closely related species of *Herichthys* (Cichlidae). Eastern Mexico, ca 25 endemic of ca 75 known species. [SK]
132	Upper Uruguay R.	Fishes	The river is relatively well known from collections made by teams of the Museu de Zoologia of the Pontifícia Universidade Católica do Rio Grande do Sul during environmental impact assessment in the area, which is destined for numerous hydroelectric power plants. The collections concern the middle and upper portions, located in Brazil. More than 130 species of fish are recorded from the middle and upper Uruguay, and the number is likely to rise to over 150 at least. About half of those may be endemic. Lucena and Kullander[73] described 11 species of *Crenicichla* from the Uruguay R., and noted that this is double the number of a similar Amazonian river. 6 of the species form a species flock originating on site and diversifying by trophic adaptation similar to cichlids of East African lakes. The lower Uruguay R., along the Argentinian-Uruguayan border, is very little studied and may have fewer endemics. [SK]
133	Western Amazonia	Fishes	Lowland Amazonian Peru, Ecuador and Colombia, and parts of Brazil, representing a large expanse of lowland Amazonia, very rich in species, but not well studied. Work in Peru and Ecuador suggest that there may be at least 1 000 species in the area, and at least half may be endemic. [SK]
134	L. Titicaca	Moll.	Gastropods: 24 species, 15 endemic. Bivalves: no data. [MSG]

Area name	Group	Remarks
135 Lower Uruguay R. and Rio de la Plata, Argentina, Uruguay, Brazil	Moll.	Gastropods: 54 species, 26 endemic. Bivalves: 39 species, 8 endemic. [MSG]
136 Paraná R.	Moll.	More than 7 species, 7 endemic, of which 3 are extinct in the wild. Bivalves: no data. [MSG]

CENTRAL AND SOUTH AMERICA

REFERENCES

1 Cumberlidge, N. 1999. *The freshwater crabs of West Africa. Family Potamonautidae.* Faune et Flore Tropicales 36. Institut de récherche pour le developpement IRD (ex ORSTOM), Paris.

2 Cumberlidge, N. 1996. A taxonomic revision of freshwater crabs (Brachyura, Potamoidea, Gecarcinucidae) from the Upper Guinea forest of West Africa. *Crustaceana* 69(6): 681-695.

3 Cumberlidge, N. 1996. On the *Globonautinae* Bott, 1969, freshwater crabs from West Africa (Brachyura: Potamoidea: Gecarcinucidae). *Crustaceana* 69(7): 809-820.

4 Türkay, M. and Cumberlidge, N. 1998. Identification of freshwater crabs from Mount Nimba, West Africa (Brachyura: Potamoidea: Potamonautidae). *Senckenbergiana biologica* 77(2).

5 Teugels, G.G. and Guégan, J.-F. 1994. Diversité biologique des poissons d'eaux douces de la Basse Guinée et de l'Afrique Centrale. In: Paradi, Teugels, G.G. and Guégan, J.-F. (eds) Biological diversity of African fresh-and brackish water. Geographical overviews. Symposium, *Ann. Mus. Roy. Afr. Centr., Zool.* 275: 67-85.

6 Teugels, G.G., Reid, G. McG. and King, R.P. 1992. Fishes of the Cross river basin (Cameroon-Nigeria): Taxonomy, zoogeography, ecology and conservation. *Ann. Mus. Roy. Afr. Centr., Zool.* 266.

7 Stiassny, M.L.J. and Raminosoa, N. 1994 The fishes of the inland waters of Madagascar. In: Teugels, G.G., Guégan, J.-F. and Albaret, J.-J. (eds) Diversité biologique des poissons des eaux douces et saumâtres d'Afrique. Synthèses géographiques. *Ann. Mus. Roy. Afr. Centr., Zool.* 275: 133-149.

8 Teugels, G.G. and Powell, C.B. 1993. Biodiversity of fishes in the eastern Niger delta, Nigeria, and species distribution in relation to white and blackwater rivers. Abstracts. Symposium International sur la diversité biologique des poissons d'eaux douces et saumâtres d'Afrique, Dakar, 15-20 November 1993, p. 62. Plus unpublished data.

9 Lévêque, C., Paugy, D. and Teugels, G.G. (eds) 1990. *Faune des poissons d'eaux douces et saumâtres de l'Afrique de l'Ouest. Tome I*, pp. 1-384. Collection Faune Tropicale XXVIII, ORSTOM, Paris/MRAC, Tervuren.

10 Lévêque, C., Paugy, D. and Teugels, G.G. (eds) 1992. *Faune des poissons d'eaux douces et saumâtres de l'Afrique de l'Ouest. Tome II*, pp. 385-902. Collection Faune Tropicale XXVIII, ORSTOM, Paris/MRAC, Tervuren.

11 Lévêque, C. 1997. *Biodiversity, dynamics and conservation. The freshwater fish of tropical Africa.* ORSTOM. Cambridge University Press. Cambridge.

12 Alcock, A. 1910. *Catalogue of the Indian decapod crustacea in the collection of the Indian Museum. Part I. Brachyura. Fasciculus II. The Indian fresh-water crabs – Potamonidae.* Calcutta.

13 Bott, R. 1970. Betrachtungen über die Entwicklungsgeschichte und Verbreitung der Süßwasser-Krabben nach der Sammlung des Naturhistorischen Museums in Genf/Schweiz. *Revue suisse de Zoologie* 77(2): 327-344.

14 Ng, P.K.L. 1988. *The freshwater crabs of Peninsular Malaysia and Singapore.* Department of Zoology, University of Singapore, Shing Lee Publishers Pte. Ltd, Singapore.

15 Ng, P.K.L. and Naiyanetr, P. 1993. New and recently described freshwater crabs (Crustacea: Decapoda: Brachyura: Potamidae, Gecarcinucidae and Parathelphusidae) from Thailand. *Zoologische Verhandelingen* 284: 1-117.

16 Ng, P.K.L. and Dudgeon, D. 1991. The Potamidae and Parathelphusidae (Crustacea: Decapoda: Brachyura) of Hong Kong. *Invertebrate Taxonomy* 6: 741-768.

17 Dai, Zhou and Peng 1995. Eight new species of the genus *Sinopotamon* from Jiangxi Province, China (Crustacea, Decapoda, Brachyura, Potamidae). *Beaufortia* 45(5): 61-76.

18 Türkay, M. and Dai, A.Y. 1997. Review of the Chinese freshwater crabs previously placed in the genus *Malayopotamon* Bott, 1968 (Crustacea: Decapoda: Brachyura: Potamidae). *Raffles Bulletin of Zoology* 45(2): 189-207.

19 Dai, A.Y. 1997. A revision of freshwater crabs of the genus *Nanhaipotamon* Bott, 1968 from China (Crustacea: Decapoda: Brachyura: Potamidae). *Raffles Bulletin of Zoology* 45(2): 209-235.

20 Dai, A.Y. and Türkay, M. 1997. Revision of the Chinese freshwater crabs previously placed in the genus *Isolapotamon* Bott, 1968 (Crustacea: Decapoda: Brachyura: Potamidae). *Raffles Bulletin of Zoology* 45(2): 237-264.

21 Bott, R. 1970. Die Süßwasserkrabben von Ceylon. *Arkiv für Zoologie* 22: 627-640.

22 Ng, P.K.L. 1995. A revision of the Sri Lankan montane crabs of the genus *Perbrinckia* Bott, 1969 (Crustacea: Decapoda: Brachyura: Parathelphusidae). *Journal of South Asian Natural History* 1(1): 129-174.

23 Kottelat, M. *et al.* 1993. *Freshwater fishes of western Indonesia and Sulawesi.* Periplus Editions, Hong Kong, with some updating.

24 Ng, P.K.L. 1994. Peat swamp fishes of Southeast Asia – Diversity under threat. *Wallaceana* 73: 1-5.

25 Ng, P.K.L., Tay, J.B. and Lim, K.K.P. 1994. Diversity and conservation of blackwater fishes in Peninsular Malaysia, particularly in the North Selangor peat swamp forest. *Hydrobiologia* 285: 203-218.

26 Kottelat, M. and Lim, K.K.P. 1995. Freshwater fishes of Sarawak and Brunei Darussalam: A preliminary annotated check-list. *Sarawak Mus. J.* 48(69): 227-256.

27 Kottelat, M. 1985. Fresh-water fishes of Kampuchea. *Hydrobiologia* 121: 249-279.

28 Kottelat, M. 1989. Zoogeography of the fishes from Indochinese inland waters with an annotated check-list. *Bull. Zool. Mus. Univ. Amsterdam* 12(1): 1-54.

29 Wu, Y.-F. and Wu, C.-Z. 1992. [*The fishes of the Qinghai-Xizang Plateau*]. Sichuan Publishing House of Science and Technology, Chengdu [Chinese, English summary].

30 Chen, Y.-R. and Yang, J.-X. 1993. A synopsis of cavefishes from China. *Proceedings of the XI International Congress of Speleology*, pp. 121-122, Beijing, updated.

31 Smith, G.R. and Todd, T.N. 1984. Evolution of species flocks of fishes in north temperate lakes. In: Echelle, A.A. and Kornfield, I.L. (eds). *Evolution of fish species flocks*, pp. 45-68. University of Maine at Orono Press, Orono.

32 Masuda, H. *et al.* 1984. *The fishes of the Japanese archipelago.* 2 vols. Tokai University Press, Tokyo,

33 Chereshnev, I.A. 1992. Rare, endemic, and endangered freshwater fishes of Northeast Asia. *J. Ichthyol.* 32(8): 110-124.

34 Chereshnev, I.A. and Skopets, M.B. 1990. *Salvethymus svetovidovi* gen. et sp. nova – a new endemic fish of the subfamily Salmoninae from Lake El'gygytgyn (Central Chukotka). *J. Ichthyol.* 30(2): 87-103.

35 Annandale, N. 1918. Fish and fisheries of the Inle Lake. *Rec. Ind. Mus.* 14: 33-64.

36 Kottelat, M. 1986. Die Fischfauna des Inlé-Sees in Burma. *Aquar. Terrar. Ztschr.* 39(9): 403-406, (10): 452-453.

37 Kottelat, M. 1990. Synopsis of the endangered Buntingi (Osteichthyes: Adrianichthyidae and Oryziidae) of Lake Poso, Central Sulawesi, Indonesia, with a new reproductive guild and description of three new species. *Ichthyol. Explor. Freshwat.* 1(1): 49-67.

38 Kottelat, M. 1990. The ricefishes (Oryziidae) of the Malili Lakes, Sulawesi, Indonesia, with description of a new species. *Ichthyol. Explor. Freshwat.* 1(2): 151-166.

39 Kottelat, M. 1990. Sailfin silversides (Pisces: Telmatherinidae) of Lakes Towuti, Mahalona and Wawontoa (Sulawesi, Indonesia) with descriptions of two new genera and two new species. *Ichthyol. Explor. Freshwat.* 1(3): 227-246.

40 Kottelat, M. 1991. Sailfin silversides (Pisces: Telmatherinidae) of Lake Matano, Sulawesi, Indonesia, with descriptions of six new species. *Ichthyol. Explor. Freshwat.* 1(4): 321-344.

41 Kottelat, M. 1997. European freshwater fishes. An heuristic checklist of the freshwater fishes of Europe (exclusive of former USSR), with an introduction for non-systematists and comments on nomenclature and conservation. *Biologia, Bratislava, Sect. Zool.* 52 (suppl. 5): 1-271.

42 Kottelat, M. and Chu, X.-L. 1988. Revision of *Yunnanilus* with descriptions of a miniature species flock and six new species from China (Cypriniformes: Homalopteridae). *Env. Biol. Fish.* 23(1-2): 65-93, updated.

43 Yang, J.-X. and Chen, Y.-R. 1995. [*The biology and resource utilization of the fishes of Fuxian Lake, Yunnan*]. Yunnan Science and Technology Press, Kunming [Chinese, English summary].

44 Li, S.-Z. 1982. [Fish fauna and its differentiation in the upland lakes of Yunnan]. *Acta Zool. Sinica* 28(2): 169-176, updated [Chinese, English abstract].

45 Kottelat, M. 1990. *Indochinese nemacheilines. A revision of nemacheiline loaches (Pisces: Cypriniformes) of Thailand, Burma, Laos, Cambodia and southern Viet Nam.* Pfeil, Munich.

46 Kottelat, M. 1998. Fishes of the Nam Theun and Xe Bangfai basins (Laos), with diagnoses of a new genus and twenty new species (Cyprinidae, Balitoridae, Cobitidae, Coiidae and Eleotrididae). *Ichthyol. Explor. Freshwat.* 9(1): 1-128.

47 Myers, G.S. 1960. The endemic fish fauna of Lake Lanao, and the evolution of higher taxonomic categories. *Evolution* 14: 323-333.

48 Kornfield, I.L. and Carpenter, K.E. 1984. The cyprinids of Lake Lanao, Philippines: Taxonomic validity, evolutionary rates and speciation scenarios. In: Echelle, A.A. and Kornfield, I.L. (eds). *Evolution of fish species flocks*, pp. 69-84. University of Maine at Orono Press, Orono.

49 Pethiyagoda, R. 1991. *Freshwater fishes of Sri Lanka*. Wildlife Heritage Trust of Sri Lanka, Colombo.

50 Pethiyagoda, R. 1994. Threats to the indigenous freshwater fishes of Sri Lanka and remarks on their conservation. *Hydrobiologia* 285: 189-201.

51 Kottelat, M. 1995. The fishes of the Mahakam River, East Borneo: An example of the limitations of zoogeographic analyses and the need for extensive fish surveys in Indonesia. *Trop. Biodiv.* 2(3): 401-426.

52 Talwar, P.K. and Jhingran, A.G. 1991. *Inland fishes of India and adjacent countries*. 2 vols, Oxford and IBH Publishing Company, New Delhi, updated.

53 Pethiyagoda, R. and Kottelat, M. 1994. Three new species of fishes of the genera *Osteochilichthys* (Cyprinidae), *Travancoria* (Balitoridae) and *Horabagrus* (Bagridae) from the Chalakudy River, Kerala, India. *J. South Asian Nat. Hist.* 1(1): 97-116.

54 Minckley *et al.* 1986. In: Hocutt, C.H. and Wiley, E.O. (eds). *The zoogeography of North American freshwater fishes*. John Wiley and Sons.

55 Bânârescu, P. 1991. *Zoogeography of freshwaters. 1. General distribution and dispersal of freshwater animals*. Aula Verlag, Wiesbaden.

56 Carlson, C.A. and Muth, R.T. 1989. In: Dodge, D.P. (ed.) Proceedings of the International Large River Symposium. *Can. Spec. Publ. Fish. Aquat. Sci.* 106.

57 Starnes and Etnier. 1986. In: Hocutt, C.H. and Wiley, E.O. (eds). *The zoogeography of North American freshwater fishes.* John Wiley and Sons.

58 Cross, F.B. *et al.* 1986. In: Hocutt, C.H. and Wiley, E.O. (eds). *The zoogeography of North American freshwater fishes.* John Wiley and Sons.

59 Smith, M.L. and Rush Miller, R. 1986. In: Hocutt, C.H. and Wiley, E.O. (eds). *The zoogeography of North American freshwater fishes.* John Wiley and Sons.

60 Swift, C.C. *et al.* 1986. In: Hocutt, C.H. and Wiley, E.O. (eds). *The zoogeography of North American freshwater fishes.* John Wiley and Sons.

61 Alvarez, F. and Villalobos, J.L. 1991. A new genus and two new species of freshwater crabs from Mexico, *Odontothelphusa toninae* and *Stygiothelphusa lopezformenti* (Crustacea: Brachyura: Pseudothelphusidae). *Proceedings of the Biological Society of Washington* 104(2): 288-294.

62 Rodriguez, G. 1986. Centers of radiation of fresh-water crabs in the neotropics. In: Gore, R.H. and Heck, K.L. (eds). Biogeography of the crustacea. *Crustacean Issues* 4: 51-67.

63 Chace, F.A. and Hobbs, H.H. 1969. The freshwater and terrestrial decapod crustaceans of the West Indies with special reference to Dominica. *United States National Museum Bulletin* 292: 1-258.

64 Rodriguez, G. and Campos, M.R. 1989. The cladistic relationships of the freshwater crabs of the tribe Strengerianini (Crustacea, Decapoda, Pseudothelphusidae) from the northern Andes, with comments on their biogeography and descriptions of new species. *Journal of Crustacean Biology* 9: 141-156.

65 Rodriguez, G. and Pereira, G. 1992. New species, cladistic relationships and biogeography of the genus *Fredius* (Decapoda: Brachyura: Pseudothelphusidae) from South America. *Journal of Crustacean Biology* 12: 298-311.

66 Rodriguez, G. and von Sternberg, R. 1998. A revision of the freshwater crabs of the family Pseudothelphusidae (Decapoda: Brachyura) from Ecuador. *Proceedings of the Biological Society of Washington* 111.

67 Rodriguez, G. 1982. *Les crabes d'eau douces d'Amérique. Famille des pseudothelphusidae.* Faune Tropicale 22, ORSTOM, Paris.

68 Rodriguez, G. 1992. T*he freshwater crabs of America. Family Trichodactylidae and supplement to the family Pseudothelphisidae.* Faune Tropicale 31, ORSTOM, Paris.

69 Magalhães, C. and Türkay, M. 1996. Taxonomy of the neotropical freshwater crab family Trichodactylidae I. The generic system with description of some new genera (Crustacea: Decapoda: Brachyura). *Senckenbergiana biologica* 75(1-2): 63-95.

70 Magalhães, C. and Türkay, M. 1996. Taxonomy of the neotropical freshwater crab family Trichodactylidae II. The genera *Forsteria, Melocarcinus, Sylviocarcinus* and *Zilchiopsis* (Crustacea: Decapoda: Brachyura). *Senckenbergiana biologica* 75(1-2): 97-130.

71 Magalhães, C. and Türkay, M. 1996. Taxonomy of the neotropical freshwater crab family Trichodactylidae III. The genera *Fredilocarcinus* and *Goyazana* (Crustacea: Decapoda: Brachyura). *Senckenbergiana biologica* 75(1-2): 131-142.

72 Parenti, L.R. 1984. A taxonomic revision of the Andean killifish genus *Orestias* (Cyprinodontiformes, Cypridontidae). *Bull. Amer. Mus. Nat. Hist.* 178: 107-214.

73 Lucena, C.A.S. de and Kullander, S.O. 1992. The *Crenicichla* species of the Uruguai River drainage in Brazil. *Ichthyol. Explor. Freshwat.* 3: 97-160.

74 WCMC 1998. *Freshwater biodiversity: A preliminary global assessment.* Groombridge, B. and Jenkins, M. World Conservation Press, Cambridge.

Index

Note: Page numbers in bold refer to figures or maps in the text; those in italics to boxed material or tables in the text or Appendices.

Note: Page numbers in bold refer to figures or maps in the text; those in italics to boxed material or tables in the text or Appendices.